Sustainable Development Goals Series

The **Sustainable Development Goals Series** is Springer Nature's inaugural cross-imprint book series that addresses and supports the United Nations' seventeen Sustainable Development Goals. The series fosters comprehensive research focused on these global targets and endeavors to address some of society's grand challenges. The SDGs are inherently multidisciplinary, and they bring people working across different fields together toward a common goal. In this spirit, the Sustainable Development Goals series is the first at Springer Nature to publish books under both the Springer and Palgrave Macmillan imprints, bringing the strengths of our imprints together.

The Sustainable Development Goals Series is organized into eighteen subseries: one subseries based around each of the seventeen respective Sustainable Development Goals, and an eighteenth subseries, "Connecting the Goals," which serves as a home for volumes addressing multiple goals or studying the SDGs as a whole. Each subseries is guided by an expert Subseries Advisor with years or decades of experience studying and addressing core components of their respective Goal.

The SDG Series has a remit as broad as the SDGs themselves, and contributions are welcome from scientists, academics, policymakers, and researchers working in fields related to any of the seventeen goals. If you are interested in contributing a monograph or curated volume to the series, please contact the Publishers: Zachary Romano [Springer; zachary.romano@springer.com] and Rachael Ballard [Palgrave Macmillan; rachael.ballard@palgrave.com].

More information about this series at https://link.springer.com/bookseries/15486

Juha I. Uitto · Geeta Batra

Editors

Transformational Change for People and the Planet

Evaluating Environment and Development

 Springer

Editors
Juha I. Uitto
Independent Evaluation Office
Global Environment Facility
Washington, DC, USA

Geeta Batra
Independent Evaluation Office
Global Environment Facility
Washington, DC, USA

ISSN 2523-3084   ISSN 2523-3092   (electronic)
Sustainable Development Goals Series
ISBN 978-3-030-78855-1   ISBN 978-3-030-78853-7   (eBook)
https://doi.org/10.1007/978-3-030-78853-7

Preface

This book is intended to provide an authoritative, interdisciplinary perspective on innovative and emerging evaluation knowledge and practice related to environment, natural resources management, climate change, and development. In recent years, evaluation has emerged as an increasingly important function in determining the worth and value of development interventions in terms of their relevance, impact, performance, effectiveness, efficiency, and sustainability. Evaluation has been formalized as a function in most development agencies on both the multilateral and bilateral sides.

We now live in the Anthropocene, a new era in which human activity has a dominant impact on planetary processes. Climate change and other environmental challenges, such as chemical pollution and the mass extinction of species, have become defining challenges of our time. The COVID-19 pandemic that began in 2020 has tragically demonstrated how closely human health and ecosystem health are intertwined. Adaptation to climate change is necessary and the impacts of the changing climate are affecting the poorest countries and regions and the most vulnerable populations in the most severe fashion. The 2030 Agenda recognizes that sustainable development depends equally on three interlinked pillars: social, economic, and environmental. All 17 Sustainable Development Goals (SDGs) incorporate each of these dimensions to a varying degree. If one of the dimensions fails, the goal is not achievable.

Evaluation must rise to the challenges of sustainable development by considering both human and natural systems and fully accounting for the environmental dimensions of development. This book explores evaluation in areas such as climate change mitigation and adaptation, agriculture, forests, and natural resources management. The chapters cover a wide range of situations, mostly drawn from real-world cases in the field in Africa, Asia, Latin America, and small island developing states. The approaches and methods of evaluation are equally wide ranging. These cases provide important lessons for advancing evaluation at the nexus of environment and development. The focus is on the role of evaluation in promoting transformational change toward a more sustainable future.

This book has its roots in the Third International Conference on Evaluating Environment and Development, held in Prague, Czechia, in October 2019, organized by the Global Environment Facility (GEF) Independent Evaluation Office (IEO) jointly with the International Development Evaluation

Association (IDEAS) and the Earth-Eval[1] community of practice. The conference brought together a large number of established and upcoming evaluators, researchers, and evaluation users from the Global North and South, representing a wide variety of organizations, to discuss the frontiers of environment and development evaluation. Following the event, the organizers identified and contacted selected participants who made key contributions at the conference and asked them to develop their ideas and papers into full-fledged book chapters according to a coherent plan. This is the outcome.

Our gratitude goes to our many partners in conceptualizing and organizing the conference, in particular Rob D. van den Berg, then president of IDEAS, and Daniel Svoboda who acted as the gracious host in Prague on behalf of the Czech Evaluation Society. We are grateful for the generous support for the preparation of this book by the Ministry of Foreign Affairs of Finland and especially the head of Development Evaluation, Anu Saxén. All our colleagues at the GEF IEO[2] contributed substantively to the conference and to this book. We would specifically like to thank Katy O'Grady, who worked with us as assistant editor, preparing the manuscript, keeping track of all the details, and ensuring that the book project was proceeding on schedule. Her professionalism and attention to detail were indispensable.

Washington, DC, USA   Juha I. Uitto

Washington, DC, USA   Geeta Batra

---

[1] https://eartheval.org/
[2] https://www.gefieo.org/

Contents

Part III   Evaluating Climate Change Mitigation and Adaptation

Part IV   Evaluation Approaches

Abbreviations and Acronyms

ADB	Asian Development Bank
AEDE	National Agency for Domestic and Energy and Environment Development
AF	Adaptation Fund
AF-TERG	Technical Evaluation Reference Group of the Adaptation Fund
AIMS	Atlantic, Indian Ocean, Mediterranean and South China Sea
API	Application programming interface
APR	Annual Performance Report
ARRI	Annual Report on Results and Impact
ASM	Artisanal and small-scale mining
ATAPED	Association of Appropriate Technology for the Protection of Environment
BEIS	Department of Business, Energy, and Industrial Strategy
BII	Biodiversity Intactness Index
CAS	Complex adaptive system
CBD	Convention on Biological Diversity
CCA	Climate change adaptation
CCM	Climate change mitigation
CDM	Clean Development Mechanism
CEE	Collaboration for Environmental Evidence
CER	Certified emissions reduction
CES	Canadian Evaluation Society
CHANS	Coupled human and natural systems
Ci-Dev	Carbon Initiative for Development
CLEAR	Centers for Learning on Evaluation and Results
Climate-KIC	Climate Knowledge and Innovation Community
CMA	Parties to the Paris Agreement
CMFP	Carbon Market Finance Programme
CMO	Context-mechanism-outcome
CMP	Parties to the Kyoto Protocol
COP	Conference of Parties
COVID-19	The disease caused by the SARS-CoV-2 coronavirus
CSO	Civil society organization
DFID	Department for International Development
DHS	Department of Health Services
EA	Evaluability assessment

EBA	Ecosystem-based adaptation
EBRD	European Bank for Reconstruction and Development
ECG	Evaluation Cooperation Group
EEGM	Energy efficiency guarantee mechanism
EO	Evaluation Office
ERPA	Emissions reduction purchase agreement
EXPAN	Expanding the Protected Area Network
FBG	Bio Guinea Foundation
FDA	Forestry Development Authority
FFI	Flora and Fauna International
FFS	Farmer field school
FSE	Special Fund for Environment
FUNBIO	Brazilian Biodiversity Fund
GEF	Global Environment Facility
GEF IEO	Global Environment Facility Independent Evaluation Office
GEI	Global Evaluation Initiative
GFW	Global Forest Watch
GHG	Greenhouse gas
GIS	Geographic information system
GIZ	Deutsche Gesellschaft fur Internationale Zusammenarbeit
GLAD	Global Land Analysis and Discovery
HELSUS	Helsinki Institute of Sustainability Science
ICF	International Climate Finance
ICMO	Intervention-context-mechanism-outcome
ICT	Information and communications technology
IDB	Inter-American Development Bank
IDEAS	International Development Evaluation Association
IDIA	International Development Innovation Alliance
IDPE	Interdisciplinary Program in Evaluation
IEG	Independent Evaluation Group
IEO	Independent Evaluation Office
IFAD	International Fund for Agricultural Development
IFI	International financial institution
IIED	International Institute for Environment and Development
IOCE	International Organization on Cooperation in Evaluation
IOE	Independent Office of Evaluation
IPBES	Intergovernmental Science-Policy Platform on Biodiversity and Ecosystem Services
IPCC	Intergovernmental Panel on Climate Change
IPDET	International Program for Development Evaluation Training
IREDA	Indian Renewable Energy Development Agency
JICA	Japan International Cooperation Agency
JNNSM	Jawaharlal Nehru National Solar Mission
J-PAL	Abdul Latif Jameel Poverty Action Lab
KYC	Know Your City
LDC	Least developed country
LLA	Locally led adaptation/action
M&E	Monitoring and evaluation

MDG	Millennium Development Goals
MEL	Monitoring, evaluation, and learning
MFE	Methodological Framework for Evaluation
MNRE	Ministry of New and Renewable Energy
NDVI	Normalized difference vegetation index
NEC	National Evaluation Capacity
NFA	National Forestry Authority
NGO	Nongovernmental organization
NGP	National Gender Policy
NOAA	National Oceanographic and Atmospheric Administration of the United States
OECD	Organisation for Economic Co-operation and Development
OECD DAC	Organisation for Economic Co-operation and Development, Development Assistance Committee
PAR	Participatory action research
PES	Payment for ecosystem services
PMO	Prime Minister's Office
PSC	Public Service Commission
QGI	Quasi-experimental geospatial interpolation
R&D	Research and development
RAC	Refrigeration and air conditioning
RBF	Results-based finance
REA	Rapid evidence assessment
S2SE	South to South Evaluation Initiative
SAMEA	South African Monitoring and Evaluation Association
SCCE	Strategic country cluster evaluation
SDG	Sustainable Development Goal
SDI	Slum/Shack Dwellers International
SDSN	Sustainable Development Solutions Network
SEI	Stockholm Environment Institute
SES	Socioecological systems
SETIG	Systems in Evaluation Topical Interest Group
SFM	Sustainable forest management
SIDS	Small island developing states
SIF	Seychelles Islands Foundation
SLM	Sustainable land management
SNA	Social network analysis
TIPC	Transformative Innovation Policy Consortium
TONC	Theory of No Change
UCEC	Union for Credit and Loan
UN	United Nations
UN OHRLLS	United Nations Office of the High Representative for the Least Developed Countries, Landlocked Developing Countries and Small Island Developing States
UNCCD	United Nations Convention to Combat Desertification
UNDP	United Nations Development Programme
UNEG	United Nations Evaluation Group
UNEP	United Nations Environment Programme

UNFCCC	United Nations Framework Convention for Climate Change
UNHCR	United Nations High Commissioner for Refugees
UNICEF	United Nations International Children's Emergency Fund
UNIDO	United Nations Industrial Development Organization
USAID	US Agency for International Development
UWA	Uganda Wildlife Authority
VIIRS	Visible Infrared Imaging Radiometer Suite
VNR	Voluntary national review
VOPEs	Voluntary organizations of professional evaluation
WRI	World Resources Institute
WTE	Waste to energy

About the Authors

Anupam Anand is an evaluation officer at the GEF Independent Evaluation Office. He has a Ph.D. in geospatial sciences from the University of Maryland, College Park.

Geeta Batra is chief evaluator and deputy director at the Global Environment Facility Independent Evaluation Office (GEF IEO). She has 27 years of experience in international development, with 10 years' experience in program development and implementation and 17 years in international development evaluation.

Dennis Bours coordinates the AF-TERG Secretariat to facilitate implementation of the evaluative work program and supports the AF-TERG Chair in the development of AF-TERG-related strategies, policies, and work plans. Bours has (co-)authored various guidance notes, evaluation reviews, and MEL manuals, and co-edited the *New Directions for Evaluation* journal special edition focusing on adaptation and resilience MEL.

Tillem Burlace is a principal consultant with NIRAS-LTS, specializing in environmental economics and monitoring and evaluation of sustainable natural resource management and climate change projects (across sustainable agriculture, forestry, energy, and land use). She has been involved in the evaluation of the Carbon Market Finance Programme from inception. Burlace is also a monitoring, evaluation, and learning advisor on FCDO Green Economic Growth Programme for Papua and has conducted cost-benefit analysis across a range of agricultural and environmental commodities. She previously worked for the Victorian Government in Australia developing and implementing market-based instruments for sustainable natural resource management. Burlace holds a masters' degree in ecological economics.

Carlo Carugi With the Global Environment Facility Independent Evaluation Office since 2009 as a senior evaluation officer, Carlo Carugi has more than 32 years' experience in environment and development, 16 of which he spent in developing countries. In the course of his career, he conducted several project, program, policy, strategy, thematic, and country-level evaluations for the European Commission, Italy, FAO, and others. Among Carugi's interests is the integration of the socioeconomic with the environmental in evaluation. He holds an M.Sc. in agricultural science from the University of Bologna and an M.Sc. in environment and development from the Imperial College at Wye, University of London.

Jindra Monique Čekan/ová, Ph.D., is a political economist with 33 years of experience in international development. She has learned from villagers and ministers in 27 countries in Africa, Latin America, and Central and South Asia. Clients have included USAID, the Bill and Melinda Gates Foundation, the Adaptation Fund, the Red Cross Movement, Catholic Relief Services, and PricewaterhouseCoopers. Since 2013, she has focused on ex-post-project evaluation and what it can teach about sustained impacts; she created Sustained and Emerging Impacts Evaluations (SEIE). Her firm, Valuing Voices at Cekan Consulting, LLC, uses mixed-methods evaluations focused on participatory development.

Dr. Ellen Fitzpatrick is an agriculture and natural resource economist who studies development policy and practice, specifically the dynamics of sustainable regional economies; food and livelihood systems; and program design, evaluation, and impact assessment. Her work includes research in Africa, Southeast Asia, the South Pacific, Latin America, and Turkey. Fitzpatrick is currently at Merrimack College in Massachusetts, serving as director of the Interdisciplinary Institute, and has held previous positions at the University of Arkansas and State University of New York – Plattsburgh. She has served as a consultant for global NGOs, US government agencies, and multilaterals.

Jeneen Garcia is an evaluation officer at the GEF IEO, where she has co-developed conceptual frameworks for assessing impact in complex adaptive systems for evaluations addressing international waters, biodiversity, multi-stakeholder partnerships, environmental and socioeconomic synergies and trade-offs, and scaling-up of environmental outcomes. She holds degrees in environmental science, marine biology, and water and coastal management, with certificates in evaluation from the International Program for Development Evaluation Training (IPDET) and the Abdul Latif Jameel Poverty Action Lab (J-PAL). She is currently program co-chair of the American Evaluation Association's Systems in Evaluation Topical Interest Group (SETIG).

Robbie Gregorowski is a monitoring, evaluation, and learning (MEL) expert specializing in climate change, systems innovation, adaptive learning, and sustainable change. In 2018, he founded Sophoi (www.sophoi.co.uk). Sophoi provides MEL expertise to help people and organizations generate new knowledge, learn, and change.

Caroline Holo is a data analyst directly supporting the implementation of evaluations and studies part of the AF-TERG evaluative work program. Before she joined the AF-TERG, she worked as MEL coordinator in Madagascar's largest marine protected area with coastal communities. She has worked on impact evaluations, MEL designs and planning, data analysis, and outreach.

Amy Jersild is an evaluator, educator, and researcher with 20 years of experience in the international development field. She is currently pursuing a Ph.D. at Western Michigan University. Her research interests include meta-evaluation, evaluability assessments, and the professionalization and internationalization of evaluation.

Lisa Keppler is a project manager in the field of environmental economics and sustainable finance for the Secretariat of the German International Climate Initiative. Between 2016 and 2019, she worked as consultant for Arepo with a focus on evaluations and assessments of energy and climate policy programs. This included the UK Government's Carbon Market Finance Programme. Keppler holds a master's degree in economics with a focus on development economics. Through her international work experience (e.g., solar sector in Kenya), she has profound knowledge of international environmental and climate change policies, development economics, and innovative financing mechanisms.

Jerry Knox is professor of agricultural water management based within the Water Science Institute at Cranfield University (UK). His international research focuses on the science, engineering, and management of water for agriculture, including assessing the relationships among water resources, abiotic stress (drought), crop yields and the environment, and the sustainability of agricultural production in the context of climate change and water-related risks.

Prashanth Kotturi works as an evaluation officer with the Independent Office of Evaluation of the International Fund for Agricultural Development. He has 10 years of experience in evaluation and research and has undertaken numerous evaluations at the project, country, and corporate levels. His geographic experience spans Africa, the Middle East, and Asia. He has a keen interest in environment and climate change, rural finance and community-driven development, and agricultural value chains.

Susan Legro, MIA, MPH, is a climate change mitigation specialist with more than two decades of experience in donor-funded mitigation and MRV projects. She has worked on global and regional initiatives to improve reporting to the UNFCCC, and her work at the country level covers more than 30 countries. She currently works as an independent consultant, primarily for UN agencies.

Ronnie MacPherson is a monitoring and evaluation specialist with over 20 years' experience in international development, working in diverse sectors and contexts including climate change, low carbon development, clean energy, funding mechanisms and aid effectiveness. Current and recent clients include the Global Environment Facility, UNCCD, Green Climate Fund, International Renewable Energy Agency, and UN Environment.

Takaaki Miyaguchi, Ph.D., is an associate professor in the Department of Global Studies at Kyoto University of Foreign Studies in Japan. Before starting an academic career in 2012, he worked at the United Nations Development Programme and other United Nations agencies for 8 years, developing and managing climate change mitigation and adaptation projects in more than 20 countries. He holds a Ph.D. in global environmental studies from Kyoto University, master of public policy from the University of Chicago, and bachelor of science from the University of Michigan, Ann Arbor.

Callum Murdoch works for NIRAS-LTS International as a senior consultant, specializing in climate finance and climate evaluation. In 2019, he took over project management of the Carbon Market Finance Programme evaluation, and led the development of the realist evaluation framework. At NIRAS-LTS, Murdoch also provides climate finance evaluation services to several multicountry portfolio evaluations, including the UK's Climate Public Private Partnership program and the Children's Investment Fund Foundation's Renewable Energy Portfolio. Prior to joining NIRAS-LTS, he worked in private finance with a leading hedge fund broker. He holds a master's degree in international law and sustainable development.

Indran A. Naidoo With 25 years of professional experience in evaluation, Indran is a global thought leader, manager, instructor, and author, renowned for reform initiatives. He is director of the Independent Office of Evaluation of the International Fund for Agricultural Development. Previously, he was director of the Independent Evaluation Office at the United Nations Development Programme, overseeing significant reform of the evaluation function, and has held senior government positions in South Africa. He has led professional networks and several conferences that advanced the role of evaluation in development. He holds a DPhil and master's degree in geography and has received several awards.

Suppiramaniam Nanthikesan is a lead evaluation officer at the Independent Office of Evaluation of the International Fund for Agricultural Development. He has more than 25 years' experience in the field of sustainable development, working in the private sector, academia, and the UN system.

Neeraj Kumar Negi is a senior evaluation officer at the Global Environment Facility (GEF) Independent Evaluation Office (IEO) and has been with the IEO since 2005. He has led several evaluations related to performance and impact of GEF policies and programs. He previously worked at Seva Mandir as a program associate. He holds a graduate diploma in forestry management from the Indian Institute of Forest Management, a master's in public affairs from Princeton School of Public and International Affairs at Princeton University, and is pursuing a Ph.D. with the School of Public Policy at the University of Maryland.

Trond Norheim, Ph.D., is a partner in Scanteam, Norway, specializing in environment and climate change, with 38 years' experience in 70+ countries.

His main competence is team leadership for design, implementation, monitoring, and evaluation. He has led multiple evaluations of GEF operations and projects focused on small island developing states, including for the EU Global Climate Change Alliance in the Pacific, the UNEP-EU Caribbean Biological Corridor, and the GEF Independent Evaluation Office to carry out strategic country cluster evaluations of the GEF SIDS portfolio in all tropical regions.

Mari Räkköläinen, Ph.D., M.Sc., is a counselor of evaluation at the Finnish Education Evaluation Centre (FINEEC), where her duties are related to methodological development and evaluation services. From 2018–2021, she was a lead evaluation expert at the Development Evaluation Unit of the Ministry for Foreign Affairs of Finland. Where she managed large-scale evaluations of strategic relevance in development policy and cooperation and was a member of the steering group for the Agenda 2030 evaluation in Finland. Prior to 2018, she was also with FINEEC, in charge of the evaluation of learning outcomes. With nearly 20 years' experience in evaluation, she has worked on many national and international evaluations. Her research has focused on the tensions between control and trust present in national-level evaluation systems, and confidence in the assessment process. Currently, she is president of the Finnish Evaluation Society.

Andy Rowe is a fellow and former president of the Canadian Evaluation Society (CES) with over 35 years in evaluation, now focusing on sustainability and climate. He has developed a new approach for evaluating impacts (Rapid Impact Evaluation) and contributes to the conceptual foundations for evaluating sustainability and use-seeking approaches for science. He initiated the CES effort to mainstream sustainability in evaluation in Canada, is a member of Footprint Evaluation and the Technical Evaluation Reference Group of the Adaptation Fund, and contributes to several ongoing evaluations of climate and sustainability. Rowe holds a Ph.D. from the London School of Economics.

Anu Saxén is director of the Development Evaluation Unit, Ministry for Foreign Affairs of Finland. Previously, she was deputy head of mission at the Finnish Embassy in Pretoria, South Africa. She has more than 30 years' experience in foreign and development policy, private business, international business development, and project management. She has held specialist, leadership, and managerial positions in donor and international financing institutions such as the European Commission and the World Bank. Anu has been a keen and consistent user of evaluations throughout her career and is acutely aware of the contribution evaluations can make to strategic planning and policy development.

Katherine Shea is a monitoring, evaluation, and learning (MEL) manager at the World Resources Institute (WRI). She led MEL for Global Forest Watch, leading two evaluations, and now supports the development of institute-wide MEL practices with the Managing for Results team. Before joining WRI, she worked on monitoring and evaluation with the World Food Program in Indonesia, Ecuador, and Uganda. She has a master's in public administration from Cornell University.

Laura Silici is an environmental economist with international field and research experience in sustainable natural resource management, agricultural innovation, and rural development. She works as a consultant to development organizations and research institutions. She previously worked as a researcher in the agroecology team at the International Institute for Environment and Development (IIED) London.

Molly Watts Sohn joined the Global Environment Facility Independent Evaluation Office in 2015 and supports the office's work on both performance and thematic evaluations. Prior to joining the World Bank, Sohn was an evaluation officer on the Monitoring, Evaluation and Learning Team at the National Democratic Institute. She holds a master's degree in Pacific International Affairs from the University of California at San Diego and a bachelor's degree from the University of California at Berkeley.

Kseniya Temnenko is a knowledge management officer at the GEF IEO and has 18 years of experience in evaluation and learning, including 12 years in knowledge management. She works on innovative synthesis studies relevant to GEF-supported programs, policies, and strategies. She also develops training materials and learning products that share lessons from evaluations.

Juha I. Uitto is director of the Global Environment Facility (GEF) Independent Evaluation Office. He has worked as an evaluator with the GEF and the United Nations Development Programme (UNDP) since 1999 and has managed and conducted numerous evaluations focusing on environment and development at the global, regional, and country levels. His work history combines positions in international development and research organizations. He has a Ph.D. in social and economic geography.

Anna Viggh As a senior evaluation officer in the GEF IEO, Anna Viggh has led thematic evaluations on adaptation to climate change, small grants, and gender mainstreaming and led country-level evaluations. She has 30 years of development and environment experience with the GEF, UNDP, the World Bank, and several NGOs. Viggh holds master's degrees from the Yale School of the Environment and George Washington University Elliott School of International Affairs.

Dr. Christine Wörlen is the founder and director of the independent research and consulting company Arepo. She is an internationally renowned expert in energy transition policy and finance and climate change evaluation. Since 2015, Wörlen has evaluated the Carbon Market Finance Programme. Before she founded Arepo, she was head of the Renewable Energies Department at the German Energy Agency and program manager at the Global Environment Facility.

Serge Eric Yakeu Djiam As a bilingual credentialed evaluator, Serge Eric Yakeu Djiam has spent more than 15 years conducting evaluations on five continents with community-based organizations, donors, governments, and United Nations agencies. He has served as visiting professor in Algeria, Cameroon, Canada, and the United Arab Emirates and is active in youth mentoring. He is past president of the African Evaluation Association, global chair of the EvalIndigenous Network, and a member of numerous evaluation associations. Yakeu Djiam holds a double international M.Sc. in rural development, an M.Sc. in research methodology and statistics, and international certificates in North/South cooperation, social science, planning and evaluation, and equity-focused evaluation.

Transformational Change for People and the Planet: Evaluating Environment and Development – Introduction

Juha I. Uitto

Abstract

The world is facing multiple crises as manifested in runaway climate change, a global pandemic, loss of ecosystems and biological species, and rapidly growing inequality. These are all closely interlinked as recognized in the 2030 Agenda for Sustainable Development. Addressing them will require broad transformational change that encompasses the economy, institutions, and how we interact with the natural environment. This chapter introduces the book that is intended to highlight how evaluation can contribute to such transformations. The chapter first reviews the state of development evaluation. It then briefly introduces the state of the global environment before discussing the implications of this context for evaluation, and how evaluation as a profession and practice must change in order to respond to the challenges of sustainability. The chapter ends by explaining the flow of the book in its four parts that focus on: transformational change, drivers of sustainability, climate change mitigation and adaptation, and evaluation approaches.

Background

The 2030 Agenda for Sustainable Development[1] is intended as a blueprint for people, the planet, and prosperity. It recognizes the interconnectedness of economic, social, and environmental development and how none of the three can succeed in the long run if any one of them fails. The 2030 Agenda is titled "Transforming Our World." Yet, despite this almost universally accepted recognition, the world is facing crises on all three fronts. Economic and social crises as expressed in continued poverty, unemployment, exclusion, and constantly increasing inequality between and especially within countries are well recognized. Climate change has similarly gained visibility as the world has witnessed increasing weather anomalies, which are no longer affecting only the developing countries, as dramatically demonstrated by the unprecedented wildfires in Australia and the West Coast of the United States. The Intergovernmental Panel on Climate Change (IPCC) has warned that if we do not limit the rise of global temperatures to 2°C above preindustrial levels, the world will face dire consequences (2018).

---

[1] https://sdgs.un.org/2030agenda

J. I. Uitto (✉)
Global Environment Facility Independent Evaluation Office, Washington, DC, USA
e-mail: juitto@thegef.org

J. I. Uitto, G. Batra (eds.), *Transformational Change for People and the Planet*, Sustainable Development Goals Series, https://doi.org/10.1007/978-3-030-78853-7_1

But a broader environmental crisis is unfolding that involves an unprecedented loss of ecosystems and biological species; places a heavy burden of chemical pollution into the oceans, land, water, and atmosphere; and poses a grave danger to human health. The COVID-19 pandemic that began in 2020 is an expression of this crisis and a direct reminder of how human health and ecosystem health are closely interlinked.

The Dasgupta Review, an authoritative report on the economics of biodiversity led by Prof. Sir Partha Dasgupta and released in February 2021, confirms that the wellbeing of every person—our livelihoods and economies—depend on the natural environment (Dasgupta 2021). It also reminds us that humans are very much part of nature—a fact that we in our technological hubris often ignore—and our economies are embedded in nature, rather than external to it. However, our current development trajectory is entirely unsustainable, which is endangering the prosperity of both current and future generations.

We therefore need to transform how we interact with nature. We need transformations of economic and financial systems, of institutions, of how we measure development, of education and how we see ourselves in relation to the rest of the planet. Such transformational change is necessary and it should be possible, but it requires knowledge and it requires alternative visions of what can be done and how. Evaluation should and can play its part in making transformational change possible.

In recent years, evaluation has emerged as an increasingly important function in determining the value of development interventions in terms of their relevance, impact, performance, effectiveness, efficiency, and sustainability. Evaluation is everywhere in public and private organizations. Many governments and government departments, notably in education, health, and social services, use evaluation to inform the approaches they take to address the issues within their mandate. Private organizations constantly evaluate their performance, whether they use the term evaluation or not. Most foundations, from the Gates Foundation to environmental actors such as the Moore Foundation, have incorporated regular evaluation, not only of their grantees but of the overall direction their funding streams take. Evaluation has been formalized as a function in most development agencies, both at the multilateral and bilateral side.

Although much progress has been made, there are still areas where evaluation has not kept up with the times. Some evaluation practice remains mechanistic and inward looking, tinkering with details rather than engaging with the big picture in the rapidly changing world. Evaluation must change to respond to challenges of sustainable development and to become an active contributor to transformational change.

That is what this book is about. It provides an authoritative, interdisciplinary perspective of innovative and emerging evaluation knowledge and practice related to environment, natural resources management, climate change, and development. It is intended to make a contribution to how evaluation can further transformation toward a more sustainable and just world.

State of Development Evaluation

What do we mean by evaluation? In their now-classic textbook on the topic, Morra Imas and Rist (2009, p. 8) define evaluation simply as the determination of the value of a project, program, or policy. They note that most of the numerous definitions include the notion of "valuing," which distinguishes evaluation from research and monitoring. The Development Assistance Committee (DAC) of the club of industrialized countries, the Organisation for Economic Co-operation and Development (OECD), has a formal definition of evaluation:

> The systematic and objective assessment of an ongoing or completed project, program or policy, its design, implementation and results. The aim is to determine the relevance and fulfillment of objectives, development efficiency, effectiveness, impact and sustainability. An evaluation should provide information that is credible and useful, enabling the incorporation of lessons learned into the decision-making process of both recipients and donors.

Evaluation also refers to the process of determining the worth or significance of an activity, policy or program. An assessment, as systematic and objective as possible, of a planned, on-going, or completed development intervention. (OECD DAC, 2010, pp. 21–22)

Evaluation is conducted for various purposes, including accountability for results achieved and for learning lessons from past experiences. According to the evaluation policy of the Global Environment Facility (GEF, 2019), the most established public funding mechanism for the global environment, the purposes of evaluation include understanding why, how, and the extent to which intended and unintended results are accrued, and their impact on stakeholders. The GEF policy (2019) emphasizes the use of evaluation, stating:

Evaluation feeds into management and decision-making processes regarding the development of policies and strategies; and the programming, implementation, and reporting of activities, projects, and programs. Thus, evaluation contributes to institutional learning and evidence-based policy making, accountability, development effectiveness, and organizational effectiveness. It informs the planning, programming, budgeting, implementation, and reporting cycle. It aims to improve the institutional relevance and achievement of results, optimize the use of resources, and maximize the impact of the contribution provided. (p. 12)

What distinguishes evaluation from related disciplines, such as monitoring and performance audit, is that these latter take the status quo as a given. They are compliance oriented with a mandate to check whether projects and programs are doing what they set out to do and moving toward the objectives set for them. Although audit and evaluation both play oversight roles in organizations, their paradigms and approaches have significant differences (Naidoo, 2020). Evaluation perspective is broader: Evaluators have the mandate to look beyond the internal intervention logic, to see how the intervention is situated in the broader context and whether it is actually making a difference to the problem it was designed to address. Evaluation may thus question the original logic and design of the intervention in light of evidence of its perfor-

mance and impact. Or so it should. But this is not always the case: Evaluators and those who commission evaluations often are not interested in challenging the fundamental assumptions on which their programs are based.

Also important to bear in mind is that plenty of evaluation takes place outside of the profession, although it may not be recognized as such. For example, many ecologists are very much concerned with the effectiveness of conservation strategies and conduct thorough studies of how to best protect ecosystems and animal and plant species in situ (e.g., Geldmann et al., 2019). This is evaluative research, although contact is often minimal between those who conduct such studies and professional evaluators, who tend to mostly be social scientists by training. Enhancing this interaction is important because both sides would benefit greatly from cross-fertilization in terms of approaches and methodologies. While evaluators ignore the natural sciences at their peril, conservationists often lack knowledge in the social sciences (Bennett et al., 2016).

Evaluation has also seen a strong trend toward professionalization of the field. The DAC has developed a set of evaluation criteria to standardize the practice among donor organizations. These influential and widely used criteria were updated in 2019 to incorporate coherence as a new criterion (OECD DAC, 2019). Professional associations, such as the International Organization on Cooperation in Evaluation (IOCE),[2] the American Evaluation Association,[3] Canadian Evaluation Society (CES),[4] and European Evaluation Society[5] have moved this agenda, sometimes establishing credentialization programs as in the case of the CES. In the field of international development, the International Development Evaluation Association (IDEAS)[6] and the EvalPartners[7] have worked to bring eval-

---

[2] https://www.ioce.net/
[3] https://www.eval.org/
[4] https://evaluationcanada.ca/
[5] https://europeanevaluation.org/
[6] https://ideas-global.org/
[7] EvalPartners (https://www.evalpartners.org/), a partnership between IOCE, UN organizations, civil society orga-

uation into the mainstream of development agendas. The United Nations Evaluation Group (UNEG)[8] and the Evaluation Cooperation Group (ECG)[9] of the international financial institutions work actively to professionalize and harmonize evaluation practice among their member organizations.

Among development organizations, evaluation capacity development is seen as a priority and is carried out through structures such as the Centers for Learning on Evaluation and Results (CLEAR) and the International Program for Development Evaluation Training (IPDET)—both under the umbrella of the new Global Evaluation Initiative (GEI)[10] established by the Independent Evaluation Group (IEG) of the World Bank Group and the United Nations Development Programme (UNDP). Focusing on national evaluation capacity in the developing countries has been a high priority for the UNDP for well over a decade. Led by the organization's Independent Evaluation Office (IEO), biannual conferences on the topic have grown significantly in scope and influence since their beginning in 2009.[11]

This is all very welcome but comes with certain risks, not least of creating an exclusive guild of evaluators closing out new or heretic ideas. Favored methods also have inevitably led to paradigm contests between the different schools of thought. Most notably, claims to a scientific method by those advocating for randomized controlled trials and other experimental techniques as a "gold standard" have drawn the derision of

others who see the "randomistas" as taking a mechanistic view of complex development problems that is culturally and socially insensitive and lacking of external validity (e.g., Bickman & Reich, 2009).

Somewhat belatedly, the randomistas—Abhijit Banerjee, Esther Duflo, and Michael Kremer—were awarded the Nobel Prize for Economics in 2019 (e.g., Banerjee & Duflo, 2011). It seems, however, that the tide is turning toward a more inclusive and comprehensive set of approaches. Experimental and quasi-experimental methods should remain in the bricolage of a wide range of tools used by evaluators (Patton, 2020a).

A desire to assign accountability through quantitative attribution of results to a specific intervention is natural. Although such attribution is appealing to many, especially intervention proponents and donors, it is particularly elusive in a complex environment. Especially when we move away from narrowly focused, targeted interventions toward more transformational efforts, credibly demonstrating the contribution of the intervention to the larger goal should suffice. Here, a well thought-through theory of change is helpful (Mayne, 2019). As Andrew Natsios, the former administrator of the American international cooperation agency, USAID, stated, already more than a decade ago: "Those development programs that are most precisely and easily measured are the least transformational, and those programs that are most transformational are the least measurable" (Natsios, 2010).

Another important attribute that distinguishes evaluation from monitoring and performance audit is the focus on learning and the ability to draw wider lessons from factors that have enabled or hampered interventions in making desired contributions.

As mentioned above, evaluation as a profession and practice is firmly anchored in social science traditions. The practice also tends to remain focused on the achievement of predetermined intervention objectives, rather than expanding its focus on the broader context in which interventions take place and their interactions. In particular, environmental aspects of interventions—including

---

nizations, and voluntary organizations of professional evaluation (VOPEs), strengthens the capacity of the VOPEs and influences policy making through promoting evaluation.

[8] UNEG (http://unevaluation.org/) brings together all units in the UN system that have evaluation as their main function, currently numbering almost 50.

[9] ECG (https://www.ecgnet.org/) has as its members 10 multilateral development banks and aims to harmonize their evaluation approaches.

[10] https://www.globalevaluationinitiative.org/

[11] The NEC Information Center (https://nec.undp.org/) functions as an online platform for knowledge on the NEC conferences, publications, and other documents and tools.

their unintended consequences—are mostly missing from evaluations. This has been confirmed in no uncertain terms by recent stocktakings of evaluation policies and practice by UNEG and CES among their respective memberships. The UNEG stocktaking found that while most member agencies consider the environment to be of medium- to high-level interest to them—and almost 60% of the agencies have environmental and social safeguards that they need to adhere to in preparing their projects and programs—environmental concerns are seldom reflected in evaluations (UNEG Working Group on Integrating Environmental and Social Impact into Evaluations, 2020). In fact, according to survey results, 84% of respondents from the UN agency evaluation units think that environmental considerations have not been adequately addressed in their evaluation guidance (the corresponding figure for social considerations was high, too, at 68%; UNEG Working Group, 2020).

The above discussion points to some persistent challenges that evaluation faces. As I outline below, the global landscape is rapidly changing and the demands for development that is environmentally sound and socially just are getting more urgent by the day. To maintain its relevance, evaluation can no longer be satisfied with ex-post assessments of whether interventions achieved what was written in their program documents. Now evaluations must include a more future-oriented, prospective dimension that provides guidance based on lessons for transformational change. Evaluation must move beyond individual interventions to systems thinking. It must embrace both social and natural sciences using the full range of appropriate approaches, methods, and data sources available.

The word *evaluation* contains the notion of value and, as evaluators, we should be clear about our values, which include respect for nature and people in an inclusive and just manner. This doesn't mean that we should abandon objectivity in our analyses, but rather that we should provide evidence-based, objective analysis of how to most effectively contribute toward development that encompasses the values that we share. In the words of Andy Rowe (2019), we must move toward sustainability-ready evaluation.

The Sustainability Context

The international development context has changed rapidly in the 2000s. Most countries of the world have signed on to the 2030 Agenda for Sustainable Development and the attendant Sustainable Development Goals (SDGs),[12] and to the Paris Agreement on Climate Change. These frameworks provide a common understanding of the universal priorities and the sense of urgency of transforming the way our societies operate. They call for a new value system that is not based only on measuring economic growth, but emphasizes sustainability and equality. They also recognize the existential threats that humankind faces due to anthropogenic climate change and environmental degradation. We have entered the Anthropocene, a new geological era in which humanity's impact on the planet overwhelms everything else.

At the same time, the dichotomy between industrialized and developing countries is blurring, in particular with the entry of large, middle-income countries like China and India on the world scene. China is poised to overtake the United States as the world's largest economy within a couple of decades. China has also been most successful in eradicating extreme poverty and lifting the living standards of millions of people. Still, according to the World Bank (Lakner et al., 2020), 689 million people—or 9.2% of the world's population—lived in extreme poverty in 2017 (using the international poverty line of less than $1.90 per day). The World Bank estimates that this number has increased by a further 88 million people (possibly going up to 115 million people) in 2020 as a result of the COVID-19 pandemic (Lakner et al., 2020). While the economic differences between countries have narrowed, inequalities within countries have grown and in significant portions of the world, fragility and conflict are increasing. At the same time, the role of non-state actors, including the private sector and civil society, in international development has also increased.

---

[12] https://sdgs.un.org/2030agenda

The 2030 Agenda recognizes that sustainable development depends equally on three interlinked pillars: social, economic, and environmental. All 17 SDGs incorporate each of these dimensions to a varying degree. If one of the dimensions fails, the goal is not achievable. However, traditional measurements of development rely almost exclusively on economic metrics and, to a lesser degree, on social indicators, while the environmental dimension is at best an afterthought. This must change and evaluation must play a role in the new thinking. Fortunately, we see some promising indications that change is on its way. The latest Human Development Report by UNDP (2020), a leading development organization, focuses on human development and the Anthropocene. The report recognizes the interlinkages between human development and the environmental challenges we face, calling for exploring new, bold paths of expanding human freedoms and easing planetary pressures. It goes on to state:

> In the face of complexity, progress must take on an adaptive learning-by-doing quality, fueled by broad innovations, anchored in deliberative shared decision making and buttressed by appropriate mixes of carrots and sticks. (UNDP, 2020, p. 5)

Climate change is now widely recognized as a defining issue of our time. The Intergovernmental Panel on Climate Change (IPCC, 2018) warns that we have about a decade to limit global warming to 1.5°C above preindustrial levels or face severe, irreversible consequences for both the people and the planet. These consequences will include increasing occurrence of extreme weather events and rising sea levels that will inundate large swaths of coastal area where the majority of major cities are located. Climate change is not something that will happen sometime in the future; its impacts are already felt around the world. According to the National Oceanographic and Atmospheric Administration of the United States (NOAA, 2021), 10 of the hottest years on record have been since 2005; the top three being 2016, 2020, and 2019. The unprecedented wildfires experienced by Australia in 2019 and the U.S. West Coast in 2020 are linked to this warming trend.

Unfortunately, as problematic as climate change is, it is not the only environmental threat we face. In essence, there are three simultaneous and interlinked crises: the climate crisis, the nature crisis, and the pollution and waste crisis.

Research by the Stockholm Resilience Center found that, specifically in areas of biosphere integrity (loss of genetic diversity) and biochemical flows (nitrogen and phosphorus pollution), we have already breached the planetary boundaries with high risk for humans (Steffen et al., 2015). The nitrogen and phosphorus that have entered the biosphere come mostly from fertilizers used for food production to feed the still increasing human population and its growing appetite.

One of the greatest challenges is the loss of habitat, ecosystem integrity, and biological diversity. We are currently facing the most rapid loss of biological species in history, earning the moniker "the sixth extinction." Research of 177 species of mammals has documented that all have lost 30% or more of their geographical range and 40% or more have experienced severe population declines (Caballos et al., 2017).

These crises are closely interlinked in that their drivers all reside in human activity. Food production is one of the main causes of environmental destruction today, as land is cleared for agriculture and cattle raising. We lose some 12 million hectares[13] of tropical forest each year primarily due to land conversion for agriculture and other economic activities. Not only does this destroy ecosystems and animal and plant species therein, it also reduces the ability of the forests to sequester carbon, thus exacerbating climate change. Other major drivers of land conversion, habitat destruction, biodiversity extinction, and climate change include urbanization and the spread of human habitat.

These are the same factors that lie behind the coronavirus pandemic that in its first year, 2020, killed almost 2 million people and devastated

---

[13] According to Global Forest Watch/University of Maryland data, the tropics lost 11.9 million hectares of forest cover in 2019 (https://www.globalforestwatch.org/).

economies around the world. The virus causing COVID-19 is zoonotic, meaning it originated in nonhuman animals before spilling over to humans. This is not the first such pandemic. Earlier examples from recent history include SARS, MERS, H1N1, Zika, Ebola, and HIV. In fact, the risk of zoonotic pandemics has constantly increased as humankind encroaches deeper into the natural world for habitation, food production, mining, transportation and other activities, thus bringing us closer to animals that act as reservoirs of viruses (UNEP, 2016; Vidal, 2020). The destruction of predators occurring as a consequence of habitat loss results in the increase of animals, such as rats and bats, that effectively transmit their viruses to humans.

The data make clear how closely related these crises are and how directly they affect humanity already today. That human health and ecosystem health are closely intertwined is obvious. Climate change continues unabated. Even if all countries lived up to their nationally determined commitments under the Paris Climate Agreement—which they mostly don't—these would be not adequate to stop global warming within the limits defined in the Agreement. With the continued warming come multiple hazards ranging from sea-level rise and weather anomalies to the spread of disease-causing mosquitoes and other vectors.

Adaptation to climate change is necessary while we continue to work on mitigation measures (Global Commission on Adaptation, 2019). The impacts of the changing climate are far from uniform across the world. They are affecting the poorest countries and regions and the most vulnerable populations in the most severe fashion. All densely populated, low-lying coastal areas from Miami to Lagos and from the Netherlands to Bangladesh will have to deal with rising sea levels and increased coastal storms, but the poorer countries and cities will have far fewer resources to do so. Small island nations face existential threats due to climate change. Although the relationship is complex, climate change combined with other factors contributing to vulnerability appears to increase the likelihood of conflict (von Uexkull & Buhaug, 2021).

As has been demonstrably the case with the COVID-19 pandemic, the people most vulnerable to environmental and health hazards are the poorest and are often minorities, indigenous peoples, people of color, and women. The environmental crises thus have a very clear social justice dimension that cannot be overlooked as we devise strategies for sustainable development.

What It Means for Evaluation

With the increasing attention given to the universally applicable SDGs in both national and international development plans, and the proliferation of international agreements and financial mechanisms focusing on the environment, the need is growing to constantly assess the effectiveness and impacts of policies, strategies, programs, and projects that are aimed to produce transformational change for the environment and human wellbeing. We must identify lessons from the past—what has produced desirable results, under what conditions, and for whom—so that we can incorporate these lessons to design better and more effective interventions for the future. Evaluation has a key role to play in this critical function, but an imperative step is furthering the approaches and methodologies for evaluating at the nexus of environment and development. Evaluation must expand its vision to encompass the coupled human and natural systems and how they interact.

All this complexity has important implications for evaluation (Bamberger et al., 2015). Evaluation must be able to provide evidence of how actions in the development sphere affect the environment and vice versa. It must be able to demonstrate the close interlinkages between ecosystem health and human health in light of evidence from the real world. Evaluation must also be able to look increasingly to the future, toward new and emerging threats and challenges, and seek solutions to them. It must broaden its accountability focus to embrace learning more broadly.

Patton (2020a, pp. 189–190) identifies 20 ways in which evaluation must transform itself to evaluate transformation. It must rise above its project mentality and start looking beyond the internal logic of the interventions that are evaluated (Feinstein, 2019; Patton, 2020a). It must embrace a systems approach toward transformational change (Magro & van den Berg, 2019). With a systems perspective, the interventions evaluated—whether they be policies, strategies, programs, or projects—must be seen as part of a landscape in which they operate and interact with other interventions.

The DAC evaluation criteria—relevance, coherence, effectiveness, efficiency, impact, and sustainability—are widely accepted, understood, and used, which provides a strong incentive to maintain them (OECD DAC, 2019). However, they have some conceptual issues. First of all, the sustainability criterion refers to the continuation of benefits from an intervention and is silent on the environmental dimension of sustainability. Making this distinction in the era of sustainable development goals is essential. Patton (2020b) has suggested the term *adaptive sustainability* to encompass ecosystem resilience and adaptability in the nexus between humans and the environment. The Scientific and Technical Advisory Panel of the GEF suggests using the term *durability* to denote the continuation of benefits, while reserving sustainability for its environmental connotation (Bierbaum & Cowie, 2018). Similarly, what seems to be missing from the DAC criteria is a sense of urgency toward transformational change. In this respect, an additional criterion would need to be introduced, whether called *transformational fidelity* (Patton, 2020b) or transformative *significance* (Feinstein, 2019).

Furthermore, evaluation must systematically search for unintended consequences that may lie outside of the immediate scope of the evaluand. We must assume that everything we do in the sphere of economic development will have unanticipated effects. These are often to the natural environment because those designing programs or projects in sectors such as energy, industry, agriculture, or infrastructure seldom take fully into account the environmental consequences. Environmental impact assessments are rarely rigorous enough to capture all the possible effects

and may be ignored if they raise inconvenient issues against a planned project. Often, too, such unanticipated results occur in the social sphere and may negatively affect vulnerable groups. Many development projects, including some related to agriculture, forestry, and mining, take place on indigenous peoples' lands where tenure rights may be less well defined and whatever environmental and social impact assessment does occur almost routinely ignores the spiritual and cultural values of a place or a resource. Even a well-meaning project for climate adaptation can have highly differentiated impacts on different groups. For instance, what may be a good solution for a commercial farmer may not be available to a small subsistence farmer whose situation may be worsened by the intervention. Evaluation must be able to capture such nuances.

Many interventions take a long time to mature and the environmental impacts are slow to materialize. An evaluation of the GEF land degradation portfolio found that measurable environmental improvements on the ground only appeared 4.5–5.5 years after the projects closed (GEF IEO, 2018a). Given the time that project preparation and implementation require, this is a decade or more after the problem was identified and the project designed. During that decade, many things will have changed and the environmental and social problems may have been exacerbated. Confining evaluation only to its summative role at the end of on intervention is not possible. Rather, we must build evaluation into the process to provide timely feedback to adaptive management. Knowledge must be extracted from the interventions as they are implemented and fed back to action to improve the practice in real time (West et al., 2019).

One of the key requirements is for multiple mixed methods that can address the issues of sustainable development and contribute to transformational change. These may involve both quantitative and qualitative approaches. Use of remote sensing and other big data show particular advantages in evaluating the environment and natural resources management (Lech et al., 2018). They must be complemented by more traditional methods, including participatory approaches that clarify the particular situations and concerns of local people and disadvantaged groups.

About This Book

The overall theme of the book is transformational change and how evaluation can contribute to it. Transformational change has been defined in numerous ways. For instance, UNDP (2011, p. 9) has defined it as "the process whereby positive development results are achieved and sustained over time by institutionalizing policies, programmes and projects within national strategies." We go beyond this definition, which focuses simply on positive development results from interventions. To be truly transformational, change must be of such scale and magnitude that it takes the object of transformation to a different place or level. Such change also must be lasting. The situation in which the world finds itself today—in terms of climate change, environmental unsustainability, and inequality—is such that mere gradual improvements will not be sufficient.

It is often said that in crises lie opportunities. Consequently, optimism still prevails that the pandemic and associated upheaval will lead to lasting societal changes. However, history teaches us that for a crisis to lead to transformation, a viable alternative to the status quo must exist that a large segment of people can align behind (Berman, 2020). We also assume implicitly that a transformation would be toward something positive, although this might not necessarily be so.[14]

Here, we use the term *transformational change* to denote change toward a more sustainable, inclusive, resilient, and environmentally sound state. Transformation is defined as a shift from the current system to a substantively new and different one (O'Connell et al., 2016, p. 19). Such a system shift is needed for humans and our non-human relatives to continue enjoying life on the planet. Evaluation can and must play a role contributing to such transformational change if we are to achieve the SDGs (Feinstein, 2019). Evaluation can help us identify factors and conditions that guide us in designing initiatives that lead to deep, sustained, large-scale impact (GEF IEO, 2018b).

The book then moves on to consider the drivers of sustainability. For decades, environmental projects and programs have had less than optimal success and their durability has been limited because they have focused on treating symptoms without addressing the root causes.

Virtually all environmental problems have their causes in economic, political, and societal factors. Terrestrial biodiversity and habitat loss are driven by deforestation and land conversion for agricultural, urban, industrial, and transportation uses. Food production and habitat for the expanding human population are fundamental drivers. Oceans are stressed by overfishing and by pollution from land and ship-based sources. Climate change is driven by fossil fuels for transportation, heating, and industry, and by intensive agriculture and deforestation.

Common threads among the factors that in turn influence all of the above. These can usually be found in the spheres of policies and economic incentives that encourage unsustainable practices of production and consumption. Therefore, we need not only to address the direct drivers but also the indirect drivers of unsustainability. And we need high-level policy engagement to transform these systems. Evaluations, too, must raise their sights toward the drivers and what influences them. Another evaluation of GEF projects found that they were able to affect enduring change when they engaged with legal, policy, and regulatory change (GEF IEO, 2018c).

The third section of the book deals with evaluating climate change action from different angles. It is obvious that efforts to slow down climate change must accelerate if we aim to get anywhere near the goals set in the Paris Agreement. The current nationally determined emissions reduction goals are nowhere near sufficient to stop warming within the 2°C, let alone 1.5°C, by the end of the century, according to the latest synthe-

---

[14]For instance, the crisis of the Weimar Republic led to the rise of Nazism, which can be described as transformational as well as disastrous.

sis report released by the UN in February 2021, calling for an urgent increase in ambition (UN Framework Convention on Climate Change, 2021). To move into this direction, we need a variety of strategies to decarbonize the economy, including financial and technological tools. Determining the efficacy of both is also important. This book provides lessons from evaluations on how to proceed on these fronts.

Climate change impacts, however, are not something for the future. They are being felt now in terms of changing climatic patterns and increased weather anomalies, storms, droughts, and wildfires. Irrespective of the success of mitigation actions, they will continue to worsen for some time. Successful adaptation to climate change—reducing people's vulnerability and building the resilience of the socioecological systems—is an important priority where we must learn rapidly from experiences. The book reviews the state of the art in the still underdeveloped field of evaluating adaptation, thus contributing to this important emerging endeavor.

The final part of the book focuses on evaluation approaches that will contribute to the quest for transformational change for the people and the planet. Our goal is not to promote one particular approach or methodology as a gold standard. On the contrary, experience shows that we do need a wide range of approaches and methods to tackle the problems of sustainable development. It is important to select the tools carefully to suit the task at hand, to answer the questions that need to be answered. Whether we choose primarily quantitative or qualitative methods should depend on the questions we wish to answer, and on the availability of data and its quality. Too often, evaluators pursue perfection in their methods and end up either adjusting their questions to suit their methods or using an extraordinary effort to hone their data. Of course, we aim for as much rigor as possible, but it is important to bear in mind the utility of the evaluation. The purpose of evaluation is to contribute to finding solutions to pending problems and to improve the performance of ongoing and future interventions. Therefore, timeliness is essential and it is most often "better to be vaguely right than exactly

wrong" (Read, 1914, p. 351[15]). We can modify the saying to also apply to questions: It is better to ask the right questions even if we cannot give exact answers.

Most of the authors in this book participated in the Third International Conference on Evaluating Environment and Development in Prague, Czechia, in October 2019, where the foundations for this book were laid. The previous conferences, held in Alexandria, Egypt, in 2008, and Washington, D.C., in 2014, also led to the publication of books that presented the state of the art at that particular time in this rapidly evolving field (van den Berg & Feinstein, 2010; Uitto et al., 2017). Although the two first conferences centered around the emerging field of evaluating climate change actions, both in terms of mitigation and adaptation, this third conference broadened the scope to cover environmental and natural resource management programs more widely, with emphasis on transformational change toward global sustainability.

The authors cover a wide range of evaluation approaches and methodologies, ranging from quantitative, including geospatial, to more qualitative and mixed methods that can be usefully applied to evaluating environment and sustainable development policies, programs, and projects at the nexus of human and natural systems.

Opening up evaluators' perspectives to these approaches is a key goal. This also requires opening up the perspectives of those who commission and use evaluations. These include donor governments and agencies, international organizations, and NGOs. They have to understand the importance of looking beyond the confines of their narrowly defined intervention. This is not necessarily easy in the face of pressures to show the effectiveness and impacts of the work that each organization is doing or funding and resistance to accept responsibility for anything outside that scope. This is why a pure focus on accountability in evaluation is dangerous. In international development, we often seem to see more interest in

---

[15]This quote or some variation of it is often attributed to John Maynard Keynes but it first appeared in the 1898 book by Carveth Read.

demonstrating to donor country citizens what their tax money achieved than in assuring that the programs and projects actually led to durable benefits to the countries and people they were intended for. Placing evaluation more in the hands of developing country partners is important so that they can ensure that the development programs contribute positively to their priorities. Even more important is empowering local people to evaluate interventions to provide downward accountability toward the claimholders and to ensure that no one is left behind.

This novel thinking—including incorporating the environmental dimension and the coupled human-natural systems, moving beyond individual projects to systems thinking, and identifying unintended consequences—does not come naturally to many evaluators trained in specific intellectual traditions and social science techniques. Yet it is necessary. This book focuses on identifying new and successful approaches and methodologies to this end. The authors share lessons gleaned from evaluations conducted by major multilateral and bilateral development agencies and financial institutions, national organizations, research and academic institutions, and the private sector. New and promising approaches are demonstrated and discussed.

References

Bamberger, M., Vaessen, J., & Raimondo, E. (2015). *Dealing with complexity in development evaluation: A practical approach.* Sage.

Banerjee, A. V., & Duflo, E. (2011). *Poor economics: A radical rethinking of the way to fight global poverty.* Public Affairs.

Bennett, N. J., Roth, R., Klain, S. C., Chan, K. M. A., Clark, D., Cullman, G., Epstein, G., Nelson, M. P., Stedman, R., Teel, T. L., Thomas, R. E. W., Wyborn, C., Curran, D., Greenberg, A., Sandlos, J., & Veríssimo, D. (2016). Mainstreaming the social sciences in conservation. *Conservation Biology, 31*(1), 56–66. https://conbio. onlinelibrary.wiley.com/doi/10.1111/cobi.12788

Berman, S. (2020, July 4). *Crises only sometimes lead to change. Here's why.* Foreign Policy. https://foreignpolicy.com/2020/07/04/ coronavirus-crisis-turning-point-change/

Bickman, L., & Reich, S. M. (2009). Randomized controlled trials: A gold standard with feet of clay? In S. I. Donaldson, C. Christie, & M. M. Mark (Eds.), *What counts as credible evidence in applied research and evaluation practice?* (pp. 51–77). Sage.

Bierbaum, R., & Cowie, A. (2018). *Integration: To solve complex environmental problems.* Scientific and Technical Advisory Panel, Global Environment Facility. https:// www.stapgef.org/resources/advisory-documents/ integration-solve-complex-environmental-problems

Caballos, G., Ehrlich, P. R., & Dirzo, R. (2017). Biological annihilation via the ongoing sixth mass extinction signaled by vertebrate population losses and declines. *Proceedings of the National Academy of Sciences USA, 114*(30), E6089–E6096. https://doi.org/10.1073/ pnas.1704949114

Dasgupta, P. (2021). *The economics of biodiversity: The Dasgupta review.* HM Treasury. https://assets.publishing.service.gov.uk/government/uploads/system/ uploads/attachment_data/file/957291/Dasgupta_ Review_-_Full_Report.pdf

Feinstein, O. (2019). Dynamic evaluation for transformational change. In R. D. van den Berg, C. Magro, & S. Salinas Mulder (Eds.), *Evaluation for transformational change: Opportunities and challenges for the sustainable development goals* (pp. 17–31). International Development Evaluation Association.

Geldmann, J., Manica, A., Burgess, N. D., Coad, L., & Balmford, A. (2019). A global level assessment of the effectiveness of protected areas at resisting anthropogenic pressures. *Proceedings of the National Academy of Sciences, 116*(46), 23209–23215. https://www. pnas.org/content/116/46/23209

Global Commission on Adaptation. (2019). *Adapt now: A global call for leadership on climate resilience.* Global Center on Adaptation and World Resources Institute. https://gca.org/reports/adapt-now-a-global- call-for-leadership-on-climate-resilience/

Global Environment Facility. (2019). *The GEF evaluation policy.* http://www.gefieo.org/sites/default/files/ieo/ evaluations/files/gef-me-policy-2019_2.pdf

Global Environment Facility Independent Evaluation Office. (2018a). *Value for money analysis for GEF land degradation projects.* http://www.gefieo.org/evaluations/value-money-analysis-gef-land-degradation- projects-2016

Global Environment Facility Independent Evaluation Office. (2018b). *Evaluation of GEF support for transformational change.* http://www. gefieo.org/evaluations/evaluation-gef-support- transformational-change-2017

Global Environment Facility Independent Evaluation Office. (2018c). *Impact of GEF support on national environmental laws and policies in selected countries.* http://www.gefieo.org/evaluations/ impact-gef-support-national-environmental-laws-and- policies-selected-countries-2017

Intergovernmental Panel on Climate Change. (2018). *Global warming of 1.5°C.* https://www.ipcc.ch/sr15/

Lakner, C., Yonzan, N., Mahler, D. G., Aguilar, R. A. C., Wu, H., & Fleury, M. (2020, October 7). *Updated estimates of the impact of COVID-19 on global poverty: The effect of new data.* World Bank. https://blogs.

worldbank.org/opendata/updated-estimates-impact-covid-19-global-poverty-effect-new-data

Lech, M., Uitto, J. I., Harten, S., Batra, G., & Anand, A. (2018). Improving international development evaluation through geospatial data and analysis. *International Journal of Geospatial and Environmental Research, 5*(2), Article 3. https://dc.uwm.edu/ijger/vol5/iss2/3

Magro, C., & van den Berg, R. D. (2019). Systems evaluations for transformational change: Challenges and opportunities. In R. D. van den Berg, C. Magro, & S. Salinas Mulder (Eds.), *Evaluation for transformational change: Opportunities and challenges for the sustainable development goals* (pp. 131–155). International Development Evaluation Association.

Mayne, J. (2019). Revisiting contribution analysis. *Canadian Journal of Program Evaluation, 34*(2), 171–191. https://doi.org/10.3138/cjpe.68004

Morra Imas, L. G., & Rist, R. C. (2009). *The road to results: Designing and conducting effective development evaluations*. World Bank. https://openknowledge.worldbank.org/handle/10986/2699

Naidoo, I. (2020). Audit and evaluation: Working collaboratively to support accountability. *Evaluation, 26*(2), 177–189. https://doi.org/10.1177/1356389019889079

National Oceanic and Atmospheric Administration. (2021). *2020 was Earth's 2nd-hottest year, just behind 2016*. https://www.noaa.gov/news/2020-was-earth-s-2nd-hottest-year-just-behind-2016

Natsios, A. (2010). *The clash of the counter-bureaucracy and development*. Center for Global Development. https://ww.cgdev.org/content/publications/detail/1424271

O'Connell, D., Abel, N., Grigg, N., Maru, Y., Butler, J., Cowie, A., Stone-Jovicich, S., Walker, B., Wise, R., Ruhweza, A., Pearson, L., Ryan, P., & Stafford Smith, M. (2016). *Designing projects in a rapidly changing world: Guidelines for embedding resilience, adaptation and transformation into sustainable development projects (Version 1.0)*. Scientific and Technical Advisory Panel, Global Environment Facility. https://www.stapgef.org/resources/advisory-documents/rapta-guidelines

Organisation for Economic Co-operation and Development, Development Assistance Committee. (2010). *Glossary of key terms in evaluation and results based management*. https://www.oecd.org/dac/evaluation/2754804.pdf

Organisation for Economic Co-operation and Development, Development Assistance Committee. (2019). *Better criteria for better evaluation. Revised evaluation criteria: Definitions and principles for use*. OECD DAC Network on Development Evaluation. http://www.oecd.org/dac/evaluation/revised-evaluation-criteria-dec-2019.pdf

Patton, M. Q. (2020a). *Blue marble evaluation: Premises and principles*. Guilford.

Patton, M. Q. (2020b). Evaluation criteria for evaluating transformation: Implications for the coronavirus pandemic and the climate emergency. *American Journal of Evaluation*. https://doi.org/10.1177/1098214020933689

Read, C. (1914). *Logic: Deductive and inductive* (4th ed.). Simpkin, Marshall, Hamilton, Kent and Co. http://www.gutenberg.org/files/18440/18440-h/18440-h.htm

Rowe, A. (2019). Sustainability-ready evaluation: A call to action. *New Directions for Evaluation, 2019*(162), 29–48. https://doi.org/10.1002/ev.20365

Steffen, W., Richardson, K., Rockström, J., Cornell, S. E., Fetzer, I., Bennett, E. M., Biggs, R., Carpenter, S. R., de Vries, W., de Wit, C. A., Folke, C., Gerten, D., Heinke, J., Mace, G. M., Persson, L. M., Ramanathan, V., Reyers, B., & Sörlin, S. (2015). Planetary boundaries: Guiding human development on a changing planet. *Science, 347*(6223), 1259855. https://science.sciencemag.org/content/347/6223/1259855

Uitto, J. I., Puri, J., & van den Berg, R. D. (Eds.). (2017). *Evaluating climate change action for sustainable development*. Springer. https://www.springer.com/us/book/9783319437019

United Nations Development Programme. (2011). *Supporting transformational change*. https://www.undp.org/publications/supporting-transformational-change

United Nations Development Programme. (2020). *Human development report 2020: The next frontier: Human development and the Anthropocene*. http://hdr.undp.org/en/2020-report

United Nations Environment Programme. (2016). *Frontiers 2016: Emerging issues of environmental concern*. https://www.unenvironment.org/resources/frontiers-2016-emerging-issues-environmental-concern

United Nations Evaluation Group Working Group on Integrating Environmental and Social Impact into Evaluations. (2020). *Stock-taking exercise on policies and guidance of UN agencies in support of evaluation of social and environmental considerations* (Vol I Main Report). United Nations Evaluation Group. http://www.unevaluation.org/document/detail/2951

United Nations Framework Convention on Climate Change. (2021). *Nationally determined contributions under the Paris Agreement: Synthesis report by the secretariat*. https://unfccc.int/sites/default/files/resource/cma2021_02_adv_0.pdf

van den Berg, R. D., & Feinstein, O. (Eds.). (2010). *Evaluating climate change and development* (World Bank Series on Development) (Vol. 8). Routledge.

Vidal, J. (2020, March 18). *Destroyed habitat creates the perfect conditions for coronavirus to emerge*. Scientific American. https://www.scientificamerican.com/article/destroyed-habitat-creates-the-perfect-conditions-for-coronavirus-to-emerge/

von Uexkull, N., & Buhaug, H. (2021). Security implications of climate change: A decade of scientific progress. *Journal of Peace Research, 58*(1), 3–17. https://doi.org/10.1177/0022343320984210

West, S., van Kerkhoff, L., & Wagenaar, H. (2019). Beyond linking "knowledge and action": Towards a practice-based approach to transdisciplinary sustainability interventions. *Policy Studies, 40*(5), 534–555. https://doi.org/10.1080/01442872.2019.1618810

Part I
Transformational Change

Evaluation for Transformational Change: Learning from Practice

Indran A. Naidoo

Abstract

The COVID-19 crisis has challenged the evaluation profession by altering the framework within which it operates. Evaluators must embrace new realities and respond to changes while not altering their principles, norms, and standards.

A review of how evaluation networks and offices have responded to changing demands showed lack of recognition for the long-term implications to the profession. Commissioners and users of evaluation now have new priorities, and nontraditional actors have entered the traditional evaluation space, offering similar expertise and meeting the demands of evaluation commissioners and users. The extensive development challenges posed by COVID-19 require a comprehensive response capacity from evaluation if it is to be transformative as a profession.

This chapter draws on national and international case studies, examining the concept of transformation from a contextual perspective and noting the relativism in the concept. It draws links between aspects, suggesting that this period is an opportunity for evaluators to learn from practice around transformation, and suggests that flexibility provides an opportunity to remain relevant and advance transformational goals.

Evaluation Must Respond to Global Signals to Be Relevant

This chapter draws on the special session on evaluation for transformation at the 2019 International Development Evaluation Association (IDEAS) conference. Since then, major changes have occurred globally, changing concepts on development and redefining its assessment. The changes in the development discourse mean that multiple reprioritizations are taking place, with major impacts on the global financial, governance, accountability, and knowledge generation systems. Operating within these contexts—and within authorizing political and social contexts—evaluation practice cannot remain detached or static. The new era has triggered new demands for evaluative knowledge and products, now met by the research sector broadly, thereby diminishing the exclusivity that evaluators once held on evaluative knowledge and outputs. Unless evaluation can demonstrate a more compelling value proposition that moves beyond serving traditional oversight and accountability needs, it may lose its privileged position. Its continued relevance will be questioned.

I. A. Naidoo (✉)
Independent Office of Evaluation, International Fund for Agricultural Development, Rome, Italy
e-mail: i.naidoo@ifad.org

J. I. Uitto, G. Batra (eds.), *Transformational Change for People and the Planet*, Sustainable Development Goals Series, https://doi.org/10.1007/978-3-030-78853-7_2

Evaluation does not occur in a vacuum but responds to demands, imperatives, and contexts, which explains both its uneven evolution and resultant variation across the globe. We have no commonly shared perspective as to what evaluation is and should achieve; with various camps of evaluation practice justifying their positions, we lack evaluation consensus and identity. Demonstrating context's effect on research demand is the COVID-19 pandemic and United Nations (UN) response, which has taken the form of supporting countries' production of socioeconomic assistance and recovery plans to ensure the most effective development interventions. Although reporting and review are key parts of the plans, they include little mention of evaluation. Such plans do not draw sufficiently from evaluative work at the country or global level. This may signal a marginalization of evaluation in favor of other forms of research during this period that requires more real-time monitoring information to support recovery than detailed and often late evaluation studies. In the era of big data, artificial intelligence, and other forms of data generation and extraction, evaluators are often not engaging with new realities.

Redefinition in the COVID-19 Crisis: Evaluators Are Not Isolated from Changes

With COVID-19 declared a global crisis in April 2020 by the Secretary-General of the United Nations (Guterres, 2020), the UN has responded to support the recovery of countries. One element of the UN response is research and analytical support to help decision making toward recovery. It has resulted in the generation of assessments on the state of development of countries around the world, with the aim of better understanding the impact of the crisis. Evaluation was affirmed for its role in guiding progress toward the UN 2030 Agenda for Sustainable Development's Sustainable Development Goals (SDGs; Mohammed, 2019). The context now, as indicated by Barbier and Burgess (2020), is one of declining resources and will require targeted

interventions to mitigate the impact of the crisis and ensure that lives and livelihoods are protected, to help rebuild in a better, more sustainable manner. Barbier and Burgess specifically highlighted that some SDGs will be sacrificed during this period while focus is on SGDs to curb the spread of the virus and tackle the immediate economic fallout. It means that addressing the SDGs with equal priority will probably not occur, even if they are referenced as important.

All of the joint UN government plans reference the SDGs, which have served to date as milestones and targets for achieving Agenda 2030. The evaluation community has been active in supporting the SDGs through providing evaluation capacity to countries, a significant contribution. This is attested to in the proceedings of the National Evaluation Capacities Conference 2019 (United Nations Development Programme, Independent Evaluation Office [UNDP IEO], 2020), where countries presented case studies of their success in using evaluation for SDG attainment. In this context and until the 2020 pandemic, the form of evaluation considered valuable was that which built measurement capacity as a basis for advancing transformation. The emphasis was for evaluation to be people-centered, shown in both the Prague Declaration and, adopted in Egypt, the Hurghada Principles (UNDP IEO, 2020), which also called for a focus on people and collectivism. The UN principle of leaving no one behind, particularly focusing on vulnerable groups such as women, has emphasized the impact of the COVID-19 crisis on the poor. The proposed UN response, apart from acknowledging a serious loss of developmental gains, is to "build back better" (United Nations, 2020). Implicit in the statement is a transformational intention. Patton (2020), in his examination of what needs to change for evaluation to be transformative, asked for an acknowledgment of the changes. Prior to the crisis era, Feinstein (2019) suggested that evaluation could be more dynamic and support transformational change. In the current era, the potential role of evaluation in generating knowledge and creating processes for a more sustainable society, and the new order will be quite different (Schwandt, 2019). Therefore, evaluation that claims to be transformative must address these demands.

Challenges to Evaluation as a Practice and Form of Transformation

Interest in evaluation has increased, reflecting on its identity as a practice with the associated research stipulations and adherence, and questioning whether this practice, if pursued in a particular manner, can be transformative. These debates will continue and Feinstein (2017) argued that evaluation will also influence knowledge management, which is critical in the era of big data, artificial intelligence, and machine learning. Classic evaluation practices and associated norms are challenged as new research formations and data producers could influence the existing oversight architecture at the country and global levels affecting evaluation demand. Efficiency considerations will be important and the old professional boundaries between audit and evaluation may not be viewed as efficient and effective (Naidoo & Soares, 2020).

The Exploratory Nature of This Chapter

This chapter draws on personal experiences to reflect on the evaluation–transformation question from an evaluation leadership perspective. The two case studies demonstrate the relativism around the concept of transformation. In Patton's (2020) discussion on Blue Marble Evaluation, he illustrated how the COVID-19 crisis has challenged the notion of the nation-state, and shown that the globe, instead, is highly interconnected with porous borders.

This may challenge evaluators who have traditionally worked within confined boundaries, departments, agencies, and country-level programs but seldom on a macro- and cross-cutting level, where issues of complexity and its multiple influences come in. Moving from units of analysis that are small and perceived as static toward addressing larger units of analysis with interconnections and influences is difficult and will be challenging in contexts of working with partners and big data, at scales larger than most evaluators deal with. This factor alone would challenge the prevalence of classic accountability evaluation, with its bias toward linear thinking and measurement. In this period, other actors will challenge the exclusive domain of judgment that evaluators have held. Very different requirements are now in place for governance in the accountability or fidelity era, as argued in Schwandt's (2019) discussion on the post-normal era. Inevitably, as resources shift, so too will governance priorities, and this will affect evaluation, irrespective of its type.

Changes to the Evaluation–Transformation Relationships over Time

My work in the field of evaluation over the last 25 years has highlighted its relationship to power and its potential and constraints to be transformational. The broad definition of positive change is making advancements toward better quality standards, and, in the process, improving transparency and accountability. This is universally applicable and, in these contexts, prioritizes elements of fidelity to assure funders and citizens that the organization performs as expected. In this context, independence is an important component for accountability obligations (Schwandt, 2019).

Evaluation by its nature is judgmental and therefore triggers a set of managerial reactions that may not always resonate with the intention of evaluators. Evaluators privilege science and assume rationality in decision-making (Schwandt, 2005). Evaluation works on the assumption that evidence is central in decision-making contexts. The independent type of evaluation generates tensions in organizations given its profile and authority but is largely accepted as a part of organizational practice. The growth in the profession, even at the level of national authorities, has produced policies that show an understanding of the relationship between independence, credibility, and utility.

The Crisis Context and Potential Loss of Judgement Proprietorship

I caution that the word "transformation" has become cliché and used broadly to describe all changes, even those that would naturally evolve over time. Evaluation is often viewed as inherently progressive and transformative, but in practice, unless there are actions to drive this goal, it tends to fail.

In the context of COVID-19, evaluative sources (research think tanks, commissioned reviews, surveys, and social media streams) also feed directly to decision makers. Monitoring information is valued due to its timely delivery, and evaluators may lose their singular propriety to performance information. Further, they may lose their direct access to decision makers as governance and accountability architecture changes. Evaluators' ability to directly access beneficiaries also will change, given the travel restrictions, and remotely generated evaluations will not carry the same level of authority as those generated from full engagement. The question is, what is the value proposition that evaluators bring into this new context? A further issue is whether the classic accountability framework for evaluation will remain dominant in the era of big data. This needs exploration.

Judging Transformation, the Challenge of Relativism

The backdrop against which an intervention takes place is important, as is whether evaluation is a practice to promote democracy. Many perspectives currently exist as to what transformative evaluation is but there is no consensus. Some scholars and practitioners identify transformation if the subject matter is inherently transformational, such as land reform or addressing discrimination. These are context specific and generally imply a form of redress, which evaluation measures and reports on. The transformative subject evaluations fall into those that promote democracy, as they generate public dialogue on performance for accountability purposes. One of the case studies I describe in this chapter relates to apartheid South Africa. In this context, evaluation was regarded as inherently transformative simply because it gave access to previously unavailable information. This may appear modest, but in such a context, against a backdrop of repression and state control, it was significant.

Context Ascribes Value and Meaning to the Concepts of Transformation

This discussion seeks to illustrate how context may attribute a higher value to change. The concept of positive transformation can be relative to how it brings about changes and is valued as such in these historical junctures. This is a value judgment and projected by a part of the evaluation community, reflecting both its diversity and differences in global growth (Naidoo, 2011, 2012).

This chapter also draws on examples from within the UN situation and highlights how aspects such as evaluation approach, methodology, and increasing evaluation outputs supported transformation. The key shift was moving from using an outsourced and consultant-driven model to a professional cadre one, which helped affect changes in learning and accountability (Naidoo, 2019).

Given the broad scope within this umbrella of political topics, as the case studies illustrate, few global standards exist to judge definitively whether or not evaluations are transformative. This chapter takes the definitional view of a transformational practice as one that brings about more fundamental changes in the sense of being able to trigger and/or sustain major changes on all fronts and meeting societal and developmental aspirations   (attainment   of   SDGs). Transformational change could also include changes in professional identity and approach, showing evaluators as change agents, or redefining evaluation decision making and governance arrangements.

Changes in Evaluation Production and Emphasis

Evaluation, irrespective of what it is termed, is one of many streams of information to influence decision making (Rist & Stame, 2006). In the era of big data, artificial intelligence, and a more active research mode on the part of commissioners and receivers of evaluation (in the form of outputs of reports, briefings, etc.), evaluation is now part of a larger flow of information at decision makers' disposal. Evaluators need to make stronger arguments about their value proposition when new scenarios, such as those caused by COVID-19, affect their work. It is too early to speculate how this will manifest, at what intensity and form, and in which countries. The new evaluation construct will probably involve very different engagement permutations than those currently in place.

The fidelity role of evaluation currently prevails and although this form has progressed and received much attention (Schwandt, 2019), it is a particular type of evaluation that on its own may not meet all of the transformative criteria of dynamic evaluation as espoused by Feinstein (2020). The fidelity type of evaluation may provide assurance of program value but may not necessarily address issues that move beyond the organizational scope or provide a foresight pitch as called for by Patton (2019). In the COVID-19 context, the classic evaluation criteria would also require another look. Ofir (2020) and Patton (2020) suggested new elements that capture the dynamic nature of changes in the context of the global pandemic. In Picciotto's (2020) discussion on renewal of evaluation, he argues that the status quo cannot remain and evaluation must be able to produce changes that are more tangible.

When undertaking evaluation for purposes of accountability, the focus is on assessing results against plans. In contexts where evaluation is not independent and focused on supporting internal audiences, its plays a more facilitative, co-creative role toward a utility focus (Patton, 2018). In both these situations, the modus operandi would differ and the actual context would influence the extent to which evaluations may have transformational purposes. Results are challenged in both contexts, and often the self-assessment undertaken by program units tends to be more favorable on ratings than assessment provided by independent evaluation units. Greater organizational dialogue is necessary to help reconcile these differences, in the spirit that evaluation is a part of organizational learning using its independent principles to improve quality.

My personal reflections from various leadership functions also indicate that the passage of time can change one's views and those of people involved at a specific time as to whether the work was truly transformational or, more modestly, contributory. The reflective approach in answering these questions is an appropriate methodology given the nascent state of development of the field of transformative evaluation.

Case Studies on the Evaluation–Transformation Nexus

South Africa National Department of Land Affairs and Public Service Commission

My work as a senior manager at the National Department of Land Affairs and the Public Service Commission (PSC) of South Africa in the post-apartheid era from 1995–2011 provided an opportunity as an evaluation professional to expand on the concept of evaluation for transformation. This was a period when the profession was still evolving as the country began establishing its own professional evaluation association. The South African Monitoring and Evaluation Association (SAMEA) was launched in 2005 and set the pace for important growth of the profession.

In my work at the National Department of Land Affairs from 1995–2000, the very function of monitoring and evaluation (M&E) did not naturally resonate with the administration. The argument was that policy was not negotiable. When assessments of the policy showed weaknesses, it took a long time for the administration to make changes. This was frustrating for evaluators who did not see adequate attention paid to evaluation. The production of basic performance information was regarded as more important than policy analysis and impact assessments.

In the still censorious but changing climate, little qualitative analysis of program worth occurred, the very modality of land reform delivery was examined, and complicated and costly processes meant rapidly rising frustrations among beneficiaries (Naidoo, 1997). Considered decades later, results show that the program continues to struggle because central issues raised at the early stage of the program were not addressed. The key question: When undertaking M&E of purportedly transformational programs, does one take the policy as a given and just assess progress, or does one have the space to question the basis of the policy?

The work of the PSC was to oversee the performance of the public service in using its powers and normative tools and measures to effect transformation. It was expressly set up as part of the democratic constitution to ensure good governance (Naidoo, 2010). This was enshrined in Chapter 10 of the constitution, which sets out nine principles and values for public administration that the PSC was to advance "without fear, favor or prejudice" (Naidoo, 2004, 2010). Its focus was on improving the capacity of the "developmental state," which was viewed as the key driver for transforming government and the country from its unequal and racially divided past. It sought to advance equity and social and economic transformation as part of the democratic era. The annual work of the PSC culminated in a State of the Public Service report that meta-assessed the organization's work against the constitution's values and principles and its demonstrated progress or lack thereof. The PSC's was a national transformation project that generated information to bring about changes to the South African public sector. However, multiple sources of information indicate governance failures persist.

The sentiment as of 2020 is that the country has not met the expectations of the developmental state, and admissions from the ruling party regarding governance deficits at multiple levels show that one organization, albeit with formal authority, cannot on its own effect transformation. It can assist, but only if those in power act. Thus, evaluation has limits as to what it can do, and its support for transformation would be modest at best in the context described. However, accumulatively with other such directed initiatives, it can make a difference.

Gaining a perspective today on whether the work of the PSC delivered a better public sector over time, as politically promised, requires attention. The setting up of commissions of enquiry to investigate corruption indicates that the ideals of a developmental state, working in the interests of its citizens, was unsuccessful. Therefore, although the PSC sought to advance good governance through its oversight work, it was not able to effect transformational change in the country. The evaluation interventions may have initiated a type of thinking and discourse based on its products, but sustaining momentum was not possible given various political and administrative leadership changes. This has been the experience of many countries around the world that have built M&E capacities only to find them being marginalized based on political appetite for candid performance results.

The Independent Evaluation Office of the United Nations Development Program: Some Strategic Choices

This case study draws from the work that resulted in the transformation of the Evaluation Office of UNDP into the Independent Evaluation Office (IEO). In this case, evaluation successfully influenced country program design toward progressive developmental agendas. This resulted in several international presentations and a UN course drawing on the experience at the

International Program for Development Evaluation Training (IPDET).

This second case study reflects on my managerial and leadership experience as director of the Independent Evaluation Office of the UNDP over an 8-year period. In this context, evaluation approaches and interventions sought to advance the notion of transformation. The movement to greater evaluation coverage created a global audience for our work, thus expanding reflection on results and making the organization more evidence based. The promotion of evaluation conversations or dialogues is better than the classic, formalistic approach to evaluation, which can be transactional (working on reacting, responses, and rebuttals in formal settings). This mode does not build collective responsibility for results and may impede organizational learning. The revised approaches included advancing the support element of evaluation, with the National Evaluation Capacity (NEC) series becoming the largest global event by national authority participation. Where stronger national evaluation capacity exists across countries, the receptiveness to evaluation becomes better with an evaluation culture always supportive of advancing discussions on development transformation. Pitching evaluation as a support for the SDGs provided added legitimacy for evaluation, and brought evaluation directly into the discussions on development. It helped integrate the ideals of the UN and imperatives of the SDGs with the priorities, aspirations, and development plans of countries.

Learning from Both Managerial Roles

In both case studies, the thrust has been that evaluation serves a reform and transformation agenda, promoting transparency and generating dialogues across the tiers and levels of functions, bringing in voices and illustrating the discrepancy between intent and outputs and outcomes. Evaluation's very nature of challenging vision feeds more into the mission of the institutions, supporting assessment of how this varies among practices. In the case of the PSC, the notion of good governance was measured by assessing and reporting on service delivery. In the case of the UNDP, the agency interventions were assessed and reported to both governing boards and countries. The practice of reflection on results is transformative, and can advance the mandate of organizations within which independent evaluation occurs.

Ensuring evaluation coverage is impactful; in the case of the PSC, the organization spanned the entire public sector with more than 140 departments across nine provinces at the national and provincial tiers of government. The impact for potential transformation increases when evaluation engages a full breadth of governance indicators. Monitoring and evaluating the nine values and principles for public administration against performance indicators allowed for rating and comparison of performance and helped legislatures and parliament hold administrators to account. The success of these measures, however, is still dependent on political will and action on results, which has not been adequately evident from the overall performance of the public sector.

At UNDP, evaluation coverage was also a key feature for increasing the evaluation critical mass. The five-fold increase in evaluation coverage that resulted from a new policy and new approaches beginning in 2013 meant that all of the 130 country programs were assessed and the results presented for action. This increased the basis for meta-synthesis and created more timely opportunities for program revision. Program countries in particular appreciated the exercise of increasing coverage and diversifying products, which provided feedback they found important for their prioritization and decision-making. Other changes to ensure consistent quality assurance through a standing evaluation advisory panel helped create a critical mass and important dialogue to assist with advancing transformation development goals. By the time of my departure, the IEO had struck the sweet spot, with the administrator being a major advocate for the function and the board pleased with the outputs and volume of evidence of UNDP performance globally, helping to justify further funding. This can be considered transformational by pushing into the public domain a sizeable portfolio of work demonstrating the development challenges of countries and bringing attention to the SDGs, the vulnerabilities of people and disadvantaged groups, the constraints, (including inherited structural impediments), and value proposition of the organization.

Some Conclusions

Challenge on the Exclusivity of Judgment

In the context of 2021, evaluators face further challenges as they may potentially lose their exclusivity on judgement ability as action research and co-creation modalities gain prominence with artificial intelligence, machine learning, big data, and reliance on streams over studies. The problem is exacerbated by the fact that assessment is difficult in complex situations (Ofir, 2020) and evaluators will encounter challenges in working in a Blue Marble context of high interconnectivity and fewer boundaries (Patton, 2020).

Reflecting on Transformation Drivers

The experiential backdrop above has laid out some reflections. Evaluation is one part of a broader administrative and political system, and leadership within this plays a role in affirming or marginalizing the process. In the case of national departments, becoming entrenched can be difficult, especially when political leaders with their own imperatives govern the administrative divide between the heads of department or permanent secretary. Further, interpretation of what constitutes administrative and political success, or even transformation, is often at odds with evidence.

How M&E negotiates this difficult terrain—often pitched toward an authority level beyond the organizational head who may not appreciate the results—requires deft leadership skills. Evaluators in this context need to recognize that their work is but one of many streams of informa-

tion that decision makers receive, consider, and eventually prioritize. The matter of independence, credibility, and use was a theme of the 2013 NEC Conference that brought this matter to the fore from the experiences of government, showing evaluation's potential and its constraining parameters. The claim of being transformational in such a context is difficult. Evaluation transforming relationships was the theme of the 2011 NEC conference, which addressed evidence-based policymaking and generated multiple publications. From the evaluator perspective, these indicate the desire for causality between outputs and change and implicit acceptance that evaluation is transformational. In reality, transformation is more complex in practice and measuring such instrumentalism is a difficult task.

The Enabling Environment for Transformation

Apart from an enabling environment to assist with transformation, certain interventions provide the platform from which transformative actions can occur. Evaluators have a tendency to focus more on the output and product rather than the related journey toward this goal. Evaluation quality is as much about the process to arrive at the product as it is about the product itself. A conducive environment for enabling transformation through evaluation will include the following elements.

Political Will and Leadership Support

The signals from the political environment are critical to open the space for evaluative conversations. Policies on their own are insufficient and evaluation or accountability advocates are important. Countries such as South Africa, which has explicitly within its constitution bodies to advance democracy and accountability, have an advantage. However, the extent to which the political masters support such institutions in the form of who they appoint, how they fund the function, and whether they take findings seriously or not is critical. On its own, evaluation policies are insufficient to create the enabling environment, and aspects such as a free press, civil society participation, and activism to hold political and administrative leaders to account are needed to bolster the evaluation function.

Evaluation capacity, including dedicated evaluation units and systems that advance a results culture, is also important. In the international system, the evaluation offices of the UN, bilaterals, and international financial institutions (IFIs) that have dedicated evaluation functions can play a role in advancing the mandate of particular agencies through promoting a results culture. The work that these offices undertake jointly with other agencies and the promotion of national or sector-specific evaluation capacity building helps create space for reflecting on results, of government and of the agencies themselves. The cumulative effect of this is helpful for advancing transformation, especially of agencies that have a normative agenda, such as gender.

The Post-Normal or COVID-19 Era

All of the UN response and recovery plans have used evidence to improve research and oversight collaboration. However, the gap remains and few functional systems feed information back to governments, partners, and citizens on the results of the various policy interventions. This means that the plans remain largely aspirational, serving a purpose of resource mobilization, but may not provide evidence of impact given the lack of M&E systems. All of the plans purport to assist the poor and marginalized, address structural and other inequalities, and build a more environmental and economically sustainable future—in essence, be transformative—but evidence of transformation cannot be known without system-wide approaches to assess the changes and report in them independently.

In this chapter, I have suggested that much relativism is present in dealing with the concept of transformation. All evaluative activity is important, and one should use caution in privileging one form over another. This becomes more important in the context of action research, where many voices and streams of information inform a more democratic and broad-based decision-making architecture. Evaluators need to move beyond only understanding and applying methods; they must recognize context and its complexity in assessment, and work with rather than apart from key players. This will enable them to enter the debate and prove value in a rapidly shifting environment that requires comprehensiveness and ability to work across sectors in a multidimensional manner. This shift certainly means that evaluators need to establish how they can be most helpful, working in teams and coalitions that are multidisciplinary and cross-cutting and working alongside a range of technological applications that support access to beneficiary voices. The ability of the profession to navigate this challenging period will alter evaluator identity and the purpose of evaluation, depending on the success of the transition.

References

Barbier, E. B., & Burgess, J. C. (2020). Sustainability and development after COVID-19. *World Development, 135*. https://doi.org/10.1016/j.worlddev.2020.105082

Feinstein, O. (2017). Trends in development evaluation and implications for knowledge management. *Knowledge Management for Development Journal, 13*(1), 31–38.

Feinstein, O. (2019). Dynamic evaluation for transformational change. In R. D. van den Berg, C. Magro, & S. S. Mulder (Eds.), *Evaluation for transformational change: Opportunities and challenges for the sustainable development goals* (pp. 17–31). IDEAS.

Feinstein, O. (2020). Development and radical uncertainty. *Development in Practice, 30*(8), 1105–1113. https://doi.org/10.1080/09614524.2020.1763258

Guterres, A. (2020). A time to save the sick and rescue the planet. *New York Times*, April 28. https://www.nytimes.com/2020/04/28/opinion/coronavirus-climate-antonio-guterres.html

Mohammed, A. J. (2019, October 20–24). *Opening remarks by the United Nations deputy secretary general*. National Evaluation Capacity Conference, Hurghada, Egypt. http://web.undp.org/evaluation/nec/nec2019_proceedings.shtml

Naidoo, I. (1997). The M&E of the Department of Land Affairs, South African Government. In A. de Villiers & W. Critchley (Eds.), *Rural land reform issues in southern Africa* (pp. 11–14). University of the North Press.

Naidoo, I. (2004). The emergence and importance of monitoring and evaluation in the public service. *Public*

Service Commission News, November/December 8–11. http://www.psc.gov.za/documents/pubs/news-letter/2004/psc_news_book_nov_2004.pdf

Naidoo, I. (2010). M&E in South Africa. Many purposes, multiple systems. In M. Segone (Ed.), *From policies to results: Developing capacities for country monitoring and evaluation systems* (pp. 303–320). UNICEF. https://www.theoryofchange.org/wp-content/uploads/toco_library/pdf/2010_-_Segone_-_From_Policy_To_Results-UNICEF.pdf

Naidoo, I. (2011, September 12–14). *South Africa: The use question – Examples and lessons from the Public Service Commission* [conference session]. National Evaluation Capacities Conference, Johannesburg, South Africa. http://web.undp.org/evaluation/documents/directors-desk/papers-presentations/NEC_2011_south_africa_paper.pdf

Naidoo, I. (2012). Management challenges in M&E: Thoughts from South Africa. *Canadian Journal of Program Evaluation, 25*(3), 103–114. https://evaluationcanada.ca/secure/25-3-103.pdf

Naidoo, I. (2019). Audit and evaluation: Working collaboratively to support accountability. *Evaluation, 26*(2), 177–189. https://doi.org/10.1177/1356389019889079

Naidoo, I., & Soares, A. (2020). Lessons learned from the assessment of UNDPs institutional effectiveness jointly conducted by the IEO and OAI. In M. Barrados & J. Lonsdale (Eds.), *Crossover of audit and evaluation practice: Challenges and opportunities* (pp. 182–193). Routledge.

Ofir, Z. (2020). *Transforming evaluations and COVID-19* [4 blog posts]. https://zendaofir.com/transforming-evaluations-and-covid-19-part-4-accelerating-change-in-practice/

Patton, M. Q. (2018). *Principle-focused evaluation: The guide*. Sage.

Patton, M. Q. (2019). Expanding furthering foresight through evaluative thinking. *World Futures Review, 11*(4), 296–307. https://doi.org/10.1177/1946756719862116

Patton, M. Q. (2020, March 23). *Evaluation implications of the Coronavirus global health pandemic emergency* [blog post]. Blue Marble Evaluation. https://bluemarbleeval.org/latest/evaluation-implications-coronavirus-global-health-pandemic-emergency

Picciotto, R. (2020). From disenchantment to renewal. *Evaluation, 26*(1), 49–60. https://doi.org/10.1177/1356389019897696

Rist, R. C., & Stame, N. (2006). *From studies to streams: Managing evaluative systems*. Transaction.

Schwandt, T. (2005). The centrality of practice to evaluation. *American Journal of Evaluation, 26*(1), 95–105. https://doi.org/10.1177/1098214004273184

Schwandt, T. (2019). Post-normal evaluation? *Evaluation, 25*(3), 317–329. https://doi.org/10.1177/1356389019855501

United Nations. (2020). *COVID-19, inequalities and building back better*. https://www.un.org/development/desa/dspd/2020/10/covid-19-inequalities-and-building-back-better/

United Nations Development Programme, Independent Evaluation Office. (2019). *Annual report on evaluation 2019*. http://web.undp.org/evaluation/annual-report/are-2019.shtml

United Nations Development Programme, Independent Evaluation Office. (2020). *Proceedings of the National Evaluation Capacity Conference 2019*. http://web.undp.org/evaluation/nec/nec2019_proceedings.shtml

Transformational Change for Achieving Scale: Lessons for a Greener Recovery

Geeta Batra, Jeneen Garcia, and Kseniya Temnenko

Abstract

Achieving transformational changes that can be then effectively scaled up requires ambition in design, a supportive policy environment, sound project design and implementation, partnerships, and multistakeholder participation. This chapter presents a framework that can be applied at the design stage to plan for change and scaling up and provides relevant lessons based on GEF interventions. Achieving change and scale can be an iterative and a continuous process until impacts are generated at the magnitude and scope of the targeted scale. Successful transformations typically adopt a systems approach and address multiple constraints to attain environmental and other socioeconomic impacts.

Introduction

COVID-19 has transformed our lives in unfathomable ways—it has altered our behaviors, cities, and the environment. It has also affirmed the inextricable link between the broader ecosystem in which we live and human health.

Land mismanagement, habitat loss, overexploitation of wildlife, and human-induced climate change have created multiple pathways for pathogens to transmit from wildlife to domestic animals and humans, affecting our health and well-being.

A recent report from the Intergovernmental Science-Policy Platform on Biodiversity and Ecosystem Services (IPBES) concluded that future pandemics will emerge more often, spread more rapidly, kill more people, and impact the global economy more than COVID-19 unless there is a transformative change to address these infectious diseases (Daszak et al., 2020). In fact, the pandemic has made it clear that solutions and future avoidance will require a transformative, systems approach to reduce the global environmental changes caused by unsustainable consumption; these changes drive biodiversity loss; climate change; pollution of oceans, land, and air; and pandemic emergence (Global Environment Facility [GEF], 2020). But the news is not all negative. As pointed out by Professor Klaus Schwab of the World Economic Forum, "The pandemic represents a rare but narrow window of opportunity to reflect, reimagine, and reset our world" (2020, para. 15). It provides a chance to develop an ambitious approach to safeguarding environmental support systems through legal and regulatory instruments, policy measures, capacity building, technological innovations, and scaling-up and replication of

G. Batra (✉) · J. Garcia · K. Temnenko
Global Environment Facility Independent Evaluation Office, Washington, DC, USA
e-mail: gbatra@thegef.org; Jgarcia2@thegef.org; KTemnenko@thegef.org

© The Author(s) 2022
J. I. Uitto, G. Batra (eds.), *Transformational Change for People and the Planet*, Sustainable Development Goals Series, https://doi.org/10.1007/978-3-030-78853-7_3

demonstrated instruments. Monitoring, evaluation, knowledge, and learning activities can play a critical role in assessing progress against initiatives implemented, informing adaptive management, and demonstrating results on environmental outcomes and on socioeconomic benefits generated. Evaluation, while generating lessons for scaling up tested approaches based on prior evidence, is also responding to this call for a systems-based approach to understanding transformation (GEF IEO, 2018; Patton, 2020; Picciotto, 2009, 2020; Uitto, 2019; van den Berg et al., 2019; World Bank Group, Independent Evaluation Group [IEG], 2016).

Addressing the linkages among biodiversity loss, climate change, and emerging diseases is imperative to preventing future pandemics. Globally, there are few funds like the Global Environment Facility, which is positioned to catalyze the transformational change in biodiversity and other environmental areas to reverse the worrisome trends in the global environment. Established in 1992, the GEF is the principal financial mechanism for the Convention on Biological Diversity and an important financial mechanism for the United Nations Framework Convention on Climate Change, the Stockholm Convention on Persistent Organic Pollutants, the United Nations Convention to Combat Desertification, and the Minamata Convention on Mercury. Working through its 18 agencies, the GEF has provided close to $20 billion in grants and mobilized an additional $107 billion in cofinancing for more than 4,700 projects in 170 countries. The GEF also funds projects in international waters and sustainable forest management that support implementation of global and regional multilateral environmental agreements. Recently, the GEF has promoted multifocal and integrated interventions that interact with broader natural and human systems, with the objective of achieving deep, systemic, and sustainable change with large-scale impact.

Over its nearly 3 decades, the GEF has designed and implemented interventions that have proven to be "transformative," with some pilot initiatives that were subsequently scaled up to achieve results at larger scale. This chapter draws on two recent evaluations investigating transformational change and scaling-up, conducted by the GEF Independent Evaluation Office (IEO), which developed systematic approaches to understand the pathways to transformational change and provide relevant lessons based on GEF interventions.

A Framework for Transformational Change and Achieving Scale

The GEF IEO evaluation to explore GEF support for transformational change defined such change as: deep, systemic, and sustainable change with large-scale impact in an area of global environmental concern (GEF IEO, 2018).[1] The underlying theory of change is that by strategically selecting projects that address global environmental concerns and are designed to support fundamental changes in key systems or markets, the GEF engages in interventions that are more likely to lead to a sustainable, large-scale impact, assuming good project design and implementation and supportive contextual conditions. The theory of change is shown in Fig. 1.

For this evaluation, the IEO selected and screened completed GEF projects along the following criteria:

1. Relevance: The intervention addresses a global environmental challenge, such as climate change, biodiversity loss, or land degradation.
2. Depth of change: The intervention causes or supports a fundamental change in a system or market identified as a root cause of an environmental concern.
3. Scale of change: The intervention causes or supports a full-scale impact at the local, national, or multicountry level.
4. Sustainability: The impact of the intervention is financially, economically, environmentally, socially, and politically sustainable in the long term, after the intervention ends.

---

[1]Evaluation team members: Andres Liebenthal, Geeta Batra, Kseniya Temnenko, Katya Verkhovsky.

Relevance
- Climate Change
- Biodiversity
- Land Degradation
- Chemicals and Waste
- International Waters
- Sustainable Forest Management

Internal Factors
- Quality of implementation
- Quality of execution
- Pre-intervention analytical and advisory activities
- Partnerships with donors

Outcome
- Depth of change
- Scale of change

TRANSFORMATIONAL MECHANISM
A mechanism to expand and sustain the impact of the intervention (through mainstreaming, demonstration, replication, or catalytic effects)

Ambition level and focus
(of intervention objectives)
- Depth of change (market and system focus)
- Scale of change

Contextual Conditions
- Government ownership and support
- Implementation capacity
- Policy environment
- NGO and community participation
- Private sector participation
- Economic and market conditions

Sustainability
- Financial
- Economic
- Environmental
- Social
- Political

Fig. 1 Theory of Change for GEF Transformational Interventions

Applying qualitative comparative analysis, the evaluation identified drivers of change in deep dives into projects, providing useful lessons for the design and implementation of future interventions (see the Appendix for a table of the projects mentioned in this chapter).

Drivers of Change

Clear Ambition in Design

The interventions that achieved transformational change aimed to address fundamental market or systemic distortions as root causes of global environmental concern. The interventions that focused on market transformation targeted the supply and demand of goods and services associated with environmental impacts of global environmental concern. The cases that aimed at system-wide transformation took a comprehensive approach to modify the functioning of components (economy, public sector, private sector, community) whose collective interaction affect the environment.

Addressing Market and System Reforms Through Policies

The policy environment had an important impact on the depth and scale of reforms promoted by transformational interventions. All cases addressed market and system changes through policies. Six of the cases helped to strengthen and implement policies to trigger and sustain transformational change, while the two remaining cases leveraged the existing enabling policy frameworks to support transformational change.

Quality of Project Design and Implementation

All interventions that achieved transformational change were well implemented in terms of quality of project design and supervision by executing agencies. Some of the salient features across all projects were: comprehensive diagnostic assessments that identified key barriers; coherent designs to address all identified barriers; the early involvement of strong executing agencies that were ready to own the project objectives; and willingness on all sides to learn and adjust the design, scope, and management of the intervention as needed to ensure its success.

Mechanisms for Financial Sustainability

The transformational interventions established mechanisms for financial sustainability by leveraging market forces and stakeholders' economic interests or by integrating changes within government budgetary systems.

Transformation does not always require large investments. Although major, multiphase, large interventions can support transformational change, relatively modest medium-sized projects, with budgets under $2 million, that target main barriers and work with key stakeholders at the right time also can have a significant impact.

Scaling-Up

Scaling-up is one mechanism for achieving transformational change and one indicator that transformational change is likely to be achieved. We define scaling-up as an increase in the magnitude of global environmental benefits and/or expansion of geographical and sectoral areas covered by those benefits, such as within a specific market or system. The mechanisms for achieving transformational change and scaling-up happen through replication, mainstreaming, linking, or catalytic effects.

Replication refers to the implementation of the same intervention multiple times, thereby increasing the number of stakeholders and/or

covering larger areas, by leveraging finance, knowledge, and policy. That is, an intervention may be implemented across a wider area either through government or other funders investing more money for this purpose, through knowledge about the intervention motivating stakeholders to implement using their own resources, through a policy requiring or encouraging stakeholders to implement an intervention, or a combination of these. In the GEF context, countries typically use replication in connection with larger financing and technical assistance provided by the multilateral development banks to reproduce successful interventions on a larger scale.

Mainstreaming involves the integration of an intervention's implementation within an institution's regular operations, usually through a policy or legal framework. While mainstreaming typically happens within a specific national or local government agency, it may also occur simultaneously through multiple government sector agencies, or in other institutions such as donors, civil society organizations, and the private sector.

Linking involves the implementation of multiple types of interventions that, by design, all contribute to the same impact at the scale of a system defined by environmental, economic, or administrative boundaries. The system could be a landscape, seascape, ecoregion, a value chain, supply chain, or a national government. Within value and supply chains, linking takes place between interventions that address causes and effects; for example, through working both in countries where deforestation or wildlife poaching occurs, and countries where demand for the forest and wildlife resources is high. Linking could also involve different interventions under a common theme or transboundary issue, such as water pollution or fisheries. Linking allows for addressing multiple environmental areas in an integrated manner within a specific geographic or ecological unit.

Large-scale catalytic effects are often associated with technological improvements whose benefits can be captured by harnessing an effective market demand. The most notable examples of a catalytic effect involve the transformation of the market or system for renewable energy development. With other types of interventions—such as those focused on biodiversity protection and land conservation—the GEF IEO found that the projects' support for cutting-edge science and technologies appeared to have faced greater challenges in capturing and monetizing the related benefits and thereby can rely only partially on market-based approaches.

Factors Influencing Transformative Change and Scaling-Up

The two GEF IEO evaluations identified internal factors that enable the achievement of transformational results, including:

- Good quality of project implementation that covers the quality of project design and supervision, including a comprehensive diagnostic assessment to identify the barriers that need to be addressed to achieve the objectives of the project
- A careful project design that reflects a coherent logical framework of activities
- A strong implementation agency that is ready to own the objectives of the project and is willing to exert the leadership and acquire the capacity and resources necessary to ensure their achievement
- A willingness to learn, adjust, and adapt the design, scope, and management of the intervention as needed to ensure its success

We found that beneficial pre-intervention analytical activities and contextual conditions include:

- Capacity building
- Building partnerships with international donor partners, which enables projects to expand their scope and scale
- Strong government ownership of and support for the project
- Implementation capacity of local institutions, especially when the activities are spread over a range of sites and local jurisdictions

- Adequacy of the policy environment to create an enabling environment for depth and scale of reforms
- Civil society and local community participation
- Private sector participation; for projects in this evaluation, the impact of private enterprises on the effectiveness of the transformational interventions was mainly defined by the extent of their (supply-side) response to the changes created by the project
- Economic and market conditions

Example 1: Transformative and Effectively Scaled Up: Lighting Africa – Market-Based Solutions for Energy Access

About 580 million people in Africa have no access to grid electricity and rely on polluting and dangerous sources of lighting such as kerosene lamps, candles, and battery-powered torches. Fuel-based lighting is generally low quality and expensive, which impedes learning and economic productivity.

Modern electric lighting products—such as solar lamps—offer an opportunity for people living in off-grid areas to replace fuel-based lamps with higher quality, safer, cleaner, and more affordable lighting devices. Despite the benefits of solar lamps, the market was not developing as quickly as expected. The market appraisal that was funded by the GEF and the International Finance Corporation/World Bank identified six barriers that inhibited market growth:

1. Consumers did not trust the available solar products because many of them were poorly made and did not work properly.
2. Consumers did not know the benefits of solar lamps, how to use them, or where to buy them. Some consumers were unaware that solar lamps existed.
3. Manufacturers and designers did not know consumer preferences for the design and function of solar lamps.

4. Supply chain entities did not know each other. Solar lamp manufacturers entering the lower-income consumer market did not have an established distribution network.
5. Lack of finance was a big problem. Designers and manufacturers, distributors and importers, and retailers needed financing to purchase and move products to the end users. Lower-income consumers needed microloans to help with the upfront cost of purchasing a solar lamp.
6. Long customs processes and import tariffs on solar lamps were a common concern for manufacturers who considered importing solar lamps to African markets.

The Lighting Africa program was created to transform the off-grid market by removing these barriers. Its goal was to help catalyze markets for quality, affordable, clean, and safe off-grid lighting. The overall approach was to demonstrate the market viability by providing market intelligence; developing a quality assurance infrastructure; facilitating business-to-business interactions; helping governments address policy barriers; providing business development services; and facilitating access to finance for manufacturers, local distributors, and consumers. The program received about $22 million in contributions from 2007 to 2013. The GEF was the largest donor, providing more than one third of the funds (World Bank IEG, 2015).

In 2014, the final evaluation of the Lighting Africa program concluded that the program had played a crucial role in transforming the market (Castalia Strategic Advisors, 2014). The key accomplishments as of 2018 are shown in Table 1.

Key factors in Lighting Africa's transformational success included:

- The program operated in areas where there was proven, strong demand for improved off-grid lighting solutions.
- It was carefully designed to simultaneously address all major market barriers. Because barriers differ from market to market, the program started with a basic program design, but

Table 1 Lighting Africa Program Impact as of June 2018

Overall Impact	
32,280,2751	**People in Africa who are currently meeting their basic electricity needs** through off-grid solar products meeting Lighting Global Quality Standards
17,920,902	**Quality-verified solar lighting products sold** through local distributorships in Africa since 2009
1,792,090	**Metric tons of GHGs avoided in Africa in the past year;** the CO_2– equivalent of taking 383,745 cars off the road for a year
Access to Finance (as of July 2016)	
$20M	**Foreign exchange credit facility established by the Development Bank of Ethiopia** with World Bank funds to support import of qualifying products, including quality-verified solar lanterns
1,000,000	**Ethiopians gained access to modern energy services** through this credit facility
800,000	**Quality-verified products imported into Ethiopia** through this credit facility
11	**MFIs** (4 in Kenya, 5 in Ethiopia, 2 in Nigeria), and **KIVA** – the crowd-funding platform, providing consumers micro-loans for quality-verified, off-grid lighting and energy products
Market Intelligence (as of January 2018)	
30	**Market Insight reports** published, facilitating entry into new markets or mobilization of investors
4	**Market Trends reports** published, analyzing the off-grid products market across Africa, including the 2018 Global Off-Grid Solar Market Trends Report
Quality Standards (as of July 2016)	
GLOBAL	**Lighting Global Quality Standards adopted** as international standard for solar lighting products by the International Electrotechnical Commission as IEC Tech Spec 62257-9-5
255	**Solar lighting and energy products** tested against the Lighting Global Quality Standards to date
101	**Solar lighting and energy products (10W–100W) currently meet** the Lighting Global Quality Standards
Partnering with Governments (as of July 2016)	

(continued)

Table 1 (continued)

Overall Impact	
8	**Countries integrated** Lighting Africa activities into their World Bank-financed energy access projects: Burkina Faso, Mali, Liberia, DRC, Uganda, Ethiopia, Tanzania, and Rwanda.

Policy (as of July 2016)	
3	**National governments (Ethiopia, Kenya, and Tanzania) and ECOWAS** have or are in the process of adopting national standards for off-grid solar products that are harmonized with Lighting Global Quality Standards
1	Institution, the UN Framework Convention on Climate Change (UNFCCC) requires solar lighting products to meet IEC Technical Specification 62257-9-5 to qualify for carbon financing (CDM).

Source: Lighting Africa (2018)
Note: Lighting Africa has contributed toward these results through its market development activities implemented in collaboration with various intermediaries across the supply chain, development partners, financial institutions, and, most important, manufacturers of solar lighting products and their distribution partners in Africa.

it tailored the components to address the specific barriers identified in the target countries.

- The program focused on market transformation. Lighting Africa did not fund solar lamps—it funded activities that created effective markets in which consumers spent their own money to buy solar lamps.

Example 2: Review, Ownership, and Partnering: Payments for Ecosystem Services in the Danube Basin

According to the International Commission for the Protection of the Danube River, about 80% of the historical floodplains in the Danube basin has been lost over the last 150 years. Among the remaining 20%, the areas along the lower Danube between Bulgaria and Romania and in the Danube delta still possess a rich and unique biological diversity that has been lost in most other European river systems. The International Commission provides multiple ecosystem services, such as biodiversity conservation, recharging of ground water, water purification, pollution reduction, flood protection, and support for socioeconomic activities such as fisheries and tourism.

The Danube PES project was launched in 2009 with the objective of demonstrating and promoting Payment for Ecosystem Services (PES) and related financing programs in the Danube River basin and other international water basins (Varty, 2012). The project was a GEF medium-size project with total GEF funding of about $1 million, cofinancing of $1.2 million from the World Wildlife Fund, and in-kind contributions from partners including government agencies, NGOs, local authorities, and private companies.[2] The project design was focused at the national levels in Bulgaria and Romania, with some outreach activities in Ukraine, Serbia, and the wider Danube river basin. It also included local-level activities where pilot PES programs were to be tested and demonstrated.

Upon completion of the project in 2014, the terminal evaluation concluded that the project had been successful in eliciting the adoption of several national-level PES concepts into national fisheries policies in Romania and Bulgaria, and their testing and implementation in four pilot programs (Stefanova, 2014). Specifically:

- The project designed and introduced a pilot program for the sustainable management and harvesting of biomass (mainly reeds) in Bulgaria's Persina Nature Park, including full cost recovery from the sale of pellets and briquettes.
- Working with the Friends of the Rusenski Lom Nature Park in Bulgaria, the project developed and helped implement a program to generate funds for the protection and maintenance of the aesthetic value and biodiversity of the reserve from the sale of postcards and other promotional materials.

---

[2]The project was also supported by a GEF project preparation grant of $25,000.

- The project established a conservation and development fund for Romania's Maramures protected area by attracting sponsorships and donations for local guesthouses and tour operators interested in repositioning the area as an ecotourism destination.
- The project mobilized public funds for the implementation of policies for the maintenance of water quality and biodiversity values in the Ciocanesti area along the lower Danube in Romania. The resulting management practices had led to improved water quality and an observed increase in the number of nesting birds.

Based on the financial, institutional, and sociopolitical support elicited by the project, the evaluation report rated the sustainability of these achievements as moderately likely. Good prospects existed for future financial commitments to sustain the project, but many of these potential resources were still unsecured, especially for the long term. The transformation was modest in scale, focusing on specific target areas within a limited geographic range. As a result of this project, four PES programs in selected wetland areas were established along the lower Danube basin.

The main factors that contributed to the project's success were:

- A timely and effective midterm review found that the project had been too ambitious in relation to its budget and time frame. On this basis, the project followed a recommended streamlining of project objectives, a refocusing on priority areas, and reduction of less important activities.
- The decision to implement the project without direct government involvement allowed the project to proceed at a time when the relevant agencies were overwhelmed with other requirements. These agencies had been involved in the design and development of the project, and actively participated in capacity building and oversight activities, establishing adequate institutional ownership that boded

well for the continued adoption, replication, and scaling-up of the piloted approaches.
- The mix of project partners was effective and efficient, with each partner making important contributions toward different aspects. Although the project introduced a very new PES concept, the good collaboration between project partners, driven by their interest in the project, was instrumental in the successful delivery of outcomes.

Not every transformative project is scaled up. Three key actions are necessary for taking impact to scale: (a) adoption of the intervention by relevant stakeholders, (b) sustained support for scaling activities, and (c) learning for adaptability and cost-effectiveness. Figure 2 includes the factors and enabling conditions that influence these three actions.

Adoption of the Intervention

Relevant stakeholders must first be willing to implement the intervention that generates impact.

Factors that contributed to stakeholders' willingness to adopt an intervention clustered into two types: those that developed a sense of ownership for the intervention, and those that made the benefits of adopting the intervention clear and salient.

Stakeholder ownership has been identified in several IEO evaluations as a key contributing factor to progress toward impact. Having ownership implies that stakeholders find a program's objectives meaningful and useful to themselves personally. Buy-in to the intervention is attributed at least in part to participatory activities or mechanisms (Garcia, 2019), such as public consultations during project preparation, village committees, and community-based natural resource management agreements.

Stakeholders are motivated to adopt the intervention because they perceive the benefits of doing so. Benefits are defined as gains or avoided losses. Gains are usually noted in the form of

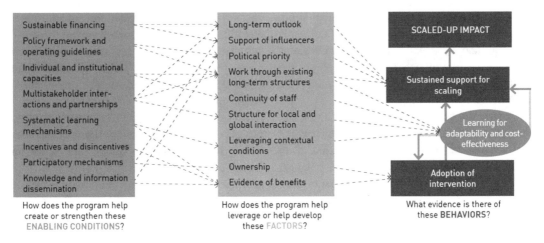

Fig. 2 Framework for Assessing the Likelihood of Scaling-Up. (Adapted from GEF IEO (2019))

higher income, cost savings, or new business opportunities; losses avoided are usually in the form of penalties, legal liabilities, or decreasing income due to a degraded natural resource base.

In some cases, adopting an intervention had the synergistic effect of both creating gains and avoiding losses. For example, in Macedonia, a cheaper alternative for PCB decontamination together with the risk of penalties for non-compliance created mutual reinforcement for private companies to decontaminate their equipment. Similarly, when farmers in China and Brazil switched to sustainable land management (Garcia, 2018), it resulted in both biodiversity protection and higher incomes, among other benefits (see project list in Appendix).

Pilot activities are sometimes not successfully scaled up because the gains are not sufficient to overcome the costs of changing the status quo. For example, a GEF project introduced the planting of buffer strips and pasture rehabilitation as part of managing nutrient pollution in the Danube River. The pilot was successful, yet did not scale in a subsequent project, in part due to state subsidies that left little incentive to include forestry activities in land management. Other components of the project that demonstrated benefits, such as reduced manure in waterways, were successfully scaled up and continue to expand without GEF support.

Sustained Support for Scaling-Up Processes

For the relevant stakeholders to implement the intervention that generates impact, supporting institutions must sustain the enabling conditions for implementation.

All successful cases of scaling-up received some form of support for longer than a typical 5-year project, mainly from their respective governments. This evaluation and other research have found that, in general, sustained support of between 10 and 20 years is necessary for scaling-up to take place.

Three factors emerged as important for ensuring long-term support for scaling-up processes: (a) scaling-up becoming a political priority, (b) gaining the support of political and economic influencers, and (c) working through existing long-term structures that depend on the appropriate choice of partner institutions.

Figure 2 highlights the necessary conditions for enabling scaling-up. First, knowledge and information dissemination, participatory processes, and incentives and disincentives are needed to motivate adoption of interventions. Second, strong institutional and individual capacities, policy framework and operating guidelines, and sustainable financing provide the resources for sustained implementation. Finally, multistakeholder interactions and partnerships and systematic learning mechanisms allow the scaling-up process to be adaptable and cost effective in the face of changing contextual conditions.

A program can support the establishment and strengthening of these enabling conditions, which can increase not just the implementation of an intervention by relevant stakeholders, but also support for scaling-up activities from institutions, in a positively reinforcing cycle. In Brazil, the GEF invested very early on in establishing FUNBIO (https://www.funbio.org.br/),[3] an organization that is now implementing scaling-up activities in the Amazon protected areas under a government mandate.

Learning for Adaptability and Cost-Effectiveness

For scaling-up processes to be sustained, supporting institutions have to learn from systematic feedback that will allow them to adapt the scaling-up process to changing contexts and make it more cost effective.

Project evaluations at midterm and closure have contributed to the scaling-up process. In many of the GEF projects we explored, learning from these evaluations typically led to reallocation of project funds. For example, in the case of Romania, the reallocation led to a shift from an expensive, concrete-based agricultural waste management platform to a cheaper and equally efficient plastic alternative, allowing more farmers to benefit from the funds. In China's termite control initiative, the learning resulted in a decision to use a more cost-effective form of integrated pest management. The

cost savings were reallocated toward additional technical training and public awareness-raising activities. Systematic learning mechanisms were usually in the form of knowledge exchange networks and regular multistakeholder meetings.

A few cases integrated adaptability into project design by allowing flexibility about which interventions to adopt and scale up based on actual contextual conditions. For example, throughout the implementation of the Rural Electrification and Renewable Energy Development project in Bangladesh, the project continuously incorporated lessons from its own pilot approaches, and—as the national demand for the solar home systems grew—the project shifted its focus to this component. The project also utilized monitoring and evaluation data from the field to incorporate new features such as LEDs to better serve lower income households, which in turn made the solar home systems more attractive to a larger population.

Conclusions

Achieving transformational changes that can be then effectively scaled up requires ambition in design, a supportive policy environment, sound project design and implementation, partnerships, and multistakeholder participation. This chapter presents a framework that can be applied at the design stage to plan for change and scaling-up and provides relevant lessons based on GEF interventions. Achieving change and scale can be an iterative and a continuous process until impacts are generated at the magnitude and scope of the targeted scale. Successful transformations typically adopt a systems approach and address multiple constraints to attain environmental and other socioeconomic impacts.

---

[3] This took place in 1991 through the Brazilian Biodiversity Fund project.

Appendix

Projects Discussed in This Chapter

Name	GEF ID	Implementing Agency	Country	Relevant Case Study
Rural Electrification and Renewable Energy Development	1209	The World Bank	Bangladesh	Bangladesh RERED
Rio de Janeiro Integrated Ecosystem Management in Production Landscapes of the North-Northwestern Fluminense	1544	The World Bank	Brazil	Brazil Rio Rural case
PRC-GEF Partnership: Land Degradation in Dryland Ecosystems: Project I-Capacity Building to Combat Land Degradation	956	Asian Development Bank	China	China IEM case
PRC-GEF Partnership: An IEM Approach to the Conservation of Biodiversity in Dryland Ecosystems - under the PRC-GEF Partnership on Land Degradation in Dryland Ecosystem Program	2369	International Fund for Agricultural Development	China	China IEM case
PRC-GEF Partnership: Forestry and Ecological Restoration in Three Northwest Provinces (formerly Silk Road Ecosystem Restoration Project)	3483	Asian Development Bank	China	China IEM case
PRC-GEF Partnership: Capacity and Management Support for Combating Land Degradation in Dryland Ecosystems	3484	Asian Development Bank	China	China IEM case
PRC-GEF Partnership: Sustainable Development in Poor Rural Areas	3608	The World Bank	China	China IEM case
PRC-GEF Partnership: Mainstreaming Biodiversity Protection within the Production Landscapes and Protected Areas of the Lake Aibi Basin	3611	The World Bank	China	China IEM case
Demonstration of Alternatives to Chlordane and Mirex in Termite Control	2359	The World Bank	China	China termite control initiative
Promoting Payments for Environmental Services (PES) and Related Sustainable Financing Schemes in the Danube Basin	2806	UNEP	Regional: Bulgaria, Romania	Danube PES Project
Lighting Africa/Lighting "The Bottom of the Pyramid"	2950	The World Bank	Regional: Ghana, Kenya	Lighting Africa
Enabling Activities to Facilitate Early Action on the Implementation of the Stockholm Convention on Persistent Organic Pollutants (POPs) in the Republic of Macedonia	1518	United Nations Industrial Development Organization	North Macedonia	Macedonia PCB case
Demonstration project for Phasing-out and Elimination of PCBs and PCB-Containing Equipment	2875	United Nations Industrial Development Organization	North Macedonia	Macedonia PCB case
DBSB: Agricultural Pollution Control Project - under WB-GEF Strategic Partnership for Nutrient Reduction in the Danube River and Black Sea	1159	The World Bank	Romania	Romania international waters case
DBSB: Integrated Nutrient Pollution Control Project-under the WB-GEF Investment Fund for Nutrient Reduction in the Danube River and Black Sea	2970	The World Bank	Romania	Romania international waters case
Brazilian Biodiversity Fund	126	The World Bank	Brazil	n/a

References

Castalia Strategic Advisors. (2014). *Evaluation of lighting Africa program – Final report*. Report to International Finance Corporation.

Daszak, P., das Neves, C., Amuasi, J., Hayman, D., Kuiken, T., Roche, B., Zambrana-Torrelio, C., Buss, P., Dundarova, H., Feferholtz, Y., Foldvari, G., Igbinosa, E., Junglen, S., Liu, Q., Suzan, G., Uhart, M., Wannous, C., Woolaston, K., Mosig Reidl, P., … Ngo, H. T. (2020). *Workshop report on biodiversity and pandemics of the Intergovernmental Platform on Biodiversity and Ecosystem Services*. IPBES Secretariat. https://doi.org/10.5281/zenodo.4147317

Garcia, J. (2018). *You win some, you lose some – Synergies and trade-offs of GEF support (Part 1)*. https://eartheval.org/blog/you-win-some-you-lose-some-%E2%80%93-synergies-and-trade-offs-gef-support-part-1

Garcia, J. (2019). *Want to scale up? Change-proof your program! Lessons from the GEF (Part 1)*. https://eartheval.org/blog/want-scale-change-proof-your-program-lessons-gef-part-1

Global Environment Facility. (2020). *White paper on a GEF COVID-19 response strategy: The complexities and imperatives of building back better.* https://www.thegef.org/sites/default/files/council-meeting-documents/EN_GEF_C.59_Inf.14_White%20Paper%20on%20a%20GEF%20COVID-19%20Response%20Strategy_.pdf

Global Environment Facility Independent Evaluation Office. (2018). *Evaluation of GEF support for transformational change*. https://www.gefieo.org/evaluations/evaluation-gef-support-transformational-change-2017

Global Environment Facility Independent Evaluation Office. (2019). *Evaluation of GEF support to scaling up impact 2019*. http://www.gefieo.org/evaluations/evaluation-gef-support-scaling-impact-2019

Lighting Africa. (2018). *Program impact as of June 2018*. https://www.lightingafrica.org/about/our-impact/

Patton, M. Q. (2020). Evaluation criteria for evaluating transformation: Implications for the coronavirus pandemic and the climate emergency. *American Journal of Evaluation*. https://doi.org/10.1177/1098214020933689

Picciotto, R. (2009). Development effectiveness: An evaluation perspective. In G. Mavrotas & M. McGillivray (Eds.), *Development aid: Studies in development economics and policy* (pp. 180–210). Palgrave Macmillan. https://doi.org/10.1057/9780230595163_8

Picciotto, R. (2020). From disenchantment to renewal. *Evaluation, 26*(1), 49–60. https://doi.org/10.1177/1356389019897696

Schwab, K. (2020, June 3). *Now is the time for a 'great reset'*. World Economic Forum. https://www.weforum.org/agenda/2020/06/now-is-the-time-for-a-great-reset/

Stefanova, V. (2014). *Terminal evaluation of the project Promoting payments for ecosystem services (PES) and related sustainable financing scheme in the Danube Basin*. United Nations Environment Programme, Evaluation Office. http://hdl.handle.net/20.500.11822/244

Uitto, J. I. (2019). Sustainable development evaluation: Understanding the nexus of natural and human systems. *New Directions for Evaluation, 2019*(162), 49–67. https://doi.org/10.1002/ev.20364

van den Berg, R. D., Magro, C., & Salinas Mulder, S. (Eds.). (2019). *Evaluation for transformational change: Opportunities and challenges for the Sustainable Development Goals*. International Development Evaluation Association. https://ideas-global.org/transformational-evaluation/

Varty, N. (2012). *Promoting payments for ecosystem services (PES) and related sustainable financing schemes in the Danube Basin. Mid-Term Review. Final Report*. United Nations Environment Programme, Evaluation Office.

World Bank Group, Independent Evaluation Group. (2015). *World Bank Group support to electricity access, FY2000-2014*. World Bank. https://ieg.worldbankgroup.org/sites/default/files/Data/reports/ElectricityAcces_v2.pdf

World Bank Group, Independent Evaluation Group. (2016). *Supporting transformational change for poverty reduction and shared prosperity: Lessons from the world bank experience*. World Bank. https://openknowledge.worldbank.org/handle/10986/24024

Part II
Drivers of Sustainability

Introduction

Neeraj Kumar Negi (✉) e-mail: nnegi1@thegef.org
Global Environment Facility Independent Evaluation Office, Washington, DC, USA

International development aid is generally provided through projects that are implemented with defined resources during a defined period. After project implementation is complete, the donor support for the project ends. Whether project benefits will continue, and whether the project will achieve its long-term objectives after implementation ends are important questions that interest donors and recipients of the support. Sustainability is an essential criterion to assess the extent to which these interventions deliver their long-term benefits. The chapters in this part discuss important aspects related to project sustainability such as observed sustainability after project completion and its drivers, concerns related to sustainability measurement, and importance of including a community-based perspective in project design and evaluation to adequately address sustainability.

The Organisation for Economic Co-operation and Development (OECD) defines sustainability as "the extent to which the net benefits of the intervention continue, or are likely to continue" (OECD/DAC Network on Development Evaluation, 2019, p. 12). When explaining its definition of sustainability, OECD notes that "depending on the timing of the evaluation, this may involve analysing the actual flow of net benefits or estimating the likelihood of net benefits continuing over the medium and long-term" (p. 12). International development organizations generally conduct project evaluations at or around implementation completion. Consequently, assessment of sustainability is mostly an estimation of likelihood of continuation of net benefits. Some of the chapters included in this part try to break this mold by bringing in evidence gathered after projects had been complete for 2 or more years.

Negi and Sohn, in "Sustainability After Project Completion: Evidence from the GEF," present findings of a desk review that covered postcompletion evaluations conducted by the Global Environment Facility Independent

Evaluation Office (GEF IEO) or independent evaluation offices of the agencies that implement GEF-funded projects. The review assessed sustainability of 62 completed projects for which results were field verified at least 2 years after implementation completion. They found that the benefit stream of most projects sustained during the postcompletion period. Although the sustainability outlook of some projects did decline during the postcompletion period, this was balanced by the instances where the outlook improved. Negi and Sohn found that multiple factors affect sustainability, including quality of project design, stakeholder buy-in, political support, financing for follow-up activities, institutional capacities, and continued support by executing partners.

In "Staying Small and Beautiful: Enhancing Sustainability in the Small Island Developing States," Batra and Norheim discuss the sustainability of 45 GEF-funded projects that were implemented in small island developing states (SIDS). Of these, results of 24 projects were field verified by the evaluators during the postcompletion period. Batra and Norheim found that project sustainability in SIDS is affected by policy and regulatory regime, country ownership, awareness, institutional capacity, and institutional partnerships. At the project level, they found that project design, adaptive management, arrangements for scaling-up and replication, and a sound exit strategy may enhance sustainability. Despite differences on specific issues, their findings are consistent with those of Negi and Sohn in the previous chapter.

Carugi and Viggh, in "From the Big Picture to Detailed Observation: The Case of the GEF IEO's Strategic Country Cluster Evaluations," present a methodological approach that they have found useful in assessing project results and sustainability. They report on their experience in assessing results and sustainability of GEF-funded projects implemented in least developed countries and two African biomes—the Sahel and the Sudan-Guinea Savanna. They argue that by proper sequencing of evaluation activities, much more may be learned about the results and sustainability of projects, and present lessons from their experience. In their work, they structured the evaluation to generate an aggregate analysis and broad trends at the front end of the evaluation, followed by field verifications for select project sites to facilitate targeted and in-depth enquiry. This approach, they found, is more useful for understanding the complexity of the evaluated interventions and the factors that are important at national and local levels.

Cekan and Legro, in "Can We Assume Sustained Impact? Verifying the Sustainability of Climate Change Mitigation Results," establish the importance of postcompletion evaluation to ensure data on actual sustainability of the projects and learning from the experience. They make two broad points. First, postcompletion evaluations are rarely conducted. Second, using GEF as an illustrative example, they argue that even when postcompletion evaluations are conducted, their methodologies may not be consistent and the evaluation reports may not be publicly available. There is little to disagree with in the broader points raised by Cekan and Legro. However, although most international development organizations do not have a postcompletion evaluation product line, most do cover postcompletion sustainability through impact and thematic evaluations that involve substantial postcompletion field verification and provide information on observed long-term sustainability of implemented activities.

In "Assessing Sustainability in Development Interventions," Fitzpatrick argues that to have a sustainable impact on intended beneficiaries, development interventions should deepen human capabilities to manage economic and social change and should be environmentally sound. Like other chapters in this part, this chapter makes a case that postcompletion evaluations are important for knowing whether project outcomes and impacts have been sustainable. Fitzpatrick emphasizes the importance of determining whether governance structures have fulfilled their functions and outcomes have been sustained. She presents a case study of an asset transfer program in Malawi to illustrate the criteria used for evaluating sustainability and to show that focus on community capacity can facilitate endogenous changes in the beneficiary communities, which in turn can enhance likelihood of sustainability of a given development intervention.

Reference

OECD/DAC Network on Development Evaluation. (2019). *Better criteria for better evaluation: Revised evaluation criteria definitions and principles for use*. Organisation for Economic Co-operation and Development. http://www.oecd.org/dac/evaluation/revised-evaluation-criteria-dec-2019.pdf

Sustainability After Project Completion: Evidence from the GEF

Neeraj Kumar Negi and Molly Watts Sohn

Abstract

This chapter examines the extent to which completed GEF projects are sustainable and the factors affecting sustainability. We considered only those projects that were covered through postcompletion evaluation at least 2 years after implementation completion, and where the evaluation reports provided adequate information related to observed sustainability during the postcompletion period. We assessed 62 projects to meet the selection criteria, then completed a desk review of the postcompletion evaluation reports and other relevant documents for these projects to assess the extent to which the project outcome was sustainable.

We found that the projects covered through postcompletion evaluations were generally sustainable, with the sustainability outlook deteriorating for some projects while improving for others. The incidence of the catalytic processes that enhance sustainability—sustaining, mainstreaming, replication, scaling-up, and market change—was higher at postcompletion evaluation, as the passage of time allows long-term project outcomes to manifest. At the project level, we observed these catalytic processes in a wider set of activities at postcompletion evaluation than at implementation completion. Factors such as financial support for follow-up, political support, follow-up by and capacities of the executing agency, stakeholder buy-in, and project design seem to play a crucial role in determining project sustainability.

Introduction

International development cooperation is aimed at helping the recipient countries address their development challenges. A significant share of this aid is provided to the recipients through a project-based modality. For projects to achieve their long-term objectives, it is important that the infrastructure created and approaches promoted by these projects are sustained. However, the extent to which this takes place typically is not ascertained because reporting on project performance usually culminates at the completion of an implementation. This chapter aims to address this gap by bringing forth evidence on postcompletion performance of GEF-supported projects.

Nina Hamilton, Ritu Kanotra, Selin Erdogan, and Spandana Battula provided research assistance support for this chapter.

N. K. Negi (✉) · M. W. Sohn
Global Environment Facility Independent Evaluation Office, Washington, DC, USA
e-mail: nnegi1@thegef.org

J. I. Uitto, G. Batra (eds.), *Transformational Change for People and the Planet*, Sustainable Development Goals Series, https://doi.org/10.1007/978-3-030-78853-7_4

International development agencies generally assess sustainability of their projects at completion of implementation. However, at that point, most of the project benefits are yet to accrue, so assessments of project sustainability generally estimate likelihood of future net benefit flows. Assessment of actual sustainability after implementation completion—say, 2 years or more after completion—is relatively rare. This is primarily because taking stock of the actual accrual of benefits during the postcompletion period is costly, relevant information may be difficult to access, and responsibility for conducting these evaluations may not be clear. As a result, little is known in terms of observed sustainability of development projects.

The evidence presented in this chapter is based on the post completion performance of projects supported by the Global Environment Facility (GEF). As of January 2021, the GEF has provided more than $21 billion through 5000 projects in 170 countries. These projects address global environment challenges related to biodiversity conservation, climate change, international waters, land degradation, sustainable forest management, and chemicals and waste.

Of the GEF-supported projects, more than 1700 have been completed, accounting for $7.5 billion in GEF grant funding. We presented a partial review of postcompletion sustainability of 53 of these projects in the GEF Independent Evaluation Office's (IEO) Annual Performance Report 2017 (GEF IEO, 2019). Since then, we have further deepened this analysis by including nine more completed projects and by reducing data gaps. Our analysis in this chapter covers 62 completed projects that were financed by the GEF and field verified 2 or more years after completion by the GEF IEO or the World Bank's Independent Evaluation Group (IEG).

We found that most of the projects predicted to be sustainable at project completion were indeed sustainable during the postcompletion period. The factors affecting actual accrual of benefits during the postcompletion period included quality of project design, availability of financing for follow-up activities, acceptance of

the project among the key stakeholders, support from the political leadership, and institutional capacities of the executing partner.

Understanding Sustainability

The Brundtland Commission in 1987 defined sustainable development as "development that meets the needs of the present without compromising the ability of future generations to meet their own needs" (World Commission on Environment and Development, 1987, p. 41). The term sustainability, however, is used with various perspectives (White, 2013). For some, most important is an economic perspective focused on intergenerational tradeoffs, and a comparison of value and costs to the society (Solow, 1993; Stavins et al., 2003). For others, conservation of ecosystems and prevention of environmental degradation is an overriding consideration (Costanza & Patten, 1995). Still others call for consideration of holistic approaches to understanding sustainability (Mebratu, 1998).

The principles of sustainable development and its application have been addressed by several scholars (Daly, 1990; Hardi, 1997; Lélé, 1991). Most consider social, economic, environmental, and institutional dimensions for sustainability assessment (Aarseth et al., 2017; Mebratu, 1998; Olsen & Fenhann, 2008; Saysel et al., 2002; Singh et al., 2009). Scholars' assumptions were related to the what and how of sustainability assessment and can have ethical and practical implications in terms of the characteristics that gain prominence versus those left out of the discussion (Gasparatos, 2010). The focus of much of the work on sustainability assessment has centered on determining the extent to which development activities avoid harm to the environment and to social, economic, and other systems affected. Such focus leaves out assessment of the durability of development interventions, especially those aimed at delivering sustainable development.

Within the context of delivering development aid through projects, sustainability may be understood as "the extent to which the net benefits of the intervention continue, or are likely to continue" (Organisation for Economic Co-operation and Development Development Assistance Committee [OECD DAC] Network on Development Evaluation, 2019, p. 12). This perspective has been used in several studies that assess sustainability of development projects. However, Patton (2020) recently criticized the criterion, arguing that the OECD DAC's narrow definition is inadequate to assess systemic transformations and does not adequately address the broader issue of sustainability.

Hoque et al. (1996), who assessed sustainability of a water, sanitation, and hygiene education project in Bangladesh, found that acceptance of the promoted practices by the beneficiaries was an important factor in ensuring sustainability 6 years after implementation completion. Pollnac and Pomeroy (2005), who studied integrated coastal management, found that a community's perception of likely benefits from an intervention affected their continued involvement in project activities and, therefore, project sustainability. Martinot et al. (2001) found that, for projects focused on solar home systems, the extent to which the promoted model ensured profitability affected its sustainability. However, these analyses are limited to a small number of projects and generalizing the findings beyond their local implementation context is usually difficult.

International development organizations such as the Asian Development Bank (ADB), European Bank for Reconstruction and Development (EBRD), United Nations Development Programme (UNDP), United Nations Environment Programme (UNEP), International Fund for Agricultural Development (IFAD), and the GEF assess likelihood of sustainability of their completed projects. This likelihood is assessed at the point of project completion and considers the results achieved and risks to accrual of future benefits. Generally, 60%–70% of these agencies' completed projects are rated in the likely range for sustainability (GEF IEO, 2019). The organizations generally take financial (and/or economic), sociopolitical, institutional, and environmental risks into account when assessing likelihood of sustainability.

Analytical Framework

In this chapter, we examine the extent to which completed projects are sustainable and the factors that affect sustainability. To assess project sustainability, we take stock of the accrued and likely benefits of projects that were completed more than 2 years prior to point of assessment. The projects covered focus on addressing environmental concerns, with benefit streams usually in the form of environmental stress reduction and improvement in environmental status, adoption of promoted technologies and approaches, changes in legal and policy environment, and improvements in institutional and individual capacities. The extent to which such expected benefits—that are demonstrably linked to project activities—accrue is a measure of the project's sustainability. Evaluators estimate the accrual of benefits and likelihood of future accrual for the time frame within which these benefits may be expected.

If the past benefit accrual and likely future accrual of a project (after accounting for risks) is close to the ex-ante projections, then the project is assessed as sustainable. Although the approach that international development organizations use to assess sustainability is analogous, our approach differs in that we make this assessment based on review of information on observed continuation of benefits gathered more than 2 years after project completion, rather than estimated likelihood of continuation of benefits at project completion. This means that assessment of project sustainability is informed more by data on actual accrual than by data on likelihood of future accrual.

The extent to which a project is sustainable may be affected by variables related to its design, implementation, and contextual factors. For example, in instances where recipient community support is important, the extent of attention to community consultations and outreach may play an important role in determining sustainability. Similarly, for a project addressing market barriers, the structure of incentives may determine how well a technology is adopted by producers and/or consumers. During implementation, attention to supervision, exit strategy, and capacity building of key institutions may affect sustainability. Timely availability of cofinancing may also enhance project sustainability. An account of such factors is provided in the project completion reports.

At project completion, the project's finances have been utilized and the project team dissolves. During the postcompletion period, project sustainability may be affected by factors such as host institutions mainstreaming project follow-up activities; support from the national government; presence of an enabling legal, policy, and regulatory environment; presence of a motivated leadership; market conditions; and general economic and political climate in the country. The account of how such factors affected a project's sustainability may be provided in the project's postcompletion evaluation report. We review these sources to gather information on the factors and mechanisms through which they affect project sustainability.

Data

The focus of our study was the project portfolio of the GEF, which provides financial support, mostly in the form of grants, for projects that address global environmental concerns. Of the 1700 completed projects, 147 received a postcompletion evaluation or a field verification by the GEF IEO or the implementing agency evaluation office. After screening the verification reports, we identified reports for 62 completed GEF projects as adequate in terms of quality of reporting on sustainability. We then reviewed the field verification reports for these projects in greater detail (see the chapter appendix for a complete list of the projects, their implementing agencies, and the countries in which the interventions took place).

The 62 projects address environmental concerns such as biodiversity conservation (23 projects), climate change (22 projects), chemicals (six projects), ozone depleting substances (four projects), international waters (four projects), and land degradation (one project); two projects addressed multiple focal areas. Fifty-three of the projects were implemented within a single country and nine were global or regional projects spanning multiple countries. In all, the projects with a national geographic scope covered 34 countries. GEF financing for these projects ranged from $0.5 million to $35 million, with an average of $8.1 million.

The 62 projects were under implementation from 2 to 12 years, with an average duration of 6 years. On average, the last field verification was conducted 6 years after project completion; final field verifications ranged from 2 to 14 years postcompletion. The World Bank was the lead implementing agency for 42 projects, UNDP for 18, and UNEP for one. Fifty-five of these projects were implemented by a single GEF Agency, with seven implemented jointly by two GEF Agencies.

Methodology

Screening and Review

We conducted this review in 2018–2020, identifying 147 completed GEF projects that had been covered through postcompletion evaluations and/or field verifications (from here on referred to as postcompletion evaluation) by the GEF IEO or the GEF Agency evaluation offices. We screened these projects to ensure that the postcompletion evaluation took place at least 2 years after the implementation was completed. This ensured that enough time had elapsed after project completion to assess actual accrual of benefits during the postcompletion period. In this 2-year period, the project execution structure financed through the project funds is generally dismantled, longer term results of the project have greater time to manifest, and risks to accrual of benefits are more likely to have materialized. We screened out 35 of the projects because their most recent post-completion evaluation took place less than 2 years after project completion.

We surveyed the postcompletion evaluations to ensure that they provided adequate information on the project's postcompletion sustainability. We considered evaluation reports prepared by the GEF IEO and the publicly available postcompletion reports by the evaluation offices of the GEF Agencies. This ensured that the evidence provided in these reports was credible; that is, provided by individuals a step removed from project implementation.

Where a project was covered through more than more than one postcompletion evaluation or field verification report, we only considered those conducted 2 years or more after completion for assessment of sufficiency of evidence by these together. After the screening process, 62 projects were retained in the pool: 42 with postcompletion evaluation reports prepared by the GEF IEO and 20 with reports prepared by the World Bank's Independent Evaluation Group.

Our desk review of available project documents included the postcompletion evaluation and field verification reports, implementation completion reports, annual progress reports, midterm reviews, and project proposal documents. We compiled the information relevant to project sustainability using an instrument that gathered numerous information sources: outcome achievements at project completion and at postcompletion field verification, projected sustainability at project completion and observed sustainability at postcompletion, mechanisms through which the projects achieved long-term impacts, barriers restraining progress, and factors driving the changes. We then organized the information gathered through the instrument in a dataset to facilitate analysis.

Assessment Approach

To assess sustainability, we reviewed the information provided in the postcompletion evaluation reports and other project documents such as the terminal evaluation, annual progress reports,

and midterm reviews. Sustainability was assessed on a 4-point scale: *sustainable*, *moderately sustainable*, *moderately unsustainable*, and *unsustainable*. In assessing performance, we considered:

- aspects such as financial, economic, social, political, and environmental sustainability
- probability and likely effect of a risk
- accrued and likely benefits
- time frame within which benefits are expected (ECG, 2012)

The assessment was both backward looking—taking account of the accrued net benefits—and forward looking—estimating the likelihood of accrual of net benefits in future.

We assessed the extent to which environmental benefits—the focus of the covered projects—had accrued at the time when the postcompletion evaluation was conducted. We also documented instances where promoted interventions were reported to be adopted outside the framework of a given project through processes such as mainstreaming, replication, scaling-up, and market change. When documenting these broader adoption processes, we identified the elements of the GEF projects that were being adopted and the scale at which they were being adopted. For some of the interventions, such as protected area management and capacity development of institutions, sustaining the momentum created by the project is also an important characteristic.

Limitations

A comparison of the performance ratings of the projects at implementation completion (see Table 1) showed that the sustainability ratings of the projects covered through postcompletion review were significantly different from the other completed projects in the GEF portfolio. This may be due to a selection bias: Projects with implementation failure due to both endogenous and exogenous reasons are generally excluded from postcompletion review and postcompletion evaluations may implicitly give more attention to

Table 1 Performance ratings at implementation completion: GEF projects rated in satisfactory/likely range as percentage of rated projects

Performance dimension (binary rating scale)	Projects with postcompletion evaluation (62)	Other projects (1644)
Outcome (Satisfactory – Unsatisfactory)	85% (61)	80% (1625)
Sustainability (Likely – Unlikely)	**78%* (59)**	**62%* (1524)**
Implementation (Satisfactory – Unsatisfactory)	83% (48)	80% (1419)
M&E (Satisfactory – Unsatisfactory)	63% (43)	65% (1455)

Source: GEF IEO (2020) dataset
Note. * = statistically significant difference

projects that provide greater opportunity to test the given project's theory of change. On performance parameters such as outcome, quality of implementation, and quality of monitoring and evaluation (M&E), the performance was not significantly different.

The postcompletion evaluations and field verifications that we reviewed were not conducted with the primary purpose of assessing sustainability. Most were completed as part of the field work to gather detailed information on a thematic area. These evaluations and verifications were conducted by different evaluators at different points in time (from 2004 to 2018). Further, the duration at which the postcompletion evaluation was conducted after project completion ranged from 2 to 14 years. This led to differences in the level of detail on issues related to sustainability in the reviewed documents.

Findings

The findings of the review indicated that, in general, the projects covered through postcompletion evaluations were sustainable. Although the sustainability outlook of some projects did deteriorate, this was balanced by improvement in the outlook for others. The review also showed that incidence of the catalytic processes of broader

adoption—sustaining, mainstreaming, replication, scaling-up, and market change—was higher at postcompletion evaluation. These catalytic processes also covered more project interventions at postcompletion evaluation than at implementation completion. The review shows that factors such as financial support for follow-up, political support, follow-up by and capacities of the executing agency, stakeholder buy-in, and project design may play a crucial role in determining project sustainability.

Sustainability During Postcompletion Period

Most projects that were assessed at implementation completion as likely to sustain were also assessed as being sustainable during the postcompletion period (see Table 2). For several projects that were assessed as unlikely to sustain, the risks did not materialize. The revised assessment based on postcompletion evaluation placed these projects in the sustainable range. For two thirds of the completed projects, the outlook on risks to sustainability improved from the point of implementation completion to the postcompletion evaluation (see Table 3). The sustainability outlook deteriorated for only about one sixth of the projects.

About one third (31%) of the projects achieved a higher level of outcome at postcompletion evaluation than at implementation completion (see Table 4), because the longer time frame since implementation facilitated greater progress. When compared to the status at implementation completion, the outcome achievement of a vast majority of projects was at the same level or higher at postcompletion evaluation.

Broader Adoption and Sustainability

Sustainability of a project is a function of whether the project's long-term effects are achieved. But seeing a project's environmental results manifest fully may take a long time. Similarly, some long-term effects may be attributed directly to a proj-

Table 2 Distribution of completed projects based on their sustainability ratings

Projected sustainability at implementation completion	Assessed performance at postcompletion evaluation		
	Sustainable range	Unsustainable range	Total
Likely range	60% (37)	15% (9)	74% (46)
Unlikely range	11% (7)	10% (6)	21% (13)
Not rated	3% (2)	2% (1)	5% (3)
Total	74% (46)	26% (16)	100% (62)

Table 3 Change in likelihood of sustainability

Projected sustainability at project completion	Change in likelihood of sustainability at postcompletion versus at completion				
	Higher	Same	Lower	Unable to assess	Total
Likely range	13% (8)	37% (23)	13% (8)	11% (7)	74% (46)
Unlikely range	5% (3)	11% (7)	2% (1)	3% (2)	21% (13)
Not rated	0% (0)	3% (2)	0% (0)	2% (1)	5% (3)
Total	18% (11)	52% (32)	15% (9)	16% (10)	100% (62)

Table 4 Change in level of project outcome achievement

Projected sustainability at project completion	Outcome achievement at postcompletion versus at completion				
	Higher	Same	Lower	Unable to assess	Total
Likely range	21% (13)	39% (24)	11% (7)	3% (2)	74% (46)
Unlikely range	8% (5)	8% (5)	3% (2)	2% (1)	21% (13)
Not rated	2% (1)	3% (2)	0% (0)	0% (0)	5% (3)
Total	31% (19)	50% (31)	15% (9)	5% (3)	100% (62)

Source for Tables 2, 3, and 4: GEF IEO (2020) dataset; review of postcompletion evaluations

ect while others may be attributed only indirectly because they involve catalytic processes and other actors. Broader adoption takes place when other stakeholders such as governments, private sector, civil society, and other donors (whether originally part of the project or not) adopt, expand, and build on initiatives through a variety of mechanisms (GEF IEO, 2019). These mechanisms include processes that sustain, mainstream, replicate, and/or scale up the supported approaches, and/or change the structure of the targeted markets. The data from the review shows that, indeed, incidence of the processes of broader adoption (such as sustaining, mainstreaming, and market change) was significantly higher at the point of the postcompletion evaluation than the point of implementation completion. Figure 1 provides a comparison of the incidence of these broader adoption processes at both points.

Although incidence of replication and scaling-up also showed nominal increase, those differences were not statistically significant. This finding is consistent with what one would expect in sustainable projects.

Higher incidence of broader adoption at post-completion evaluation was also evident in several instances at a more granular level. Our review tracked the extent to which other actors were adopting interventions related to:

- technology dissemination
- governance arrangements (including development of legal and policy measures)
- management approaches (including development of management plans and strategies)
- development of institutional capacities (through training, awareness, and support for operational infrastructure)

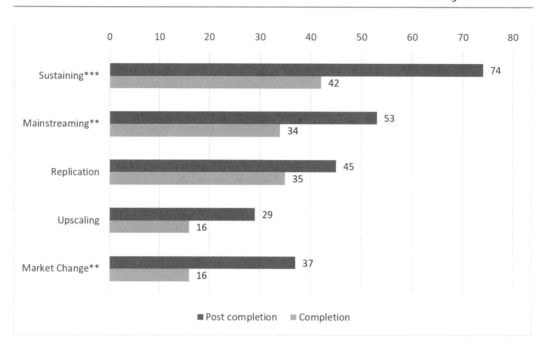

Fig. 1 Incidence of Broader Adoption Processes at Implementation Completion and at Postcompletion

Note. Graph shows percentage of projects studied ($N = 62$); ** = $p < .001$; *** = $p < .005$

Table 5 Broader adoption processes and the elements adopted

	Sustaining		Mainstreaming		Replication		Scaling-up		Market change	
	PC	C	PC	C	PC	C	PC	C	PC	C
Technology dissemination	27*	11*	5	5	27	16	11	5	31*	15*
Governance arrangements	39*	23*	44*	29*	3	2	10	11	16	6
Management approaches	40**	18**	5	11	11	16	10	8	3	2
Institutional capacities	39	37	16	15	23	21	13	6	11	5

Note. Figures indicate percentage of projects studied ($N = 62$); *PC* postcompletion, *C* project completion; * = statistically significant difference at 90% confidence, ** = at 95%

Table 5, which presents the findings of this tracking, shows that several project elements were adopted by other actors for a higher percentage of projects at postcompletion evaluation than at implementation completion. Although for several project-supported interventions the difference in incidence of adoption was not statistically significant, the direction of change in general was consistent with an increase in broader adoption.

The review found that environmental status change and broader adoption was taking place at postcompletion for a nominally higher percentage of projects and at a higher scale. We assessed the scale at which environmental stress

reduction and/or environmental status change (insignificant, local, large scale, or no change), and broader adoption, was taking place both at postcompletion evaluation and at implementation completion. Table 6 presents the findings of this assessment. It shows that environmental status change/stress reduction and broader adoption were taking place for a nominally higher percentage of projects, and at a large scale, at postcompletion evaluation than at implementation completion. Although the difference in incidence was not statistically significant, the direction of change was consistent with most projects being sustainable at project completion (see Table 2).

Table 6 Environmental Status Change/Stress Reduction and Broader Adoption

	At postcompletion	At completion
Environmental status change/stress reduction		
Yes	69%	61%
At large scale	39%	34%
At local scale	31%	27%
No	31%	39%
Insignificant scale	13%	15%
No evidence	18%	24%
Broader adoption		
Yes	84%	75%
At large scale	45%	35%
At local scale	39%	40%
No	16%	25%
Insignificant scale	5%	21%
No evidence	11%	3%

Note. Figures indicate percentage of projects studied ($N = 62$)

Factors that Facilitate Sustainability

The narratives of the postcompletion evaluation reports discussed the factors that affected project outcomes and sustainability. These include financial support for follow-up, political support, follow-up by and capacities of the executing agency, stakeholder buy-in, and project design weaknesses.

Financial Support for Follow-Up

Availability of financial support for follow-up activities is an important factor in a project's sustainability. Projects for which key stakeholders (including national and local governments, development agencies, NGOs, and private sector organizations) provided support for follow-up faced reduced risks and were able to progress well toward achieving their long-term outcomes.

Of the 19 projects for which we assessed outcome achievement at postcompletion at a higher level than at implementation completion, availability of financial support for follow-up was a key factor in 12 projects (63%). For example, the GEF-supported Mekong River Basin Water Utilization Project, implemented by the World Bank in Cambodia, Lao PDR, Thailand, and Vietnam, developed the procedures and guidelines for the Mekong River Commission for management of the basin. These procedures and guidelines have been implemented through a series of follow-up projects funded through World Bank loans. In Mexico, the GEF-funded Introduction of Climate Friendly Measures in Transport project provided support for the development of the first bus rapid transit line in Mexico City. Thanks to sustained financial support by the government, more metrobus lines were subsequently added, leading to expansion of the low-carbon public transit system.

Availability of financial support for follow-up also reduces risks to sustenance of the progress made by the project. For the Renewable Energy Development project in China, sustained government financial support and regulations have stabilized the changes in the renewable energy market and reduced the risk of losing the gains related to market transformation. In case of the Lewa Wildlife Conservancy project in Kenya, the risks to sustainability decreased because of an increase in the fundraising capacity of the conservancy and continued support from the government.

In contrast, lack of financial support for follow-up activities can adversely affect a project's ability to achieve its long-term outcomes. For example, the Caribbean Planning for Adaptation to Global Climate Change project established 18 stations to monitor sea level rise. However, postcompletion evaluation found that none of these stations were transmitting data consistently after project completion because network maintenance had not been funded and continuity in capacity-building efforts was lacking. Similarly, in Ethiopia, the Conservation and Sustainable Use of Medicinal Plants project aimed at supporting in situ conservation of medicinal plants in the Bale Mountains National Park. However, the park was under resourced and an expected follow-up project did not materialize. This affected implementation of the plans and guidelines developed as part of the project and its replication to other areas in the country.

Political Support

Political support for a project and its follow-up is another important factor that affects project sustainability. The support is especially important for projects that aim to influence the legal, policy, and regulatory framework of a country. In other projects, continued support from the political leadership may help government agencies and departments prioritize follow-up to a given project. Evidence from the postcompletion evaluations showed political support (or lack thereof) had a critical effect on project sustainability in several instances. Of the 19 projects with outcome achievement at a higher level at postcompletion evaluation than at implementation completion, strong political support was a key factor for 11 projects (58%).

The China Renewable Energy Development project progressed well toward achieving its catalytic effects because the Chinese government adopted the Renewable Energy Law of 2006. Similarly, the India Ecodevelopment project, which piloted a financing mechanism in a protected area, received support from the political leadership that amended the nation's Wildlife Act to mandate that similar mechanisms be established in all tiger reserves. In Bulgaria, a high level of political support for the Ozone Depleting Substances Phase-out project led to sustained efforts for development of appropriate legislation; establishment of procedures to permit, record, and monitor production of ozone-depleting substances; and implementation of measures to address illegal trade of these substances.

Follow-Up by, and Capacities of, Executing Partner

The support provided by an international development agency through a project generally ends at implementation completion. Typically, activities are implemented on the ground by an executing agency that has a track record or mandate to address concerns that are the focus of the project. We found that, after implementation completion, the follow-up by the executing agency—and its capacities to follow up—seems to affect sustainability. This was a key factor in seven (37%) of the 19 projects for which outcome achievement

was higher at postcompletion evaluation than at implementation completion.

Several examples illustrate the role of executing agencies in facilitating project sustainability. The World Bank-implemented Alternate Energy project in India aimed at development of the renewable energy sector through support for small hydro projects, solar photovoltaic, and wind energy. One project component was to enhance the capacities of its executing partner, the Indian Renewable Energy Development Agency (IREDA), through technical support and training, and through support for enhancing its operational capacity. After completion of the project, IREDA has continued supporting renewable energy development projects and is able to carry out its mandate more effectively due to its enhanced capacities. IREDA's leadership has helped in enhancing the sustainability of the project. The Lewa Wildlife Conservancy Project in Kenya was executed by the Kenya Wildlife Conservancy, an NGO. After implementation completion, the conservancy has continued to create and manage new community conservancies that support wildlife populations, and this has enhanced the sustainability of the GEF-supported project outcomes.

Stakeholder Buy-In

Buy-in on the part of key stakeholders—or lack of it—appears important in determining project sustainability. Of the 19 projects for which outcome achievement was assessed to be at a higher level at postcompletion evaluation than at implementation completion, for six (32%), strong stakeholder buy-in was a key factor in facilitating progress. Strong stakeholder involvement in the Ozone Depleting Substances Phase-out project in Bulgaria led to sustained efforts by the participating enterprises in maintaining equipment, which enhanced project sustainability. The Lewa Wildlife Conservancy project generated sustained support from local communities and national government by providing representation to the national and local government on the Lewa Wildlife Conservancy board. That sustained political support helped in replication of community conservancies in the region, and buy-in from local communities facilitated the efficient creation and management of new community conservancies, which have contributed to the outcomes of stable and improving wildlife populations.

Project Design

Each project is expected to be well designed so that the scarce resources are used for activities that are relevant, effective, and efficient. The project design should address key risks to the project and incorporate measures to mitigate risks that may jeopardize progress. Identifying examples where appropriate design made a project relevant and effective is difficult, but weaknesses in project design that limit a project's ability to achieve its long-term outcomes are more apparent.

In several projects, weaknesses in project design negatively affected progress and sustainability. The design of the Caribbean Planning for Adaptation to Global Climate Change project did not give attention to maintenance of the sea level rise monitoring stations created by the project. This affected functioning of these stations: Three years after project completion, none of the stations were transmitting data consistently. In Romania, the progress made by the Danube Delta Biodiversity project was jeopardized because the project did not adequately consider livelihood concerns of the residents of the Danube Delta area, making it difficult to sustain these communities' interest in conservation. The design of the Ship-Generated Waste Management project, which covered countries in the Caribbean islands, did not anticipate that ship-generated waste at sea (driven by cruise ships) would be a substantial contributor to pollution compared to waste that enters the land-based system. Instead of contributing to reduction of ship-generated waste, the project focused more on the threat from marine and coastal solid waste pollution to the land-based system. After project completion,

this was assessed to be a major missed opportunity for the project and the gains from the project were limited because the main concern was not addressed.

Conclusion

Although the sustainability outlook of some projects changed from sustainable to unsustainable (or the reverse) with time, the review found little difference in the percentage of projects that were sustainable at the postcompletion evaluation. Passage of time allowed several long-term outcomes of the projects to manifest. It also allowed catalytic processes of broader adoption to take root—in several instances, interventions supported by the projects were being sustained, mainstreamed, replicated, scaled up, and/or leading to market change.

The completed projects covered in the review were somewhat higher achieving projects to begin with. Therefore, the performance of an average approved project would be somewhat lower because the average would include projects that experienced implementation failure (these are generally excluded from postcompletion evaluations) or unsatisfactory outcomes (generally underrepresented in postcompletion evaluations). Nonetheless, results do show that in most instances, projects that are assessed to have performed well at implementation completion are able to sustain their performance. This allays a major concern in the development community that gains made up to implementation completion may be lost during the postcompletion period.

Our review showed that numerous factors may affect sustainability, including availability of financial support for follow-up, political support for the project, follow-up by and capacities of the executing partner, stakeholder buy-in, and shortcomings in project design. These factors should be given attention to improve likelihood of sustainability of development projects.

Appendix: Completed GEF-funded Projects with Postcompletion Evaluation

GEF ID	Project Name	Implementing Agency	Country
15	Programme for Phasing Out Ozone Depleting Substances	UNDP/ UNEP	Tajikistan
18	Kenya - Lewa Wildlife Conservancy	World Bank	Kenya
49	Coastal Wetlands Management	World Bank	Ghana
50	Kenya - Conservation of the Tana River Primate National Reserve	World Bank	Kenya
54	Bwindi Impenetrable National Park and Mgahinga Gorilla National Park Conservation	World Bank	Uganda
57	Biodiversity Conservation	World Bank	Bolivia
59	Regional - OECS Ship-Generated Waste Management	World Bank	Regional (Antigua and Barbuda, Dominica, Grenada, St. Kitts and Nevis, St. Lucia, St. Vincent and Grenadines)
64	Demand Side Management Demonstration	World Bank	Jamaica
69	Danube Delta Biodiversity	World Bank	Romania
71	In-Situ Conservation of Genetic Biodiversity	World Bank	Turkey
74	Ozone Depleting Substance Consumption Phase-out (first tranche)	World Bank	Russian Federation
76	Alternate Energy	World Bank	India
84	India - Ecodevelopment	World Bank/ UNDP	India
90	Russia Biodiversity Conservation Project	World Bank	Russian Federation
93	Ozone Depleting Substances Phase-out Project	World Bank	Bulgaria
94	Technical Support and Investment Project for the Phaseout of Ozone Depleting Substances	World Bank	Hungary
100	Danube Delta Biodiversity	World Bank	Ukraine
105	Caribbean Planning for Adaptation to Global Climate Change (CARICOM)	World Bank	Regional (Antigua and Barbuda, Barbados, Bahamas, Belize, Dominica, Grenada, Guyana, Jamaica, St. Kitts and Nevis, St. Lucia, Trinidad and Tobago, St. Vincent and Grenadines)
107	Ukraine - Ozone Depleting Substances Phaseout	World Bank	Ukraine
112	Photovoltaic Market Transformation Initiative	World Bank/ IFC	Global (Kenya, India, Morocco)
114	Russian Federation - Ozone Depleting Substance Consumption Phaseout Project	World Bank	Russian Federation
115	Phaseout of Ozone Depleting Substances	World Bank	Poland
134	South Africa - Cape Peninsula Biodiversity Conservation Project	World Bank	South Africa
192	Bhutan Integrated Management of Jigme Dorji National Park (JDNP)	UNDP	Bhutan
292	Russian Federation - Capacity Building to Reduce Key Barriers to Energy Efficiency in Russian Residential Buildings and Heat Supply	UNDP	Russian Federation
325	Coal Bed Methane Capture and Commercial Utilization	UNDP	India

(continued)

GEF ID	Project Name	Implementing Agency	Country
344	Lithuania Phase Out of Ozone Depleting Substances	UNDP/ UNEP	Lithuania
351	Ethiopia - A Dynamic Farmer-Based Approach to the Conservation of Plant Genetic Resources	UNDP	Ethiopia
358	Sustainable Development and Management of Biologically Diverse Coastal Resources	UNDP	Belize
370	India - Development of High-Rate Biomethanation Processes as Means of Reducing Greenhouse Gas Emissions	UNDP	India
386	India - Optimizing Development of Small Hydel Resourcces in the Hilly Regions of India	UNDP	India
404	Energy Efficiency	World Bank	India
445	Barrier Removal for the Widespread Commercialization of Energy-Efficient CFC-Free Refrigerators in China	UNDP	China
446	Renewable Energy Development	World Bank	China
593	Programme for Phasing Out Ozone Depleting Substances	UNDP/ UNEP	Turkmenistan
615	Mekong River Basin Water Utilization Project	World Bank	Regional (Cambodia, Lao PDR, Thailand, Vietnam)
631	Conservation and Sustainable Use of Medicinal Plants	World Bank	Ethiopia
643	Renewable Energy for Agriculture	World Bank	Mexico
769	Programme for Phasing Out Ozone Depleting Substances	UNDP/ UNEP	Kazakhstan
778	Indigenous and Community Biodiversity Conservation (COINBIO)	World Bank	Mexico
784	Methane Capture and Use (Landfill Demonstration Project)	World Bank	Mexico
818	Conservation of Globally Threatened Species in the Rainforests of Southwest Sri Lanka	UNDP	Sri Lanka
837	Conservation and Sustainable Use of the Mesoamerican Barrier Reef	World Bank	Regional (Belize, Guatemala, Honduras, Mexico)
878	Protected Area Management and Wildlife Conservation	World Bank/ ADB	Sri Lanka
885	Reversing Environmental Degradation Trends in the South China Sea and Gulf of Thailand	UNEP	Regional (Cambodia, China, Indonesia, Malaysia, Philippines, Thailand, Vietnam)
941	China – Demonstration for Fuel-Cell Bus Commercialization	UNDP	China
945	National Protected Areas System	World Bank	Ecuador
1058	Pacific Islands Renewable Energy Programme (PIREP)	UNDP	Regional (Cook Islands, Fiji, Micronesia, Kiribati, Marshall Islands, Nauru, Niue, Papua New Guinea, Palau, Samoa, Solomon Islands, Tonga, Tuvalu, Vanuatu)
1079	Off-Grid Rural Electrification for Development (PCH / PERZA)	UNDP/World Bank	Nicaragua

(continued)

GEF ID	Project Name	Implementing Agency	Country
1084	Mainstreaming Adaptation to Climate Change Project (MACC)	World Bank	Regional (Antigua and Barbuda, Barbados, Bahamas, Belize, Dominica, Grenada, Guyana, Jamaica, St. Kitts and Nevis, St. Lucia, Trinidad and Tobago, St. Vincent and Grenadines)
1124	Integrated Participatory Ecosystem Management in and Around Protected Areas, Phase I	UNDP	Cabo Verde
1155	Introduction of Climate Friendly Measures in Transport	World Bank	Mexico
1356	Forest Sector Development Project	World Bank	Vietnam
1544	Rio de Janeiro Integrated Ecosystem Management in Production Landscapes of the North-Northwestern Fluminense	World Bank	Brazil
1682	Facilitating and Strengthening the Conservation Initiatives of Traditional Landholders and Their Communities to Achieve Biodiversity Conservation Objectives	UNDP	Vanuatu
1872	Community Agriculture and Watershed Management	World Bank	Tajikistan
2767	LAC Regional Sustainable Transport and Air Quality Project	World Bank	Regional (Argentina, Brazil, Mexico)
2947	Renewable Energy and Rural Electricity Access (RERA)	World Bank	Mongolia
2952	Thermal Power Efficiency	World Bank	China
3148	DBSB Agricultural Pollution Control Project - under the Strategic Partnership Investment Fund for Nutrient Reduction in the Danube River and Black Sea	World Bank	Croatia
3510	LDC/SIDS Portfolio Project: Capacity Building for Sustainable Land Management in Sierra Leone	UNDP	Sierra Leone
3973	Armenia Energy Efficiency Project	World Bank	Armenia

References

Aarseth, W., Ahola, T., Aaltonen, K., Økland, A., & Andersen, B. (2017). Project sustainability strategies: A systematic literature review. *International Journal of Project Management, 35*(6), 1071–1083. https://doi.org/10.1016/j.ijproman.2016.11.006.

Costanza, R., & Patten, B. C. (1995). Defining and predicting sustainability. *Ecological Economics, 15*(3), 193–196.

Daly, H. E. (1990). Toward some operational principles of sustainable development. *Ecological Economics, 2*(1), 1–6.

Evaluation Cooperation Group. (2012). *Big book on evaluation good practice standards*. Author. https://www.ecgnet.org/documents/4792/download https://ecgnet.org/document/ecg-big-book-good-practice-standards

Gasparatos, A. (2010). Embedded value systems in sustainability assessment tools and their implications. *Journal of Environmental Management, 91*(8), 1613–1622. https://doi.org/10.1016/j.jenvman.2010.03.014.

Global Environment Facility Independent Evaluation Office. (2019). *GEF annual performance report 2017* (Evaluation report No. 136). https://www.gefieo.org/evaluations/annual-performance-report-apr-2017

Global Environment Facility Independent Evaluation Office. (2020). Annual performance report 2020. https://www.gefieo.org/evaluations/annual-performance-report-apr-2020

Hardi, P. (1997). *Assessing sustainable development: Principles in practice* (Vol. 26). International Institute for Sustainable Development.

Hoque, B. A., Juncker, T., Sack, R. B., Ali, M., & Aziz, K. M. (1996). Sustainability of a water, sanitation and hygiene education project in rural Bangladesh: A 5-year follow-up. *Bulletin of the World Health Organization, 74*(4), 431.

Lélé, S. M. (1991). Sustainable development: A critical review. *World Development, 19*(6), 607–621.

Martinot, E., Cabraal, A., & Mathur, S. (2001). World Bank/GEF solar home system projects: Experiences and lessons learned 1993–2000. *Renewable and Sustainable Energy Reviews, 5*(1), 39–57.

Mebratu, D. (1998). Sustainability and sustainable development: Historical and conceptual review. *Environmental Impact Assessment Review, 18*(6), 493–520.

OECD DAC Network on Development Evaluation. (2019). *Better criteria for better evaluation: Revised evaluation criteria definitions and principles for use.* Organisation for Economic Co-operation and Development. http://www.oecd.org/dac/evaluation/revised-evaluation-criteria-dec-2019.pdf

Olsen, K. H., & Fenhann, J. (2008). Sustainable development benefits of clean development mechanism projects: A new methodology for sustainability assessment based on text analysis of the project design documents submitted for validation. *Energy Policy, 36*(8), 2819–2830. https://doi.org/10.1016/j.enpol.2008.02.039.

Patton, M. Q. (2020). Evaluation criteria for evaluating transformation: Implications for the coronavirus pandemic and the global climate emergency. *American Journal of Evaluation, 2020.* https://doi.org/10.1177/1098214020933689.

Pollnac, R. B., & Pomeroy, R. S. (2005). Factors influencing the sustainability of integrated coastal management projects in the Philippines and Indonesia. *Ocean & Coastal Management, 48*(3–6), 233–251.

Saysel, A. K., Barlas, Y., & Yenigün, O. (2002). Environmental sustainability in an agricultural development project: A system dynamics approach. *Journal of Environmental Management, 64*(3), 247–260.

Singh, R. K., Murty, H. R., Gupta, S. K., & Dikshit, A. K. (2009). An overview of sustainability assessment methodologies. *Ecological Indicators, 9*(2), 189–212. https://doi.org/10.1016/j.ecolind.2011.01.007.

Solow, R. (1993). An almost practical step toward sustainability. *Resources Policy, 19*(3), 162–172. https://doi.org/10.1016/0301-4207(93)90001-4.

Stavins, R. N., Wagner, A. F., & Wagner, G. (2003). Interpreting sustainability in economic terms: dynamic efficiency plus intergenerational equity. *Economics Letters, 79*(3), 339–343. https://doi.org/10.1016/S0165-1765(03)00036-3.

White, M. A. (2013). Sustainability: I know it when I see it. *Ecological Economics, 86*, 213–217. https://doi.org/10.1016/j.ecolecon.2012.12.020.

World Commission on Environment and Development. (1987). *Report of the World Commission on Environment and Development: Our common future.* https://sustainabledevelopment.un.org/content/documents/5987our-common-future.pdf

From the Big Picture to Detailed Observation: The Case of the GEF IEO's Strategic Country Cluster Evaluations

Carlo Carugi and Anna Viggh

Abstract

This chapter introduces strategic country cluster evaluations (SCCEs), a concrete example of how the Independent Evaluation Office (IEO) of the Global Environment Facility (GEF) has dealt with the increasing complexity of GEF programming. This complexity reflects the interconnectedness—in terms of both synergies and trade-offs—between socioeconomic development priorities and environment conservation imperatives that is typical of many country settings in which GEF projects and programs are implemented, such as least developed countries and small island developing states. SCCEs address this complexity by applying a purposive evaluative inquiry approach that starts from aggregate analyses designed to provide trends and identify cases of positive, neutral, or negative change, and proceeds to in-depth data gathering aimed at identifying the specific factors underlying the observed change in those specific cases. By establishing the interconnectedness and sequencing of the various evaluation components, rather than conducting these in parallel, SCCEs provide an opportunity to focus on a limited set of purposively selected issues that are common in clusters of countries and/or portfolios. This enables a comprehensive understanding of the factors at play in complex national and local settings.

Introduction

Since the early 1990s, evaluations have been conducted according to the five criteria of relevance, effectiveness, efficiency, impact, and sustainability (Organisation for Economic Co-operation and Development, Development Assistance Committee [OECD DAC], 1991, 2019). This approach has worked rather well, especially at the project level where most of the evaluation body of work was being done. Standard evaluation methods have included the review of project documentation, portfolio analysis, interviews at agencies' headquarters, and field observations in a selection of project sites assessed using specialized technical expertise.

The introduction of more complex delivery modalities that started in the 2000s—sector approaches, budget support modalities, and programs—and the advent of the Millennium Development Goals (MDGs), recently replaced by the Sustainable Development Goals (SDGs), brought about a corresponding increased complexity in evaluation. The SDGs take an inte-

C. Carugi (✉) · A. Viggh
Global Environment Facility Independent Evaluation Office,
Washington, DC, USA
e-mail: ccarugi@thegef.org; aviggh@thegef.org

J. I. Uitto, G. Batra (eds.), *Transformational Change for People and the Planet*, Sustainable Development Goals Series, https://doi.org/10.1007/978-3-030-78853-7_5

grated approach that links the three pillars of sustainability: social, economic, and environmental (United Nations Department of Economic and Social Affairs, 2015). Although such integration is necessary to move toward sustainable development, it undeniably poses significant challenges in terms of identifying suitable metrics and indicators to assess achievements and results in a way that breaks the "data silos" performance measurement approach that was typical of the MDGs era (ICLEI, 2015).

The Global Environment Facility (GEF), a partnership set up as a result of the 1992 Rio Earth Summit, underwent a similar evolution. From a project-based delivery institution focusing on the environment, the GEF is increasingly moving toward more complex, programmatic, interconnected, and synergetic delivery modalities that consider the environmental with the social and economic dimensions. These GEF integrated programming modalities aim at tackling the main drivers of environmental degradation and achieving impact at scale (GEF Independent Evaluation Office [IEO], 2018a). The GEF has designed these strategies because many of these drivers extend their influence beyond national boundaries. To participate in integrated, multiple-country initiatives, governments need to find a balance between their national sustainable development priorities and their commitments to contribute to the global goals of international environmental conventions.

In the GEF, project and program evaluations are conducted by GEF partner Agencies. The GEF IEO conducts complex evaluations at levels higher than projects (GEF IEO, 2019a). To better capture the successes and challenges the GEF has faced in its move toward more complex, integrated programming, IEO evaluations increasingly consider innovative ways to address the complexity of assessing the environmental with the social and economic, including how these three dimensions play out at the national and local levels. The way GEF support is operationalized at the country level is increasingly a key IEO area of enquiry.

Challenges and Opportunities in IEO Complex Evaluations

Complex evaluations typically use mixed methods involving both quantitative and qualitative tools and analyses. In mixed-methods research, methods sequence and dominance are central concepts. The rationale for the mixed-method explanatory sequential design is often that the quantitative analysis provides a general understanding of the main research results while the qualitative data and their analysis refine and explain those results (Walker & Baxter, 2019). When this approach is applied in evaluation, aggregate quantitative analysis can also inform subsequent qualitative deep dives in specific projects/project sites to explain the main trends and provide additional insights. This is the usual approach in academic research, which, unlike evaluation, does not usually face tight deadlines to serve decision makers' specific information needs.

In practice, tight timelines make for difficulty in applying a coherent sequencing in conducting the various quantitative and qualitative components of a typical complex, higher level evaluation. A long time is needed for process issues, and the tasks that take the longest usually are (from most to least time consuming): (a) contracting the various firms and individual experts; (b) getting in touch and agreeing on the field mission dates and modalities with GEF national stakeholders in countries chosen for field data gathering; (c) setting up stakeholder engagement mechanisms such as peer review panels and reference groups, and the functioning of those mechanisms; and (d) arranging the mission logistics while complying with security procedures of the institution (which in the GEF case is the World Bank). Afterwards, when the time comes to bring it all together, the evaluators must triangulate the different sets of qualitative and quantitative data and information, looking for coherence and connectedness between the various pieces of evidence.

To address this challenge, a few years ago the IEO developed a systematic approach to triangulate evidence and identify key findings in country portfolio evaluations (Carugi, 2016). This approach ensures the systematic use and analysis of all the data and information gathered, while respecting tight deadlines. Systematic triangulation can also help in addressing common challenges in evaluation, such as the scarcity or unreliability of data, or the complexities of comparing and cross-checking evidence from diverse disciplines. Although comprehensive, systematic triangulation does not allow evaluators to purposively dive deeply on a limited set of selected key themes that are common to multiple country or portfolio settings.

The Strategic Country Cluster Evaluation Concept

A way to address the challenge of assessing complex environmental and development interventions that require comparing and cross-checking evidence from diverse disciplines is to apply a sequenced, purposive approach in the conduct of an evaluation. That is what the IEO has done with strategic country cluster evaluations (SCCEs). SCCEs focus on a limited set of common themes across clusters of countries and/or portfolios that involve a critical mass of GEF investments toward comparable or shared environmental challenges and that have gained substantial experience with GEF programming over the years. Starting from aggregate portfolio analysis to identify trends and cases of positive and absent or negative change, SCCEs are designed to dive deeply into those themes and unpack them through purposive evaluative inquiry. SCCE design is based on a conceptual analysis framework, an approach the GEF IEO developed earlier at the country level,[1] to enable comparison of findings across geographic regions and/or portfolios. In addition to the aggregate portfolio analysis, SCCEs use geospatial analysis to identify

change on key environmental outcome indicators over time. Targeted field verifications follow in specific hot spots selected based on the findings of the geospatial and portfolio analyses. The purpose of field verifications is to identify and understand the determinants of the observed change or lack thereof.

The identification of factors hindering and/or enabling the sustainability of GEF outcomes was one of the main themes selected by the GEF IEO for deep-dive investigation in SCCEs. In 2017, the IEO completed a desk study on the sustainability of GEF project outcomes (GEF IEO, 2019b).[2] The study analyzed the IEO datasets of terminal evaluation ratings to assess correlations among sustainability, outcomes, implementation, broader adoption, project design features, country characteristics, and other variables. The analysis took stock of projects for which field verifications were conducted by the IEO at least 2 years after project completion. According to the study, the following contributing factors were at play in those cases where past outcomes were not sustained: (a) lack of financial support for the maintenance of infrastructure or follow-up, (b) lack of sustained efforts from the national executing agency, (c) inadequate political support including limited progress on the adoption of legal and regulatory measures, (d) low institutional capacities of key agencies, (e) low levels of stakeholder buy-in, and (f) inadequate project design characterized by flaws in the theory of change of projects.

The IEO further explored these issues by applying the new SCCE purposive evaluative enquiry approach to three different clusters of country portfolios. The SCCEs' main objectives were: (a) to provide a deeper understanding of the determinants of the sustainability of the outcomes of GEF support and (b) to assess the relevance and performance/impact of the GEF toward the main environmental challenges from the countries' perspective. Gender, climate resilience, private sector, and GEF operations in fragile situations were also assessed as cross-cutting issues.

---

[1] From 2006 to 2016, the GEF IEO conducted 26 country portfolio evaluations and studies that used the country as the unit of analysis to examine the totality of GEF support across all GEF Agencies and programs. The new strategic country cluster evaluations build on this experience.

[2] Negi and Sohn's chapter in this book updates this review.

A unique area of SCCE research was the environment vs. socioeconomic development nexus, a concept that is central to sustainable development. This nexus is too often neglected in development interventions, both by donors and developing countries alike (GEF IEO, 2020). Efforts to integrate socioeconomic development with environment conservation/sustainable use both at national and local levels depend on the interest of country governments. Many governments in the least developed countries (LDCs) believe that achieving both at the same time is difficult, and perceive, rather than a nexus, that major trade-offs exist between environment and socioeconomic/livelihoods objectives. Countries differ on: (a) reliance on natural resources, (b) susceptibility to natural disasters, (c) the poor's dependence on the environment, and (d) the government's economic development and other priorities. SCCEs investigated if and how the existence of a nexus between socioeconomic development needs and environmental conservation priorities (or lack thereof) contributed to or hindered the observed sustainability of project outcomes.

Applications of the SCCE Approach

The approach discussed in the previous sections has been applied to three clusters of countries, one covering the GEF portfolio of projects and programs in two biomes,[3] one covering LDCs, and one covering the small island developing states (SIDS) portfolios.[4] The African biomes covered by the first SCCE were the Sahel and the Sudan-Guinea Savanna. Selection of these two biomes was based on the countries' comparable land-based environmental challenges. These countries also face challenges related to governance, demographics, migration, conflict, and fragility, which work as drivers for the environmental issues at hand. Most countries in the two selected biomes are LDCs, and half are fragile (World Bank, 2020).

The LDCs SCCE covered 47 countries that are currently designated by the United Nations as LDCs.[5] Focus on LDCs was based on these countries' greater challenges related to sustainability of outcomes over several GEF periods (GEF IEO, 2019b) and related economic, social, and environmental challenges. Most LDCs are characterized by a low level of socioeconomic development. They have weak human and institutional capacities, low and unequally distributed income, gender inequality, and scarce domestic financial resources. LDCs often suffer from governance crisis, political instability, and, in some cases, internal and external conflicts. Twenty-eight of the 47 LDCs are fragile (World Bank, 2018). The SIDS SCCE covered 39 small island developing states in the AIMS (Atlantic, Indian Ocean, Mediterranean, and South China Sea), Caribbean, and Pacific regions. The choice to evaluate the SIDS as a strategic country cluster was based on their shared geophysical constraints that result in disproportionately large economic, social, and environmental challenges.

Methodological Considerations

Selection of case study countries in the three SCCEs drew upon sustainability cohorts composed of national and regional projects completed between 2007 and 2014 and having Annual Performance Report (APR) ratings (GEF IEO, 2018b, 2019b, c) to allow for observation of the actual sustainability of outcomes 4–5 years after

---

[3] A biome is an ecological zone sharing similar habitats or vegetation types. Its uniformity is defined by the type of plant life in relation to temperature and rainfall patterns. Each biome consists of several terrestrial ecoregions (a smaller class). An ecoregion covers a realm of land/water having geographically distinctive communities, sharing the same environmental conditions and ecological dynamics (Data Basin, 2011).

[4] Because the SIDS SCCE is discussed at length in a separate chapter, this chapter gives detailed examples of site visits only for the African biomes and the LDCs SCCEs.

[5] For more information on the United Nations definition of LDCs, see https://www.un.org/development/desa/dpad/least-developed-country-category/creation-of-the-ldc-category-and-timeline-of-changes-to-ldc-membership-and-criteria.html

project completion. Projects in the African biomes and LDC cohorts were classified as: (a) having both outcome and sustainability ratings in the positive range (i.e., highly satisfactory, satisfactory, or moderately satisfactory); (b) having both outcomes and likely sustainability ratings in the negative range (i.e., highly unsatisfactory, unsatisfactory, or moderately unsatisfactory); (c) having either positive outcome and negative likely sustainability ratings, or the inverse; and (d) not having either outcome or sustainability ratings, or both (see Tables 1, 2, and 3).

Also informing the selection of country case studies were trends over time of key environmental outcome indicators at geolocated project sites, with the aim of identifying cases of positive and absent or negative change. Country case study selection started with the identification of the main environmental challenges faced by the countries covered by the respective SCCE. These challenges were classified by biome in the case of the African biomes SCCE and by geographic country category in the case of the LDCs and SIDS SCCEs. Projects with both positive and negative outcome and sustainability ratings in each portfolio were tagged to each environmental challenge.

Guided by the mapping of countries and projects to environmental challenges, the IEO selected countries with the largest number of national and regional projects with positive and negative outcome and sustainability ratings. This method ensured the largest number of observable data points and coverage of possible factors affecting sustainability. The countries selected also included those in which projects addressed the most commonly shared environmental challenges. In the African biomes SCCE, these were deforestation and land degradation, threats to biodiversity, and desertification. In the LDCs

Table 1 Outcome and sustainability ratings matrix

Change		Outcome rating					
		Highly satisfactory	*Satisfactory*	*Moderately satisfactory*	*Moderately unsatisfactory*	*Unsatisfactory*	*Highly unsatisfactory*
Sustainability rating	*Likely*	Positive			Neutral		
	Moderately likely						
	Moderately unlikely	Neutral			Negative		
	Unlikely						

Table 2 African biomes SCCE: Selection of countries based on APR ratings prior to missions

Project	Outcome and sustainability ratings				Total
	Both positive	Both negative	Neutral[a]	No ratings[b]	
Country	10	16	16	4	46
Regional	7	4	4	7	22
Total	**17**	**20**	**20**	**11**	**68**

[a]Positive outcome and negative sustainability, or negative outcome and positive sustainability
[b]Projects without either outcome rating, sustainability rating, or both

Table 3 LDCs SCCE: Selection of countries based on APR ratings prior to missions

Project	Outcome and sustainability ratings				Total
	Both positive	Both negative	Neutral[a]	No ratings[b]	
Country	25	21	29	12	87
Regional	14	7	10	9	40
Total	**39**	**28**	**51**	**21**	**127**

[a]Positive outcome and negative sustainability, or negative outcomes and positive sustainability
[b]Projects without either outcome rating, sustainability rating, or both

SCCE, these were deforestation and land degradation, and biodiversity loss. Water-related challenges were also important and included water quality and quantity, threats to marine resources, and coastal and coral reef degradation.

The application of the pre-mission selection process based on outcomes and sustainability ratings was accompanied by typical logistics and organizational considerations such as site accessibility and seasonality. In Bhutan, evaluators did not make the final selection of project sites to visit until after discussion upon arrival in the country with stakeholders in the Gross National Happiness Commission, relevant line ministries, and technical agencies such as the National Soil Services Center. For example, in the case of sustainable land management (SLM), these discussions resulted in the LDC SCCEs evaluation team visiting a site in Zhemgang District, selected out of three possible sites to logistically coordinate with site visits to the other projects in the sustainability cohort and in consideration of road conditions in the mountainous country. This choice was made because the SLM project sites are located in areas of high incidence of land degradation that are inhabited by most of the country's poorest and most vulnerable communities. Although the terminal evaluation had rated the project's outcomes in the positive and sustainability in the negative range, the evaluation team could verify that the SLM measures introduced

by the project were still in operation 5 years after the project was completed. Selecting Zhemgang District for a site visit allowed the evaluation team to observe, 5 years postcompletion, the main sustainability factors fostering positive SLM results in mountainous ecosystems alongside unforeseen hindering factors. The team could verify the status of SLM measures introduced by the project and directly collect information on their continued use and maintenance from the remote rural communities living in those highly degraded lands. Photos 1 and 2 show meetings to finalize selection of sites and interview rural communities in Bhutan.

For the African biomes SCCE, once the evaluation team had selected the countries and projects based on the pre-mission selection process described above, they prepared geospatial maps for each project site prior to the missions to the country. Once in the country, evaluators used these maps to select the sites to visit in the field verification mission (see Fig. 1). This ensured the conduct of field observations in specific project locations selected both in highly degraded areas and in areas where vegetation had actually increased.

The evaluation team shared these maps with stakeholders (on a laptop/smartphone in the field, or on paper in local offices) to stimulate discussions and identification of the key factors at play driving the change observed in the map—see Photos 3 and 4. Local technicians, locally elected

Photos 1 and 2 LDCs SCCE – Finalizing sites selection and interviewing rural communities in Bhutan
Photo 1: Meeting the Gross National Happiness Commission (Thimphu, March 2019)

Photo 2: Discussing SLM measures with farmers during site visits (Zhemgang, March 2019)

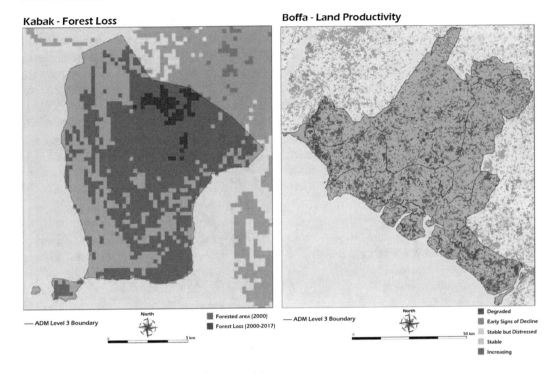

Fig. 1 African biomes SCCE – project sites geospatial maps

Photos 3 and 4 African Biomes SCCE – Discussing environmental change with local stakeholders
Photo 3: Field visit in Kaback Commune (Guinea, March 2019)

Photo 4: Kyenjojo District technical staffs reviewing Albertine Rift forest loss maps (Uganda, May 2019)

representatives, and community members all confirmed the environmental changes in the areas indicated in the maps and provided additional insights on when, how, and why those changes occurred. For example, in Tolo (Guinea), areas of increasing vegetation were subject to intense afforestation efforts accompanied by strict enforcement measures by local government for-

estry technicians. In Kaback, the anti-salt dikes built with GEF support in highly degraded coastal areas were insufficient in both height and width to withstand water intrusion. Attempts were made in Konimodouya and Katonko to change the approach by building a more robust dike, but these too could not withstand the rising sea-level pressure.

The purposive selection processes described above allowed for an in-depth, more granular, and comprehensive understanding of which specific factors have influenced the observed sustainability following project completion. In Tolo, the watershed identified for relocating farmers from the Bafing Lake had insufficient water for irrigation, a case of poor project design (see Box 1). In Zhemgang, the quality of project design led to the highly positive observed sustainability postcompletion (see Box 2).

Box 1: Field Visit in a Site Selected Based on Pre-mission Analysis: Tolo, Bafing Lake (Guinea)

The GEF project applied a coherent ecosystem approach to the whole watershed, working with all the stakeholders involved. Evaluators selected two sites to visit in Tolo: The first was on protection measure to rehabilitate the Bafing Lake banks, and the second involved community-based farming in the adjacent watershed. The lake is a source for 50% of the water going to the Senegal river. Around the lake is a community village. One of the project objectives was to reduce deforestation around the lake that leads to erosion and water loss from the lake basin. Deforestation is due to land clearing for slash-and-burn, itinerant agriculture. The local forest department enforces a forest-cutting ban around the lake. The project relocated the farmer community around the lake to a watershed 2 km from the village, where communities could practice horticulture. This delocalization measure was informed by a socioeconomic study followed by intensive participatory activities and negotiations, which provided a management arrangement for the distribution of land in the watershed and included granting some compensation measures to the farmers.

Years after the delocalization of the activity from the lake, the ecosystem of the lake banks has been slowly rehabilitated through intense reforestation measures (see Photo 5). The area has become green, with no agricultural activities around the lake, favoring the settling in of a small microclimate that benefits the whole ecosystem. It was reported that years ago, one could cross the lake by foot in April due to damage from deforestation. The banks around the lake, once degraded from unsustainable agriculture activities, are now green.

Access to water remains the key impediment for agriculture in the Mamou region. The two hectares of watershed where the farmers have been delocalized has an irrigation system with canals that allows water to be spread on the field and six groundwater wells, all of which are thanks to the Community Land Management Project investments. The mission found this area underused. Farmers reported that despite the investments made, they can have enough irrigation water only for 6 months in a year (see Photo 6).

Geospatial Analysis Following Project Field Visits

The IEO conducted targeted geospatial analysis once teams returned from the missions, using the geographic coordinates collected with GPS tracking software apps installed in the team members' smartphones and the information gathered during field observations. Both in Bhutan and Guinea, this analysis showed increased vegetation despite lower precipitations in the project sites visited, providing complementary data to shed more light on the observed changes (see Figs. 2 and 3). The reforested areas evidenced by the satellite photos taken in 2012 and 2019 on the Bafing lake basin in Guinea (Fig. 4) are the result of GEF-induced farmer relocation and afforestation activities accompanied by strict government enforcement. This temporary project success depends on the farmers continuing to practice horticulture in the watershed where they have been relocated.

Box 2: Field Visit to a Site Selected Based on Pre-mission Analysis: Zhemgang District, (Bhutan)

The project aimed to strengthen institutional and community capacity for anticipating and managing land degradation. SLM practices were piloted in three *geogs* (groups of villages), where farmers were trained in SLM techniques. The project sites were in areas of high incidence of land degradation that were inhabited by the country's poorest and most vulnerable communities. The project resulted in an increase in farmers practicing SLM techniques, a reduction in sediment flows in selected watersheds, regeneration of degraded forest land, and improved grazing land in the pilot *geogs*. The postcompletion site visit to a pilot *geog* in a remote area in Zhemgang noted continued practice of SLM techniques such as land terracing, hedgerows, fruit orchards, tree plantations, and irrigation systems. Selling produce both in the district and in Gelephu on the border with India has provided increased income for residents. Villagers confirmed in interviews that more land is under cultivation, and 60% of households continue using SLM techniques learned from the project. The remainder of the households discontinued using SLM due to shortages of water and losses caused by wildlife such as bears and wild boars. The government has provided some electric fencing, but it is not sufficient. The continued practice of SLM techniques has also helped improve and retain soil and convert shifting land cultivation to sustainable land cover (see Photo 7).

Among the project outcomes were the preparation and implementation of the 2007 Land Policy Act that incorporated SLM principles in programs and policies including the National Land Policy, the Forestry Policy, the National Adaptation Program of Action, and the National Biodiversity Action Plan. SLM principles have been incorporated in the government's 12th five-year plan (2018–2023) and in plans on poverty reduction and increased food security.

Key factors driving postcompletion sustainability were good project design and government support, including highly relevant objectives in line with government priorities and relevant activities to achieve the stated objectives. The project design was guided by a bottom-up approach with participatory planning that focused on community priorities, phased implementation allowing for adjustment throughout implementation based on learning from pilots, decentralization to strengthen the role of communities and local authorities, use of knowledge and information on farmer incentives, and an integrated multisectoral approach. Before the completion of the project, institutional, financial, technical, and policy arrangements were made for sustaining its outcomes.

In Zhemgang, both forest and vegetation cover in pastures have increased since the onset of the project. In Fig. 5, the 2010 image clearly shows large areas of relatively bare ground, which are subsequently covered by vegetation in 2018. The findings of the post-mission geospatial analysis of the SLM project confirmed the field visit finding of improved sustainability of outcomes years after project completion.

Lessons from the SCCE Experience

Using the selection process described in this chapter and a combination of geospatial analysis prior to and after field visits to targeted project sites, the SCCEs revealed that most of the field-verified projects maintained or sustained their outcomes postcompletion. This was the case for 87% of the projects field verified in the African biomes SCCE (16 projects), 81% in the

Photo 5 Reforestation around Bafing Lake

Photo 6 Watershed relocation land

SIDS SCCE (24 projects) and 70% in the LDCs SCCE (25 projects). More important, the selection of projects with a combination of outcome and sustainability ratings in both the positive and negative range and tagged to the main environmental challenges faced by the country led to a diverse group of projects selected for deep-dive analysis into which specific factors contributed to these improvements in the observed postcompletion sustainability. Enhanced learning led to a better understanding of how the environment and development nexus (or lack thereof) played out in contributing to or hindering the observed sustainability. This would not

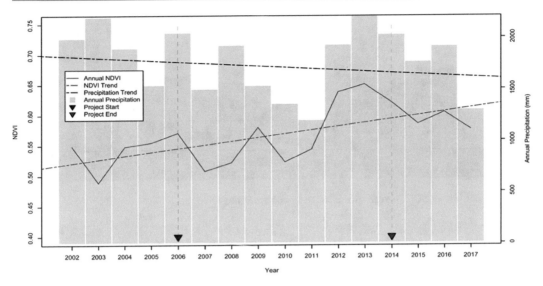

Fig. 2 African biomes SCCE – vegetation increase vs. lowering annual rainfall in the Bafing region, 2012 and 2019

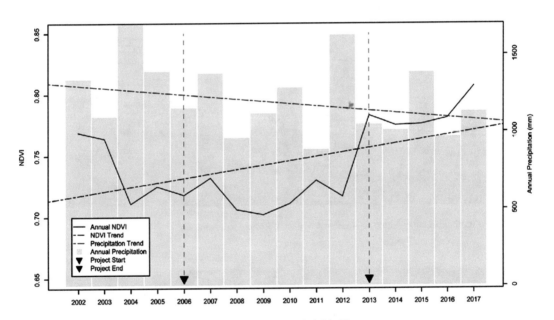

Fig. 3 LDCs SCCE – vegetation increase vs. lowering annual rainfall in Zhemgang

have been possible to achieve with the same granularity through the usual randomized approaches to country, project, and site selections for field verification applied in parallel to the conduct of aggregate analyses in previous IEO evaluations.

A second important lesson that informs the preparation of future IEO work plans is that applying the described sequencing approach from aggregate analysis to detailed observation took a long time. This investigation was possible because the three SCCEs were conducted in the

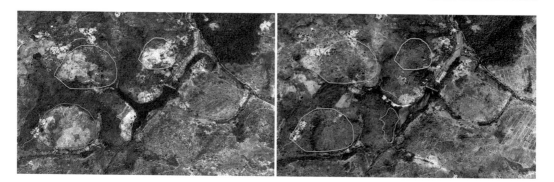

Fig. 4 African biomes SCCE – vegetation increase around the Bafing Lake

Photo 7 Fruit orchards contributing to soil conservation, observed in Zhemgang

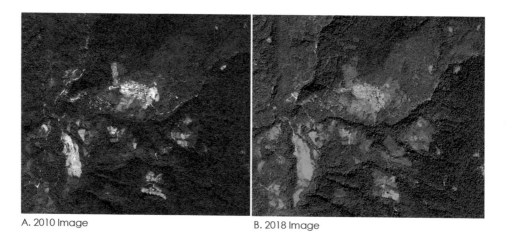

A. 2010 Image

B. 2018 Image

Fig. 5 Vegetation increase in Zhemgang

2 years following the completion of the Sixth Comprehensive Evaluation of the GEF, corresponding to a slightly lower intensity in the GEF decision makers' demand for evaluative evidence from the office. To minimize the long timeframes that may result from sequencing, the most time-demanding activities should be conducted first. In the case of SCCEs, aggregate geospatial analysis was the most time-consuming and complex component, followed by making the arrangements for the missions in selected countries.

At times, pre-mission analysis needs adjustment to account for country-specific logistics and other organizational considerations influencing the final site visit selections. In the case of the LDCs SCCE, site selection had to account for the remoteness and challenges of traveling to several sites during a visit in a mountainous country. When this happens, care should be taken in adjusting site selection to allow as much compliance as possible with the results of the pre-mission aggregate analysis, while accounting for variation due to the site changes in the final deep-dive analysis, as was done in Bhutan.

Applying a purposive evaluative enquiry approach to evaluation encompasses sequencing the evaluation data-gathering and analysis components so that each component informs the following one. This approach has the potential to produce a deeper, more granular and comprehensive understanding of the issues being evaluated. This was achieved by introducing the new SCCE approach, in which evaluators used geospatial analysis with aggregate portfolio analysis and review of project documentation to design the case studies' deep dives in terms of issues to focus on, data and information to gather, and exact locations for gathering those data.

The project selection method based on projects' positive and negative outcomes and sustainability ratings was very useful for new discoveries. Among these, field visits to 36 completed projects in 12 LDCs by the three SCCEs found that 25 projects sustained or progressed further in achievement of their outcomes after project completion. Teams found that these improvements were mainly attributed to two factors: the quality of project design and positive changes in the con-

text taking place postcompletion. Although previous analyses already indicated the importance of good project design for fostering the sustainability of project outcomes (GEF IEO, 2019b), less was known about the different ways in which various contextual factors come progressively into play 4–5 years after a project is completed. This understanding sheds new light on how to best take advantage of the country- and site-specific context factors that enable the sustainability of GEF interventions, a lesson that further contributes to improving project design.

References

Carugi, C. (2016). Experiences with systematic triangulation at the Global Environment Facility. *Evaluation and Program Planning, 55*(4), 55–66. https://doi.org/10.1016/j.evalprogplan.2015.12.001

Data Basin. (2011). *Terrestrial ecoregions of the world* [Data set]. https://databasin.org/datasets/68635d7c77f1475f9b6c1d1dbe0a4c4c/

Global Environment Facility Independent Evaluation Office. (2018a). *Evaluation of programmatic approaches in the GEF* (Evaluation Report No. 113). http://www.gefieo.org/evaluations/evaluation-programmatic-approaches-gef

Global Environment Facility Independent Evaluation Office. (2018b). *Strategic country cluster evaluation (SCCE): Sahel and Sudan-Guinea savanna biomes selection of case study countries.* http://www.gefieo.org/sites/default/files/ieo/documents/files/scce-biomes-2018-case-study-countries.pdf

Global Environment Facility Independent Evaluation Office. (2019a). *The GEF evaluation policy.* https://www.gefieo.org/sites/default/files/ieo/council-documents/files/c-56-me-02-Rev.01.pdf

Global Environment Facility Independent Evaluation Office. (2019b). *GEF annual performance report 2017* (Evaluation Report No. 136). https://www.gefieo.org/sites/default/files/ieo/evaluations/files/apr-2017_0.pdf

Global Environment Facility Independent Evaluation Office. (2019c). *Strategic country cluster evaluation (SCCE): Least developed countries selection of case study countries.* https://www.gefieo.org/sites/default/files/ieo/documents/files/scce-ldc-2018-case-study-countries.pdf

Global Environment Facility Independent Evaluation Office. (2020). *Strategic country cluster evaluation (SCCE): Sahel and Sudan-Guinea Savanna biomes Volume 2 – technical documents.* https://www.gefieo.org/sites/default/files/ieo/evaluations/files/scce-biomes-2018-v2.pdf

ICLEI Local Governments for Sustainability. (2015). *Measuring, monitoring and evaluating the SDGs*

(ICLEI Briefing Sheet, Urban Issues, No. 06). https://www.local2030.org/library/236/ICLEI-SDGs-Briefing-Sheets-06-Measuring-Monitoring-and-Evaluating-the-SDGs.pdf

OECD DAC Network on Development Evaluation. (2019). *Better criteria for better evaluation: Revised evaluation criteria definitions and principles for use*. Organisation for Economic Co-operation and Development. http://www.oecd.org/dac/evaluation/revised-evaluation-criteria-dec-2019.pdf

Organisation for Economic Co-operation and Development, Development Assistance Committee. (1991). *DAC criteria for evaluating development assistance* [Factsheet]. Organisation for Economic Co-operation and Development. https://www.oecd.org/dac/evaluation/49756382.pdf

United Nations Department of Economic and Social Affairs. (2015). *Social development for sustainable development*. https://www.un.org/development/desa/dspd/2030agenda-sdgs.html

Walker, C., & Baxter, J. (2019). Method sequence and dominance in mixed methods research: A case study of the social acceptance of wind energy literature. *International Journal of Qualitative Methods, 18.* https://doi.org/10.1177/1609406919834379

World Bank. (2018). *Harmonized list of fragile situations FY 18.* http://pubdocs.worldbank.org/en/189701503418416651/FY18FCSLIST-Final-July-2017.pdf

World Bank. (2020). *Classification of fragile and conflict-affected situations.* https://www.worldbank.org/en/topic/fragilityconflictviolence/brief/harmonized-list-of-fragile-situations

Staying Small and Beautiful: Enhancing Sustainability in the Small Island Developing States

Geeta Batra and Trond Norheim

Abstract

Spread over the ocean regions of the Caribbean, the Pacific and Atlantic, the Indian Ocean, the Mediterranean, and the South China Sea, the small island developing states (SIDS) are a distinct group of developing countries often known for their rich biological diversity, oceans, tourism, and fisheries. The pressures on these and other natural resources is most immediate in the islands where the high vulnerability to the impacts of climate change, limited land and water resources, often unsustainable natural resource use, and other particular economic vulnerabilities are disrupting livelihoods. The COVID-19 pandemic has further exacerbated the SIDS economies and livelihoods. Over the past 25 years the Global Environment Facility (GEF) has supported interventions in SIDS through $578 million in financing, in critical areas such as biodiversity protection, climate resilience, and energy access through renewable energy. But how effective and sustainable have these interventions been? What factors influencing the sustainability of GEF interventions can pro-vide insights for future project design and implementation? This chapter draws on findings from a recent country cluster evaluation on SIDS conducted by the Independent Evaluation Office (IEO) of the GEF. It presents the main environmental challenges in SIDS, the evidence on the relevance and effectiveness of GEF interventions in addressing these challenges, and the main risks to sustainability of outcomes. Important contextual factors that affect sustainability in SIDS include good policies and legal and regulatory frameworks, national ownership of projects, environmental awareness, institutional capacity, and strategic institutional partnerships. Project-related factors including good project design and adaptive project management, scaling-up and replication based on lessons learned, and a good exit strategy are also important for sustainability.

Keywords

GEF · Sustainability · SIDS · Biodiversity · Climate change

G. Batra (✉)
Global Environment Facility Independent Evaluation
Office, Washington, DC, USA
e-mail: gbatra@thegef.org

T. Norheim
Scanteam, Oslo, Norway

Introduction

Spread over the ocean regions of the Caribbean, the Pacific and Atlantic, the Indian Ocean, the Mediterranean, and the South China Sea, the

© The Author(s) 2022
J. I. Uitto, G. Batra (eds.), *Transformational Change for People and the Planet*, Sustainable Development Goals Series, https://doi.org/10.1007/978-3-030-78853-7_6

small island developing states (SIDS) are a distinct group of developing countries often known for their rich biological diversity, oceans, tourism, and fisheries. The pressures on these and other natural resources is most immediate in the islands where the high vulnerability to the impacts of climate change, limited land and water resources, often unsustainable natural resource use, and other particular economic vulnerabilities are disrupting livelihoods. The COVID-19 pandemic has further exacerbated the SIDS economies and livelihoods. Over the past 25 years the Global Environment Facility (GEF) has supported interventions in the SIDS through $578 million in financing, in critical areas such as biodiversity protection, climate resilience, and energy access through renewable energy. But how effective and sustainable have these interventions been? What factors influencing the sustainability of GEF interventions can provide insights for future project design and implementation?

Despite many regional and national differences indicative of the heterogeneity across SIDS, with context-specific environmental and socioeconomic development challenges (United Nations Office of the High Representative for the Least Developed Countries, Landlocked Developing Countries and Small Island Developing States [UN OHRLLS], 2015; United Nations Environment Programme [UNEP], 1999, 2008, 2010, 2013; World Bank, 2009, 2015), these nations share certain geophysical constraints, environmental challenges, and economic vulnerabilities due to their small size, geographic remoteness, and fragile environments. Their resource base is limited with a predominant focus on natural resources and tourism, domestic markets are typically small, and remoteness results in high costs for energy, infrastructure, and transportation, and a heavy dependence on a few markets for exports. Their openness makes them particularly vulnerable to economic shocks and their growth has been sluggish (OECD, 2018). SIDS are also highly vulnerable to climate change and natural disasters. Climate change is causing sea-level rise, beach erosion, coral bleaching, more invasive alien species, and is fundamentally adversely impacting the main economic sectors of agriculture, fishing, and tourism.

The GEF Independent Evaluation Office (IEO) conducted a strategic country cluster evaluation (SCCE) of SIDS in 2019–2020, evaluating the relevance and effectiveness of GEF interventions in countries in the Atlantic, Indian Ocean, Caribbean, and the Pacific. The overarching objectives of the SCCE were:

1. To assess the relevance of GEF support in addressing the main environmental challenges in SIDS.
2. To provide a deeper understanding of the determinants of sustainability of outcomes for future design and implementation.

To address these questions, we analyzed GEF SIDS projects completed between 2007 and 2014 for sustainability of outcomes; these date parameters provided sufficient time after project completion to observe early trends towards sustainability of outcomes. To further explore the determinants of sustainability, we undertook country case studies of Kiribati and Vanuatu in the Pacific; Comoros, Maldives, Mauritius, and Seychelles in the Indian Ocean; Guinea-Bissau in West Africa/Atlantic; and Belize, Dominican Republic, Jamaica, and St. Lucia in the Caribbean. The evaluation also examined cross-cutting issues on gender, vulnerability/resilience, and private sector engagement in relation to their role in achieving sustainable outcomes.

Environmental Challenges in SIDS

SIDS confront many severe challenges, especially climate change that results in sea-level rise, the increased impact of natural disasters and invasive alien species, problems relating to nonsustainable use of land and water affecting the productive sectors, and issues with the governance of the natural resources (UN OHRLLS, 2015; UNEP, 1999, 2008, 2010, 2013; World Bank, 2009, 2015).

According to an Intergovernmental Panel on Climate Change (2019) special report, *The Ocean and Cryosphere in a Changing Climate*, the sea level is likely to rise 0.61–1.10 m by 2100 if global greenhouse gas emissions are not mitigated. However, a rise of 2 meters or more cannot be ruled out. Even if efforts to mitigate emissions are effective, extreme sea level events will become common before 2100, and probably by 2050 in many locations. Without ambitious adaptation, the combined impact of hazards such as coastal storms and high tides will drastically increase the frequency and severity of flooding and land erosion in low-lying SIDS (OECD, 2018). Particularly at risk from rising sea levels are the Bahamas, Kiribati, Maldives, the Marshall Islands, and Tuvalu, where between 30% and 55% of the land is less than 5 m above sea level (World Bank, n.d.).

Beach erosion is another common problem in SIDS, and has increased due to climate change. The coral reefs around many islands are also severely affected by global warming, which is causing ocean acidification, reef degeneration, and more frequent coral bleaching. Coastal tourism-related development and an influx of tourists put pressure on coastal areas and feed into coral reef degradation. More than 70% of Antigua and Barbuda's coral reef is threatened by coastal development; in St. Vincent and the Grenadines, the coral reefs around Tobago Cays are under threat of further deterioration due to the anchoring of cruise ships. The development of marinas, hotels, and other tourism-related facilities has also put pressure on mangroves and wetlands and reduced important fish breeding habitats.

The primary sectors of agriculture, agroforestry, fisheries, and tourism are important in most SIDS. In atoll countries, soils are mostly infertile and not conducive for agriculture. Limited freshwater resources combined with excessive drainage in these islands makes agriculture even more difficult, with the result that annual crops often are produced only in the rainy season. Climate change and unusual weather variability have made agricultural production planning increasingly difficult. The volcanic islands often have fertile soils and a large number of crops can be produced at different altitudes. However, the soils are often degraded due to deforestation and overexploitation by a relatively high population, and strong tropical rainfalls cause erosion and landslides. Poor land management practices such as slash-and-burn agriculture, uncontrolled livestock grazing on fragile lands, poor road construction, and unplanned or poorly planned settlements in landslide-prone areas have further exacerbated land degradation (Food and Agricultural Organization of the United Nations, 2017). In SIDS in Latin America and the Caribbean, land degradation costs an estimated $4.8 billion dollars annually, and impacts approximately 125 million people within the region (UNEP, 2014). It directly impacts human livelihoods and survival, with significant negative implications for the most vulnerable groups in society.

Many SIDS see themselves as large ocean states, as their ocean territories are approximately 20.7 times greater than their land area, and many are promoting sustainable use of ocean resources while generating economic growth, building social and financial inclusion, and preserving and restoring ocean ecosystems (Meddeb, 2020). The oceanic and coastal fishing industry represents an important source of nutrition and revenue for SIDS populations. However, unsustainable commercial fishing has put pressure on marine resources. In Nauru, Palau, and Tonga, commercial fishing accounts for 50%–70% of total fishery activity, and although the number of tons produced per year is rather small, it does have an impact on fish stocks. The top three fish-exporting SIDS—Fiji, Kiribati, and Papua New Guinea—have lower rates of commercial fishing, ranging from 10.0% to 28.6% of the country's respective total fishery activity. Marine resources here are also threatened by natural disasters, mainly cyclones, damaging fishing grounds and fish breeding habitats, and seabed mining is a critical issue in Papua New Guinea. In all SIDS, illegal, unreported, and unregulated fishing; harmful fishing subsidies; pollution; habitat degradation; governance structures; and a lack of policies and their enforcement pose threats to marine resources.

The isolated nature of SIDS makes for small populations and restricted habitats, leading in turn to often unique but also extremely fragile biodiversity, where species often lack the ability to adapt to rapid changes. Countries that currently face immediate threats to their flora and fauna include Cabo Verde, Cook Islands, Guinea-Bissau, Kiribati, Palau, São Tomé and Príncipe, the Solomon Islands, and Vanuatu.

Invasive alien species are the primary cause of species extinctions in island ecosystems. If left unchecked, these species can degrade critical ecosystem services on islands, such as the provision of water and the productivity of coastal areas. Large numbers of invasive alien plants often cause problems in the agricultural sector and forest areas, and in freshwater bodies. Invasive animal species are also a big problem, such as the small Indian mongoose (*Herpestes auropunctatus*) and rats that prey on native animals and eat bird eggs. In the ocean, invasive alien species have been less frequently reported, but the lionfish (*Pterois volitans*) that is native to the Indo-Pacific is a problem in the Caribbean, where its toxic spikes are a threat to biodiversity and tourists.

Another challenging issue in SIDS is waste management, due to lack of space and deficient waste-handling systems. Solid waste is frequently burned or discarded in the sea or in nearby mangroves. Large amounts of solid waste are accumulated on land, then often flow into the ocean. The substantial number of tourists and tourist facilities in many SIDS increase the amount of waste produced. In St. Vincent and the Grenadines, wastewater from tourist yachts has severely polluted the eastern coasts. Solid and liquid waste make their way to the coastal areas, contaminating beaches and marine ecosystems. Sewage water most often goes directly into the sea without any treatment. Permeation of aquifers by wastewater, including contaminated water from agricultural production (fertilizers, pesticides), also reduces water quality.

Many SIDS have rich but currently untapped repositories of mineral resources. Extraction is an important source of foreign capital and government revenue, and a source of jobs, but is also associated with negative environmental effects. Mining takes a toll on the environment in several SIDS. For example, some of Guyana's and Suriname's extractive processes for gold use cyanide and mercury, which are both highly toxic. Impacts from mining include soil contamination, deforestation, removal of soil surface, and biodiversity loss. In the Americas, SIDS particularly at risk from the environmental impacts of mining are Cuba, the Dominican Republic, Guyana, Haiti, Jamaica, and Suriname. In the Pacific, phosphate mining in Nauru has a major impact on natural resources. For many SIDS, sand mining and seabed mining are practices that have a major impact on the integrity and sustainability of local ecosystems.

The discussion above highlights the fact that environmental issues in SIDS are clearly interrelated and impacted by economic constraints such as limited diversification; small markets; high levels of indebtedness; high costs of energy, infrastructure, communication, and transportation; limited institutional capacity; and brain drain. Growing recognition of the vital importance of the oceans to the economies and livelihoods in SIDS has increased calls for integrated "blue economy" approaches, the "sustainable use of ocean resources for economic growth, improved livelihoods and jobs, and ocean ecosystem health" (World Bank, 2017). At the same time, SIDS face fundamental challenges that must be tackled immediately—especially their high vulnerability to the impacts of climate change, reflected in the need for sustainable management of natural resources on land and in the ocean, and the need to convert to renewable and less costly energy sources. Adaptation measures are complicated by limited land and water resources, lack of awareness, and long-standing traditions of unsustainable exploitation of resources. As such, appropriate environmental interventions would require an integrated approach to land, water, forest, biodiversity, and coastal resource management, which in turn would have an impact on economic livelihoods. In the next section, we highlight some of the main SIDS interventions of the GEF that aim to address these complex and systemic challenges.

GEF Interventions in SIDS

The GEF has a mandate to protect the global environmental commons—the biodiversity, water, oceans, healthy forests, land, and stable climate on which the planet and human health depend. The pressures on these resources are immediate in SIDS, in view of their unique biodiversity and vulnerability. Although the GEF does not have an official strategy for SIDS, it has for more than 25 years supported projects in critical areas for SIDS such as biodiversity protection on land and in the ocean, resilience to climate change and related disaster risk management, increased energy access through renewable energy and energy efficiency, halting and reversing land degradation, cooperation on international waters, and improved chemicals management. In total, between 2006 and 2018, the GEF has invested $1.37 billion in SIDS through 337 interventions, 219 of which were at the country level and the others at a regional level. Recently, the GEF has planned an additional $233 million commitment through 2022. Of GEF allocations in SIDS, 43% are in Asia, 37% in Latin America and the Caribbean, and 20% in the Atlantic, Indian Ocean, Mediterranean and South China Sea (AIMS). The GEF implements projects through 18 implementing agencies. The United Nations Development Program (UNDP) has implemented more than half the GEF projects in SIDS; together, the UNDP, UNEP, and World Bank have implemented more than 85% of the GEF SIDS portfolio.

This evaluation included a desk review of the portfolio of 286 GEF projects in the 39 SIDS. Most GEF projects reviewed are focused on climate change mitigation and adaptation, including energy, followed by biodiversity (31%) and international waters. Table 1 presents the environmental domains addressed in the GEF SIDS projects. Fifteen percent of the portfolio's projects address more than one area. Many projects cover watershed management from an integrated natural resource management perspective, sometimes with a ridge to reef approach, and establish alliances with the agricultural sector in conservation

Table 1 Environmental domains in SIDS GEF projects

Environmental domains	Projects	%
Threats to terrestrial biodiversity	71	24.91
Deforestation and land degradation	48	16.84
Climate change mitigation—emission reduction	45	15.79
Climate change adaptation—sea-level rise	42	14.74
Water quality and quantity	39	13.68
General capacity building	39	13.68
Climate change mitigation—renewable energy and energy efficiency	35	12.28
Threats to marine resources	34	11.93
Coastal and coral reef degradation	30	10.53
Threats to freshwater fishery resources	17	5.96
Waste management	14	4.91
Climate change adaptation—natural disasters	5	1.75

of soil, water, and biodiversity.[1] Many projects under one focal area generate co-benefits in other areas, especially between the areas of biodiversity and climate change, but these co-benefits are often not measured. The chapter appendix provides a list of all projects discussed in this chapter, their implementing agencies, and the countries in which the interventions took place.

We found that all projects reviewed had a satisfactory rating for relevance to the national environmental challenges and were relevant for the environmental priorities in relation to national priorities.[2] This was further reflected in the governments' interest in employing GEF funding to confront their challenges. GEF-financed projects

---

[1] Two examples are A Ridge to Reef Approach for the Integrated Management of Marine, Coastal and Terrestrial Ecosystems in the Seychelles; and Conserving Biodiversity and Reducing Land Degradation Using a Ridge to Reef Approach in St. Vincent and the Grenadines.

[2] Some examples include Support to the Alignment of Jamaica's National Action Programme to the UNCCD 10 Year Strategy; Mainstreaming Global Environmental Priorities into National Policies and Programmes; Renewable Energy Technology Development and Application, which supported the Maldives national strategy in the area of renewable energy; and Sustainable Management of POPs in Mauritius, which was designed to comply with the priorities in the Mauritius National Implementation Plan on hazardous waste.

are also well aligned with the GEF and convention strategies in climate change, biodiversity, sustainable forest management, and hazardous waste. The ministers of environment and other government officials interviewed highlighted that the GEF is an important source of funding that fits into their priorities and planning. This is also reflected in the country programs of GEF Agencies that have a national presence, including the UNDP, the World Bank, and regional development banks (Inter-American Development Bank, African Development Bank, and Asian Development Bank). Discussions around formulation and identification of priority areas take place between the GEF focal point in a country's government and relevant ministries and agencies, with consultation with relevant GEF Agencies.

Climate Resilience

The GEF adaptation projects support investment, policy, and capacity-building measures in a range of sectors that are vulnerable to climate risk, including agriculture, fisheries, water resources, health, and urban and coastal settlements. To improve climate resilience and reduce disaster risks, the GEF supports land use planning with an integrated and sustainable natural resources management approach and disaster risk management focused especially on prevention and mitigation of natural disasters. Through its two adaptation funds, the Least Developed Countries Fund and the Special Climate Change Fund, the GEF has built an active portfolio of projects across SIDS in Africa, the Indian Ocean, Asia-Pacific, and Latin America and the Caribbean. Recent GEF support has focused on:

- Disaster preparedness and resilience, including mapping of disaster-prone areas and establishment of local early warning systems, as well as ecosystem-based approaches
- Innovative tools to manage disaster risk such as risk insurance facilities, risk pooling, risk transfer, and supportive policy and capacities
- Win-win solutions that can deliver both adaptation and global environmental benefits, such as improved access to drinking water (including rainwater harvesting), improved access to clean and resilient energy, more climate-resilient smallholder food systems, and integrated semiurban and urban planning

Building the capacity of the private sector to engage in climate change adaptation and mainstreaming community and gender considerations are also important aspects.

For example, the Kiribati Adaptation Program focused on climate resilience and disaster risk management, including the design of seawalls to protect against sea-level rise and coastal erosion. The subsequent phases continued the process, strengthening climate resilience based on the strategies and designs developed, and improved the seawall designs based on lessons learned during the previous phase (see Fig. 1).

Fig. 1 Seawall models used in the World Bankv Kiribati Adaptation Program (KAP)
From left: Failed eroded KAP II sandbag seawall, KAP III seawall with cement sandbags, and KAP III rock seawall using imported rocks. (2019 photos courtesy T. Norheim)

Integrated Resource Management Through Ridge to Reef

Thirty percent of the GEF projects in SIDS consider integrated approaches such as ridge to reef, whole island approach, or blue economy. The GEF is supporting SIDS countries in implementing such approaches to sustainably manage soil, water, and biodiversity while also considering renewable energy resources and productive sectors such as agriculture, forestry, fisheries, and tourism. "Ridge to reef" is an integrated watershed management approach in which the planning area starts at the top of the island and ends at the coral reef. The approach is designed to reverse the degradation of coastal resources by finding ways to reduce the flow of untreated wastewater, chemicals, nutrients, and sediments from land-based economic activities and cities into deltas, coastal zones, and oceans. Two ecosystems are specifically important for the resilience and economic viability of the coastal zones: the mangroves and the coral reef. Ridge to reef is one important measure to help defend these ecosystems that protect human settlements against natural disasters and are also important for productivity of fisheries. Consequently, this approach employs integrated water resource management and integrated coastal management plans that come together into long-term sustainable use of natural resources, while limiting the impact on fragile environments.

Blue Economy

Another priority area of the GEF is strengthening national blue economy opportunities through a combination of national and regional investments. GEF support aims to sustain healthy coastal and marine ecosystems, catalyze sustainable fisheries management, and address pollution reduction in marine environments. The GEF assists SIDS in identifying sustainable public and private national investments through funding of collective management of coastal and marine systems and implementation of integrated ocean policies and legal and institutional reforms. This support is often channeled through regional GEF programs, which also encourage South-South knowledge transfer. Examples from the various regions are the Pacific's Strategic Action Program, Addressing Land-Based Activities in the Western Indian Ocean, and Catalyzing Implementation of the Strategic Action Programme for the Sustainable Management of Shared Living Marine Resources in the Caribbean and North Brazil Shelf Large Marine Ecosystems.

Protected Areas

GEF assistance has included the establishment of new protected areas, building capacity for planning and effective area management including co-management with local stakeholders, and establishment of protected area funds and other mechanisms for sustainable financing. The GEF supports strategies to reduce the negative impacts of tourism, fisheries, and agriculture, while at the same time allowing traditional communities situated in and around the areas to carry out sustainable income-generating activities from fruit, nuts, fish, eco-tourism, etc., based on the ecosystems' carrying capacity.

Land Use Management

The GEF's work in land degradation—specifically deforestation and desertification—has emphasized the need to take an integrated approach to sustainable land management while ensuring the sustainability of livelihoods. The GEF has now expanded this approach to include the United Nations Convention to Combat Desertification (UNCCD, 2017) guiding principle of land degradation neutrality. The GEF's support to SIDS has evolved in the same way, seeking to ultimately halt and reverse land degradation, restore degraded ecosystems, and sustainably manage the resources.

The many environmental challenges on land and in the ocean are interconnected, and GEF projects to confront these challenges recognize this. Addressing a single challenge separately is not possible, because management of soil, water, and waste impacts the ocean, and thereby human economic activities, especially fisheries. Examples of projects demonstrating this include the regional program Combating Living Resource Depletion and Coastal Area Degradation in the Guinea Current LME Through Ecosystem-Based Regional Actions; Integrated Ecological Planning and Sustainable Land Management in Coastal Ecosystems of Comoros; and Integrated Management of the Yallahs River and Hope River Watersheds.

Invasive Alien Species

Invasive alien species are one of the main causes of ecosystem degradation and species extinctions in SIDS. Many SIDS have been geographically isolated for thousands of years and are therefore more vulnerable to the effects of alien species. The GEF continues to support the implementation of comprehensive prevention, early detection, control, and management, while emphasizing a risk management approach that focuses on the highest risk invasion pathways.

Chemicals and Waste

Toxic chemicals, other hazardous waste, and waste arriving from the ocean present acute challenges to the fragile ecosystems in SIDS and their coastal areas. GEF programs seek to address the sound management of chemicals and waste through strengthening the capacity of subnational, national, and regional institutions and strengthening the enabling policy and regulatory framework in these countries.

Renewable Energy and Energy Efficiency

Several SIDS have a huge potential of untapped renewable energy resources from solar, wind, hydroelectric, tidal, geothermal, and biomass resources, but continue to meet a high percentage of their energy needs by burning fossil fuels. The GEF supports SIDS to strengthen national energy security, develop clean energy policies, catalyze private investments in the renewable energy sector, and facilitate the use of advanced renewable energy and energy efficiency technologies in agriculture and urban and rural development, with co-benefits to health, community development, poverty eradication, and women's empowerment.

Within each of these areas, the GEF must assure support to achieve global environmental benefits. This evaluation showed that, consistent with the challenges SIDS confront, the most important areas include maintaining biodiversity goods and services (36.8%) and support for low-emission development (35.1%), followed by enhancement of the countries' capacity to implement multilateral environmental agreements and mainstream them into national and subnational policy, planning, financial, and legal frameworks (26%).

In all the different focal areas, GEF interventions have mostly focused on strategy implementation and institutional capacity development, and on various aspects of knowledge management. Infrastructure investment is included in only a few projects, usually at a small scale. Institutional strengthening, including training, continues to be important, not only for SIDS governments, but also for effectiveness and efficiency in all GEF projects (see Table 2). These issues are especially significant for the least developed countries (LDCs) that have fewer resources for the public sector. SIDS are also in favor of regional projects with South-South sharing of knowledge, which is yielding important benefits for the smallest and poorest countries. The evaluation found support for regional programs especially in the Indian Ocean, but also in the Caribbean and Pacific, where countries are in favor of regional programs if they include a strong national component (for pilot projects), and transfer of knowledge/lessons learned that especially benefit the smallest countries.

Table 2 GEF contribution areas

Area	Sub area	Projects	%
Strategy implementation	Technologies and approaches	120	42.11
	Implementing mechanisms and bodies	81	28.42
	Financial mechanisms for implementation and sustainability	62	21.75
Institutional capacity development	Policy, legal, and regulatory frameworks	172	60.35
	Governance structures and arrangements	66	23.16
	Informal processes for trust building and conflict resolution	1	0.35
Knowledge management	Knowledge generation	125	43.86
	Information sharing and access	92	32.28
	Awareness raising	73	25.61
	Skills building	152	53.33
	Monitoring and evaluation	73	25.61

Table 3 Positive environmental outcomes mentioned in the terminal evaluation reports in SIDS

Area of positive environmental outcome	Projects	%
Threats to terrestrial biodiversity	18	51.43
Deforestation and land degradation, including SLM	13	37.14
Water quality and quantity	10	28.57
Waste management	8	22.86
Threats to marine resources	7	20.00
Coastal and coral reef degradation	5	14.29
Climate change mitigation, emission reduction	5	14.29
Renewable energy and energy efficiency	5	14.29
Climate change; sea level rise	2	5.71
Other	1	2.86

Note: The total number of projects is higher than the number of the projects with terminal evaluations (45) because several projects had more than one environmental outcome

Table 4 Areas of positive changes in building institutional capacity/governance in GEF projects in SIDS

Area of capacity building, institutional development, or improved governance	Projects	%
Capacity and skills development	38	86.36
Awareness raising	32	72.73
Development of plans, policies, codes, covenants, laws, and regulations	25	56.82
Knowledge management, information-sharing, and knowledge systems	24	54.55
Institutional and decision-making processes, structures, and systems	17	38.64
Environmental monitoring systems	14	31.82
Decision-makers' information and access to information	6	13.64
Trust-building and conflict resolution	2	4.55
Other	1	2.27

Note: The total number of projects is higher than the number of the projects with terminal evaluations (45) because several projects had more than one area of positive change

Performance and Sustainability of GEF Projects in SIDS

Based on a detailed review of 45 closed SIDS projects with terminal evaluation reports prepared at closure, we observed positive environmental institutional capacity building and socioeconomic outcomes in more than 75% of the projects (for example, see Box 1). The findings were further validated through in-country visits to these projects in 2018. The main positive environmental impacts were in the areas of biodi-

versity, deforestation/land degradation, and water quality/quantity (see Table 3). Socioeconomic outcomes were observed in the areas of income generation/diversification, private sector engagement, and civil society engagement. All the projects except one (97.78%) reported improvements in institutional capacity or governance (see Table 4).

Overall, the SIDS portfolio performance was slightly lower than that of the overall GEF portfolio. Factors contributing to this include limited project preparation time, particularly for projects cutting across various environmental areas; the relative complication of GEF projects, compared to those of other funding agencies, and the projects' additional burden on existing limited capacity; and weak national institutional capacity for procurement. Of note, nearly all GEF projects are implemented during a single phase with a duration of 4–5 years. New projects with similar or complementary goals are often approved without designing a coherent next phase based on results and lessons learned. Monitoring information, including the availability of baseline data, continues to be a challenge.

Sustainability

Sustainable development is defined as "development that meets the needs of the present without compromising the ability of future generations to meet their own needs" (World Commission on Environment and Development, 1983). For a project, sustainability is understood as the likelihood of continuation of the benefits after completion of project implementation. Donors are increasingly interested in ensuring that benefits continue past interventions to ensure longer term outcomes and impacts.

Box 1: Case Study: Geospatial Analyses on the Outcomes of the Iyanola—Natural Resource Management Project in the NE Coast of St. Lucia

This case study demonstrates the relevance and effectiveness of GEF interventions using geospatial analysis. The $7.3 million Iyanola—Natural Resource Management of the NE Coast project was launched in 2015 to improve the effective management and sustainable use of the natural resource base of the northeast coast of Saint Lucia and generate multiple global environmental benefits. The region hosts Iyanola dry forests that are classified as the key biodiversity areas and as important bird areas. These dry forests are unique to the region and an important habitat for a combination of rare and endemic flora and fauna species, with ecosystems rich in biodiversity and unique dry scrub forests and pristine beaches (see Figs. 2 and 3). The forest region is also endowed with a variety of environmental resources that form an important and potential socioeconomic and cultural asset base of the island's national economy.

The Iyanola dry forests area is threatened mainly by agriculture expansion, logging, and forest fire due to slash-and-burn practices. To address these threats, the GEF project adopted a cross-sectoral, strategic approach to integrated landscape management involving forest, coastal, and land use management. The main activities included developing a regulatory framework, enhancing capacity to produce biodiversity-friendly goods and services, restoration, and piloting land use plans. Time series forest loss data (see Fig. 4) shows an increase in forest loss in the protected area before the project implementation started in 2015, and a slight decrease during the project period. At the 2018 data point, the percent loss had further decreased to 0.05% in the protected area and about 0.04% in the buffer areas.

As a result of the GEF interventions, vegetation cover increased between 2015 and 2016 in restoration sites (see Fig. 5). The average normalized difference vegetation index (NDVI) at the three restoration sites increased by 20% between 2015 and 2018; the productivity tapered down in 2018 compared to the previous 2 years, perhaps due to a decrease in precipitation. The plantation of native and nonnative trees together with the understory led to increased vegetation productivity, also verified during site visits.

Fig. 2 Iyanola dry forests, St. Lucia

Fig. 3 Location of the
Iyanola project sites

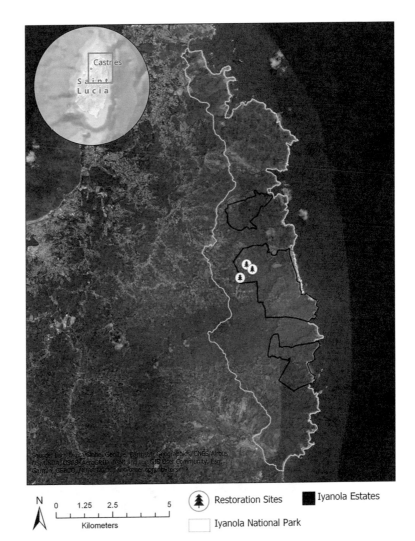

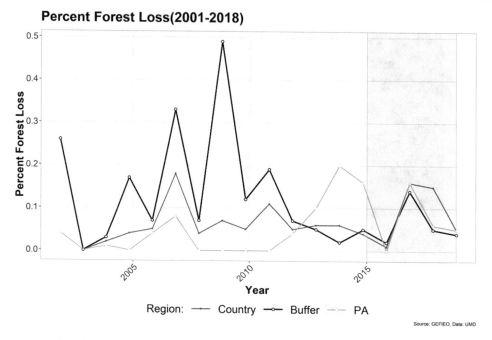

Fig. 4 Percent forest loss, 2001–2018. (Source: GEF IEO, 2019)

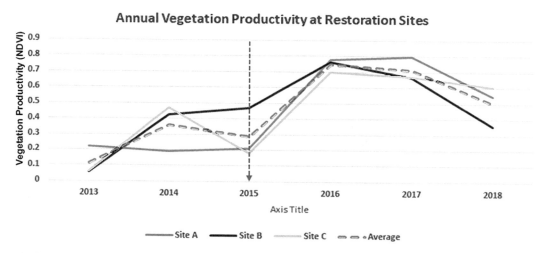

Fig. 5 Landsat-derived vegetation productivity at the restoration sites; NDVI before and during the project. (Source: GEF IEO, 2019)

The GEF SIDS SCCE examined 45 projects which were rated for their likely sustainability at completion. Half the projects had an overall sustainability rating of likely or moderately likely (see Table 5). We found relatively small differences between the different dimensions of sustainability, with political sustainability being most likely.

Factors Affecting the Sustainability of Outcomes in SIDS

The main contextual and project-related factors that affect sustainability are summarized in Table 6. These were developed based on field visits and further confirm results from our desk analysis of the 45 projects on national and regional

Table 5 Ratings on four dimensions of project sustainability in SIDS

Sustainability dimension	Likely/moderately likely	Moderately unlikely/unlikely	Not rated/NA
Financial	53	18	28.88
Political	62	9	28.89
Institutional	51	22	26.67
Environmental	49	9	42.22

Table 6 Observed contributing and hindering factors influencing the sustainability of outcomes

Sustainability	Contributing factors	Hindering factors
Context related	• Legal and institutional framework for environment and protected areas • Government policies supporting environmental conservation, climate change mitigation, and adaptation • National ownership of projects, reflected in government support and budget allocation • Strategic institutional partnerships • Public–private partnerships in the key sectors • Sustainable national financing mechanisms, e.g. environmental funds, to cofinance projects • General institutional capacity, especially in the public sector	• Low institutional capacity, especially in the relatively smaller countries, with low ownership, little institutional memory, high turnover, and brain drain • Unfavorable political conditions and events in some countries (coup d'etat, corruption, civil protests) • Often weak national and local environmental NGOs with low technical capacity and limited influence on decision making and low capacity on local level to implement planned activities • Low level of environmental awareness, reflected in the public's attitude to waste and to renewable energy sources • Pressure from the agricultural and tourism sectors to exploit sensitive areas, from a land, coastal, and marine environment perspective • Natural disasters and unfavorable environmental conditions (hurricane, drought, earthquake, tsunami) • Infrastructure constraints that make transport and communication across islands difficult, impacting learning and knowledge sharing
Project related	• Training and institutional capacity building, including introduction of new technology and new techniques • Buy-in and sense of ownership among key project stakeholders • Adaptive project management • Strength of project teams and engagement of steering committees • Strategic institutional partnerships • Replication and scaling-up based on lessons learned, including small-scale local investments financed by GEF-SGP, NGO/CSO, and the private sector	• Project design that does not consider previous projects in the sector and lessons learned • Little consideration of impact and sustainability in the project design • Insufficient involvement of main stakeholders during design and implementation • Weak project monitoring and risk management • Insufficient national and local capacity building to assure continuation of activities • Lack of exit strategy and future financing to sustain the projects' momentum

interventions that were completed between 2007 and 2014, allowing time after project completion to observe the long-term sustainability of outcomes. Below, we discuss these factors with examples from GEF projects.

Project-related factors that influence sustainability include training and building institutional capacity, good project design and adaptive project management, an engaged project steering committee, building strategic institutional partnerships, scaling up and replication based on lessons learned, and a good exit strategy.

We found that the most important project-related hindering factor was the quality of project design, which sometimes gave little consideration to long-term impact and sustainability. Many of these SIDS projects had a short time horizon for planned outcomes and impact, and the issue of sustainability was often considered only from a financial point of view. Not enough consideration was given to previous projects in the same sector (e.g., biodiversity, energy) and even though the project documents always list preceding projects, seldom did deep analysis occur of lessons learned that could help avoid repeating errors from the past. Project plans from international consultants often provided a theoretical approach without on-the-ground technical and social knowledge; at the same time, most SIDS have less specialized capacity for project design. Therefore, collaboration between national specialists and international counterparts is necessary. Another challenge for many SIDS is that GEF projects must include a high percentage of cofinancing.

Important contextual factors that affect sustainability in SIDS were found to be national policies and legal and regulatory frameworks, national ownership of projects, national environmental funds, environmental awareness, institutional capacity, and strategic institutional partnerships.

The most important context-related factor was the national-level legal and regulatory framework for environment and protected areas, and the extent to which the laws were enforced. For instance, in Comoros, unsustainable forest and agricultural practices, including slash-and-burn and overexploitation for firewood and timber, have greatly reduced the possibility for regeneration of natural forest ecosystems. The government has developed policies and incentives to promote agricultural production and self-sufficiency of food products.

National ownership of the projects is an important contributing factor for sustainability, as reflected in local stakeholder participation and government support and budget allocation. Ownership by national institutions was clearly demonstrated in the case of the Partnership for Marine Protected Areas in Mauritius, in which various departments of the government have provided for sustained budgeting for the conservation of marine resources and biodiversity since the early 1990s. After the project closed in 2012,

the total annual budget for marine conservation was estimated at $5.2 million and increased up to an average of approximately $9.9 million from 2013 to 2018, with a peak budget of $12.4 million in 2016 due to the construction of the Blue Bay Marine Park Centre.

The establishment of national environmental funds is important for sustainable development financing as demonstrated in Guinea-Bissau through the Biodiversity Conservation Trust Fund, which was able to achieve sustainable results, particularly in capacity building and institutional strengthening. Its most important result was the creation of the Bio Guinea Foundation (FBG) with an initial government funding of 1 million Euros. It is a public fund but managed autonomously with its own board, with the goal of covering the costs of the protected area system and supporting other biodiversity conservation initiatives. The Bio Guinea Foundation is to be capitalized with $1.7 million, including $0.9 million from the GEF's framework strengthening project.

Strategic institutional partnerships, including public–private partnerships, have been another key contributing factor in project sustainability. Long-term partnerships with national NGOs for protected area management have been fundamental for social, environmental, and financial sustainability of protected areas. In Seychelles, the protected area site Vallée de Mai is situated within the Praslin National Park, managed by the National Parks Authority, but is managed separately by the NGO Seychelles Islands Foundation (SIF). The site has the highest concentration of the endemic coco-de-mer palm (*Lodoicea maldivica*), found only on the islands of Praslin and Curieuse. The entrance fees from tourists visiting the site are used to cofinance the UNESCO World Heritage Sites of the Aldabra Atoll more than 1000 km away, where income from tourism is not so easy to manage.

Institutional Capacity, Environmental Awareness, and Economic Pressure

Our evaluation found that low levels of institutional capacity, lack of environmental awareness, and pressure from economic sectors had negative effects on projects' sustainability.

Overall low institutional capacity causes problems especially in the poorest SIDS countries, and brain drain has been an issue in the Pacific islands and the Caribbean. High turnover at the national level is another issue in many SIDS, especially in countries with a dynamic private sector such as Mauritius and the Dominican Republic. Some SIDS, such as Guinea-Bissau, Comoros, and Maldives, have passed through periods of political instability and coup d'etat, which not only affected ongoing projects but also cut off financing from many development agencies for long periods.

Low technical capacity and limited direct influence on decision making of national and local environmental NGOs was another issue noted in Comoros, Kiribati, and Mauritius, where it has limited opportunities for national dialogue on sustainable development and reduced opportunities for partnerships such as those supporting local communities. In contrast, we found that the environmental NGOs in Jamaica and Seychelles were technically strong and had significant influence on political decisions.

Low levels of environmental awareness are reflected in the public's attitudes toward waste and renewable energy sources. One example is attitude regarding the use of disposable plastic. During country visits, evaluators observed huge amounts of solid waste along the coast line, along roads, and even in protected areas. However, communities that act a certain way based on short-term self-interest, mostly due to incentives, differ from communities that act based on awareness that their actions will benefit them and their livelihoods in the future. Awareness raising is a slow process, especially in countries with high poverty rates.

In Guinea-Bissau, the lack of public awareness was demonstrated in large amounts of garbage directly in front of schools established inside national parks, even though the schools have environmental education on the curriculum.[3] The general lack of environmental awareness was also clearly shown in Comoros, where solid waste was found all over the country and waste collection and handling is limited even in urban areas and tourist resorts. However, some governments, including Mauritius, Samoa, and Seychelles, have taken effective measures to forbid single-use plastic bags and conduct awareness campaigns through public media. The national component of the project in Mauritius installed waste incinerators and grids in the four main streams to prevent solid waste from entering the port waters in the Municipality of Port Louis.[4]

Another common challenge is pressure from economic sectors such as agriculture and tourism to exploit environmentally sensitive areas. Deforestation in SIDS due to the advance of the "agricultural frontier" has mostly been limited by lack of road infrastructure in the island interior, steep areas on the volcanic islands, and poor soils on the atoll islands. On the other hand, natural habitats such as mangroves and wetlands in coastal areas have often been eliminated due to shrimp farming and construction of coastal tourist resorts.

Sustainability changes over time as circumstances change. Two thirds of the 24 projects that were subject to field verification had positive (moderately likely or likely) sustainability ratings at completion, while the observed sustainability rating for the same projects was higher after the passage of time, at 81.25%.[5] For example, in Guinea-Bissau, sustainability ratings improved with the political situation after the coup d'etat ended. Another factor is the GEF's project funding timeline: The GEF normally finances only one project phase, with the expectation that the results will be achieved within that period. This evaluation, however, found that just one project intervention is often not sufficient to achieve sustainability. Multi-phase projects, such as the Kiribati Adaptation Program presented

---

[3] The problem is being addressed through local and national awareness raising through the project Strengthening the Financial and Operational Framework of the National Protected Areas System in Guinea-Bissau.

[4] Projects such as Addressing Land-Based Activities in the Western Indian Ocean have helped reduce the threats of waste to the health of marine and coastal ecosystems.

[5] The evaluation noted differences in sustainability ratings over time in 12 projects. In nine, sustainability improved over time; two projects had a decline in the sustainability rating; and in one project in St Lucia, the changes were mixed.

earlier, have a higher likelihood of sustainability.[6] As an alternative to several phases, replication and scaling-up of project activities can strengthen the sustainability of outcomes. These follow-up interventions may be financed through other national or international sources.

GEF's Overall Additionality in SIDS

The GEF's strongest areas of additionality in SIDS are strengthening institutions and assistance with legal and regulatory frameworks, which, as the discussion above highlights, are very important for sustainability of outcomes (GEF IEO, 2018). Projects across SIDS have achieved results in other areas of additionality to varying degrees, with the weakest area being accessing private sector financing (see Table 7).

Table 7 GEF's main areas of additionality in SIDS

Additionality elements	Project design	Results achieved
Innovation additionality		
Focus on solar technology	✓	✓
Ridge to reef approach	✓	✓
IAS	✓	✓
Socioeconomic additionality		
Encouraging of local solutions	✓	✓
Social inclusiveness	✓	○
Social and economic benefits	✓	○
Institutional/governance additionality		
Strengthening of institutions	✓	✓
Environmental governance	✓	✓
Financial additionality		
Access to private sector financing	✓	○
Policy/regulatory additionality		
Strengthening of the policy and regulatory environment	✓	✓
Environmental additionality		
Adaptation	✓	✓

The evidence is also limited on projects achieving socioeconomic co-benefits and social inclusion. In terms of broadening and ensuring sustainable impact, the most important mechanism is mainstreaming activities in biodiversity and climate change through policies, strategies, and activities of the countries. The second significant channel is sustaining progress in environmental outcomes through attention to the project and contextual factors presented.

Conclusions

Despite the heterogeneity across SIDS, they confront many common and severe challenges: climate change that results in sea-level rise, the increased impact of natural disasters and invasive alien species, problems relating to nonsustainable use of land and water affecting the productive sectors, and issues with the governance of natural resources. These are further impacted by common economic constraints such as limited diversification; small markets; high levels of indebtedness; high costs of energy, infrastructure, communication and transportation; limited institutional capacity; and brain drain. The COVID-19 pandemic has further exacerbated the situation, impacting the tourism industry that is an integral part of these economies. However, drawing on evaluative evidence and lessons from GEF projects, this chapter highlights two important points for sustainability of interventions. First, investment in proven integrated interventions, such as blue economy and ridge to reef approaches, is necessary. Expanding marine and coastal activities could help diversify these economies that are heavily reliant on the tourism sector. Second, attention to contextual and project-related factors is very important. Putting these economies on a path to sustainability and a greener recovery will require investments in sound policy and regulatory frameworks, institutional strengthening, financing from the public and private sectors, and innovations in locally driven solutions that generate economic benefits and are socially inclusive.

---

[6]The Preparation Phase KAP I was implemented 2003–2005 and the Pilot Implementation Phase KAP-II during 2006–2011 followed by the Expansion Phase KAP III. The program's expected outcome is strengthened climate resilience for Kiribati, especially on the main islands.

Appendix: Projects Discussed in Chap. 7

Project name	GEF ID	Country	Implementing agency
SIP: Integrated Ecological Planning and Sustainable Land Management in Coastal Ecosystems of the Comoros in the Three Island of (Grand Comore, Anjouan, and Moheli)	3363	Comoros	IFAD
SPWA-BD: Guinea Bissau Biodiversity Conservation Trust Fund Project	3817	Guinea-Bissau	World Bank
Strengthening the Financial and Operational Framework of the National PA System in Guinea-Bissau	5368	Guinea-Bissau	UNDP
Integrated Management of the Yallahs River and Hope River Watersheds	4454	Jamaica	IABD
Support to the Alignment of Jamaica's National Action Programme to the UNCCD 10 Year Strategy and Preparation of the Reporting and Review Process	5893	Jamaica	UNEP
Kiribati Adaptation Program—Pilot Implementation Phase (KAP-II)	2543	Kiribati	World Bank
Renewable Energy Technology Development and Application Project (RETDAP)	1029	Maldives	UNDP
Partnerships for Marine Protected Areas in Mauritius	1246	Mauritius	UNDP
Sustainable Management of POPs in Mauritius	3205	Mauritius	UNDP
A Ridge-to-Reef Approach for the Integrated Management of Marine, Coastal and Terrestrial Ecosystems in the Seychelles	9431	Seychelles	UNDP
Iyanola—Natural Resource Management of the NE Coast	5057	St. Lucia	UNEP
Conserving Biodiversity and Reducing Land Degradation Using a Ridge-to-Reef Approach	9580	St. Vincent and the Grenadines	UNDP
Mainstreaming Global Environmental Priorities into National Policies and Programmes	5655	Vanuatu	UNDP
Combating Living Resource Depletion and Coastal Area Degradation in the Guinea Current LME through Ecosystem-based Regional Actions	1188	Regional: Angola, Benin, Congo, Cote d'Ivoire, Cameroon, Gabon, Ghana, Equatorial Guinea, Guinea-Bissau, Liberia, Nigeria, Sierra Leone, São Tomé and Principe, Togo, Congo DR	UNDP
Catalyzing Implementation of the Strategic Action Programme for the Sustainable Management of Shared Living Marine Resources in the Caribbean and North Brazil Shelf Large Marine Ecosystems (CMLE+)	5542	Regional: Antigua and Barbuda, Barbados, Brazil, Belize, Colombia, Costa Rica, Dominica, Dominican Republic, Grenada, Guatemala, Guyana, Honduras, Haiti, Jamaica, St. Kitts and Nevis, St. Lucia, Mexico, Panama, Suriname, Trinidad and Tobago, St. Vincent and Grenadines	UNDP
Strategic Action Programme (SAP) of the Pacific Small Island Developing States	530	Regional: Cook Islands, Fiji, Micronesia, Kiribati, Marshall Islands, Nauru, Niue, Papua New Guinea, Solomon Islands, Tonga, Tuvalu, Vanuatu, Samoa	UNDP
Addressing Land-based Activities in the Western Indian Ocean (WIO-LaB)	1247	Regional: Kenya, Comoros, Madagascar, Mauritius, Mozambique, Seychelles, Tanzania, South Africa	UNEP

References

Food and Agricultural Organization of the United Nations. (2017). *Land degradation assessment in small island developing states.* http://www.fao.org/3/a-i7744e.pdf

Global Environment Facility Independent Evaluation Office. (2018). *An evaluative approach to assessing GEF's additionality.* https://www.gefieo.org/council-documents/evaluative-approach-assessing-gef%E2%80%99s-additionality

Global Environment Facility Independent Evaluation Office. (2019). *Small island developing states (SIDS) strategic country cluster evaluation (SCCE).* https://www.gefieo.org/evaluations/small-island-developing-states-sids-strategic-country-cluster-evaluation-scce

Intergovernmental Panel on Climate Change. (2019). Summary for policymakers. In H.-O. Pörtner, D. C. Roberts, V. Masson-Delmotte, P. Zhai, M. Tignor, E. Poloczanska, K. Mintenbeck, M. Nicolai, A. Okem, J. Petzold, B. Rama, & N. Weyer (Eds.), *IPCC special report on the ocean and cryosphere in a changing climate.* https://www.ipcc.ch/srocc/chapter/summary-for-policymakers/

Meddeb, R. (2020, December 2). *Small island developing states do not have the luxury of time.* United Nations Development Programme. https://www.undp.org/content/undp/en/home/blog/2020/small-island-developing-states-do-not-have-the-luxury-of-time-.html

Organisation for Economic Co-operation and Development. (2018). *Geographical distribution of financial flows to developing countries 2018: Disbursements, commitments, country indicators.* Author. doi:https://doi.org/10.1787/fin_flows_dev-2018-en-fr

United Nations Convention to Combat Desertification. (2017). *Scientific conceptual framework for land degradation neutrality. A report of the Science-Policy Interface.* https://www.unccd.int/publications/scientific-conceptual-framework-land-degradation-neutrality-report-science-policy

United Nations Environment Programme. (1999). *Pacific islands environmental outlook.* UNEP Regional Office for Latin America and the Caribbean.

United Nations Environment Programme. (2008). *Africa: Atlas of our changing environment.* UNEP Division of Early Warning and Assessment (DEWA). https://www.unenvironment.org/resources/report/africa-atlas-our-changing-environment

United Nations Environment Programme. (2010). *Latin America and the Caribbean: Atlas of our changing environment.* UNEP Regional Office for Latin America and the Caribbean.

United Nations Environment Programme. (2013). *Arab region atlas of our changing environment.* http://hdl.handle.net/20.500.11822/8620

United Nations Environment Programme. (2014). *Emerging issues for small island developing states.* https://www.unenvironment.org/resources/report/emerging-issues-small-island-developing-states

United Nations Office of the High Representative for the Least Developed Countries, Landlocked Developing Countries and Small Island Developing States (2015). *Small island developing states in numbers: Climate change edition 2015.* https://sustainabledevelopment.un.org/content/documents/2189SIDS-IN-NUMBERS-CLIMATE-CHANGE-EDITION_2015.pdf

World Bank. (2009). *Timor-Leste: Country environmental analysis.* https://openknowledge.worldbank.org/handle/10986/28126

World Bank. (2015). *Maldives: Identifying opportunities and constraints to ending poverty and promoting shared prosperity.* https://openknowledge.worldbank.org/handle/10986/23099

World Bank. (2017). *What is the blue economy?* [Infographic]. https://www.worldbank.org/en/news/infographic/2017/06/06/blue-economy

World Bank. (n.d.) *DataBank: World development indicators.* https://databank.worldbank.org/source/world-development-indicators

World Commission on Environment and Development. (1983). *Our common future.* United Nations.

Resources

CBD Country profile for all countries visited: https://cbd.int

Database of GEF projects: www.theget.org/projects

GEF–UNDP Small Grants Program (SGP): www.sgp.undp.org

IUCN Country Profiles: https://www.iucn.org/regions

The Global Register of Introduced and Invasive Species: www.griis.org

The IUCN Red List of Threatened Species: https://www.iucnredlist.org

World Bank Country Profiles: https://www.worldbank.org/en/country

Assessing Sustainability in Development Interventions

Ellen Fitzpatrick

Abstract

Sustainability is often claimed as an impact in development interventions although there is rarely a shared understanding of what it means, how to design for it, and especially how to assess the likelihood that intended streams of benefits will continue. This chapter asserts that to design and later to evaluate an intervention with sustainable impacts, the intervention must deepen indigenous capabilities to manage the program, to solve problems, and to innovate. The design and implementation also must operate within environmental boundaries, not extracting resources beyond the ability to regenerate or degrading environmental services—that is, design and implementation must incorporate the primacy of the environment. A postprogram evaluation 3–10 years after a program has ended provides evidence on whether the program is likely to have sustainable impacts. A case study of an asset transfer program in Malawi highlights the criteria for evaluating sustainability: deepened capabilities and social capital, reinvestment in program activities, and the development of backward and forward linkages catalyzing growing economic opportunities.

Keywords

Sustainable impact · Capabilities · Environment · Design · Evaluation · Malawi

This chapter presents a framework for transforming sustainability from a broad, amorphous idea to a concrete and trackable impact that can be used in the design and evaluation of programs and projects. This framework suggests that if NGO and government programs are to have sustainable impacts, they must deepen human capabilities to manage economic and social change and to operate within environmental limits. To effectively evaluate whether a program is likely to have sustainable outcomes and impacts, a postprogram assessment is essential.

The discussion begins with a description of the problem: an industry born out of a colonial world where NGOs, governments, and foundations have sought to reduce poverty through technology transfers, education, and infrastructure with little attention to the impacts on the environment and the long-term work of collaboration and partnership. The second section presents a framework that defines key components for the design and evaluation of programs for sustainability. The third section presents a case

E. Fitzpatrick (✉)
Interdisciplinary Institute at Merrimack College, North Andover, MA, USA
e-mail: fitzpatricke@merrimack.edu

J. I. Uitto, G. Batra (eds.), *Transformational Change for People and the Planet*, Sustainable Development Goals Series, https://doi.org/10.1007/978-3-030-78853-7_7

study of a postprogram evaluation that provides evidence of sustainable outcomes, including deepened community capabilities and improved livelihoods, and missed opportunities for environmental stewardship. The last section summarizes the argument and makes recommendations for evaluation practices that encourage a more deliberate approach to sustainability.

The Problem

Sustainability as a concept was first introduced in 1987 in response to a United Nations concern about the challenges of economic and social development in a world with increasing ecological degradation. The UN World Commission on Environment and Development produced the study *Our Common Future*, also known as the Brundtland Report (1987). This report defined sustainable development as "development that meets the needs of the present generation without compromising the ability of future generations to meet their own needs" (p. 16). This perspective on sustainability has been critiqued as being anthropocentric and situating other species and environmental services in service to humans. Concern has also been raised that "meeting needs" may sanction over-consumption, especially in the Global North, and justify the continuous extraction of resources from the Global South (Farley & Smith, 2014). The challenge for NGOs, governments, foundations, and academics is that because the term *sustainability* has never been well defined, or perhaps understood, it has been co-opted so that it means whatever the user wants it to mean. Corporations use it to market their products or to secure customer loyalty, multilaterals use it to justify export promotion, and NGOs use the term in program planning, grant seeking, and fund raising to signal a sensitivity to environmental issues.

A shared understanding of the core elements of sustainability and a commitment to assess these elements after a program closes is important for the development community for three reasons. First, when a new program begins, whether focused on improved livelihoods, electricity, clean water, or improved health, it raises communities' expectations. This poses ethical issues: Is it fair to raise expectations when it is unclear whether the program benefits will continue? If we are not reasonably sure that a program will yield long-term streams of benefits, is it fair to ask communities to commit their time, resources, and trust to this activity? Second, do we as development practitioners also have an obligation to do no harm? Are we confident that the proposed activity won't make participants more vulnerable? For example, if a program is proposed that increases income in the short term but degrades an environmental resource, have our actions limited future access to environmental services and hence damaged intergenerational equity? Finally, a postprogram evaluation provides an opportunity to learn about the successes and failures of the program's design and implementation, and provides important information about the likelihood that outcomes and impacts will continue into the future.

Although the Brundtland Report began the serious discussion of biophysical limits of our shared environment, has the popularity of the word sustainability or sustainable development brought us any closer to meeting basic needs for this generation or securing the ability of the earth to meet the needs of future generations? NGOs and multilaterals spend an estimated $150 billion each year in development assistance (World Bank, 2020). Funders and evaluators may want to know whether the work of this large industry has made a difference in terms of well-being, security of the marginalized, and protection of our environment.

Global poverty rates are variable across regions but from 2015 to 2017, approximately 52 million people rose out of poverty. This is 0.5% per year, a decline from 1% per year that was achieved between 1990 and 2015. This decline makes achieving the 2030 goal of less than 3% of the population living in poverty more unlikely

(World Bank, 2021).[1] Poverty rates remain high in many global south countries, especially those affected by political turmoil. The World Bank suggests the region most challenged (where extreme poverty will likely remain in the double digits past 2030) is in sub-Saharan Africa, which had a poverty rate of 41.1% in 2015 (World Bank, 2018). We have spent large sums of money but have made only discouraging progress in meeting basic needs for the current generation.

Our influence on the global environment is less sanguine. Although global environmental problems are largely driven by consumption in the global north countries, the resource extraction and waste burden has taken its toll. The Living Planet Index documents a 68% decline in mammals, birds, amphibians, retiles, and fish from 1970 to 2016 (Almond et al., 2020), a troubling reflection of the health of our ecosystem. The Biodiversity Intactness Index (BII), which examines how much of the earth's original biodiversity remains, shows a global BII at 79%, significantly below the recommended "safe level" of 90% (Almond et al., 2020). So, we have an opportunity to make significant improvement in the practice of sustainability. To do this, we need a shared understanding of what sustainability means and then the willingness to transfer this shared meaning to the design and evaluation of interventions that aspire to contribute to sustainably improving lives.

As we think about the sustainability of programs, we face two related questions: Is the program environmentally sustainable; that is, does the program explicitly account for the resources and environmental services used? Second, are the outcomes and impacts sustainable? While specific outcomes will likely change as the program evolves, do those outcomes contribute to the sustainability of program impacts?

The Environment as a Closed System

To answer the first question, we must acknowledge that we live in a closed system, where the integrity of our environment sets the limits on our economic and social opportunities. As early as 1966, Ken Boulding encouraged us to integrate growth with limits into our thinking when he contrasted the cowboy economy to a spaceship economy; that is, to treat how we grow not as if we have unending resources at our disposal but as if we are all on a spaceship, where what we use and dispose is limited (Boulding, 1966). This requires that the design of programs articulate how the economic, social, and environmental systems are interacting.

Systems thinking can help trace the connection between economic/social change and environmental limits by allowing us to identify feedback loops and causal relationships. Evaluators often use systems thinking as they reproduce a program's theory of change. This approach can also be used to embed a theory of sustainability into program theory. Just as we use linking hypotheses to demonstrate how activities lead to outcomes and outcomes to impact, we can use a theory of sustainability to trace out how the design of a program can lead to sustainable outcomes and impacts and how interventions affect limited environmental resources and services.

To assess whether program activities adhere to the boundaries (limits) of the environment, we can draw from ecological sciences and the role of biodiversity and its importance for resilience. Although resilience is rapidly becoming an overused word, if we understand resilience to mean the ability of a system to bounce back after a shock, then it is an important component of sustainability. Protecting and maintaining diverse ecosystems absorbs some of the service loss that may occur in the attempts to improve human well-being. An example of a diverse food system comes from a visit to the Huicholi who live in the highlands of the Pacific coast of Mexico. They relayed a story of a government official who approached the community to plant hybrid maize, arguing that the yields of the improved variety

---

[1] Poverty rates increased in 2020 in part due to the COVID-19 pandemic. Although the pandemic has been global in its reach, the vulnerable have become more so. This reveals how tenuous the climb out of poverty has been and how little resilience has been created by seven decades of international aid.

would provide more food for the community. The elders replied: "If we take your seed and replace ours, we put our community in danger. Your seeds may produce much more in a good year, but we plant many kinds of maize seed. If the rains don't come, we have maize that will grow. If there is too much rain, some of our maize will grow. The diversity of what we plant ensures that we will eat; can you promise the same with your seeds?" When a biodiverse ecosystem receives a shock, it is more likely that the system can maintain its processes, absorb the disturbance, and retain its function and structure. If biodiversity declines, the system will have decreased ecological resilience and be in danger. Therefore, an intervention will not be sustainable if it diminishes biodiversity health, ecological resilience, and ecosystem vitality. The ecological services provided by the environment create possibilities in the economic and social sphere. Without these possibilities, economic and social systems would collapse. As we design for changes in economic and social spheres, we must also understand the nature and limits of our use. The environment is a closed system; it can't expand as we modify economic and social systems to improve well-being. It is incumbent on us to design for improvements in well-being within this closed system or we will compromise intergenerational equity.

Catalyzing Capabilities to Ensure Sustainable Outcomes and Impacts

Different types of development interventions—whether livelihoods; food production; infrastructure; or water, sanitation, and hygiene—require that we draw on environmental resources and services that have an impact on the ecosystem. For example, when we introduce livestock into a community, we are concerned about how this increased population will affect land use, natural flora, and water supply. We may also be concerned about the introduction of disease, invasive species, and dependencies on external inputs. All of these influence the sustainability and resilience of the ecosystem. Another important component—perhaps a precondition—of sustainability for any

intervention is twofold; first, it must respond to a need or value expressed by the community, what evaluators call relevance, and second, the intervention must be designed to extend the community's capabilities.

Catalyzing or extending communities' capabilities requires that communities are cocreators and become innovators, problem solvers, and managers. Communities are not just collaborators but lead actors (Mog, 2004). If the program is to be sustainable, the community must take over the roles of the NGO. Further, the program should create change by melding community assets with new knowledge, information, or external assets (Mog, 2004).

If a program is implemented without a commitment to this process, outcomes are likely to dissipate when external program management or funding ends. Elements of process that enable sustainable outcomes and impacts require that communities participate in identification of the problem, collaborate in the design of the program, and develop decision-making structures. Another important element is that participants understand and are a part of any new technology, techniques, or methods so that they are able to solve problems and adapt to changing circumstances. Woolcock (2000) extends capabilities to also include the assessment of programs. He suggests that instead of focusing on externally imposed performance indicators, assessments focus on how the community sees the program evolving over time. Evaluators can contribute to this process by working with the community to illustrate how capacity building has led to outcomes and impacts over time, and also to diagnose where failures have occurred and how to learn from them (Woolcock, 2000).

Learning, adaptation, and innovation are more than one-time changes. For example, a common objective for many NGOs is to improve livelihoods. From an economist's lens, the binding constraint may be a lack of physical assets, so an asset such as dairy cattle is introduced. If no accompanying investment is made to further organizational and technical expertise, or to innovate in response to changing situations, then the cows will likely be consumed in a few years and no long-term change in livelihoods will occur. The ability of participants to learn, adapt, and innovate is the lynchpin to sustainable outcomes.

Marginalized communities are not without assets, both human and physical. A theory of sustainability should include the identification and use of local assets. Social capital is an important local asset in most global south communities, especially rural communities. Lin defined social capital as resources embedded in a social structure that can be accessed and/or mobilized in purposive action (Lin, 2008). These social relations have an economic value in that households may rely on social networks to exchange knowledge, provide safety nets, and create economic opportunities (Hartmann, 2014). Networks contain resources that increase capabilities—the ability to access resources and combine those resources in new ways. Networks influence norms and behavior such as adoption of new techniques or technologies, or participation in collective activities (Fitzpatrick & Akgungor, 2019). Program activities that deepen and extend these networks contribute to sustainable outcomes by endogenizing capabilities within the communities.

Postprogram Evaluation

The third evaluative component of sustainability is the simplest but frequently is not carried out by NGOs and governments. This is the practice of revisiting communities several years after external funding and management of program activities has ended. One reason this is rarely done is structural: Funding agencies often require a midterm and final evaluation but not a postprogram evaluation. Furthermore, budgets for monitoring, midterm, and final evaluations are usually integrated into project proposals, but rarely is there a budget for postprogram evaluation. Final evaluations are summative, determining if the program was delivered as designed and if the outcomes detailed in the program's theory of change were met. A postprogram evaluation, which usually takes place 2–7 years after the close of the program, examines outcomes and unexpected streams of benefits (and costs), and establishes their relationship to program activities. Impacts, while projected in program documents, usually are long-run phenomena that occur because of planned and unplanned sustained outcomes. The program evaluation provides an opportunity to test the efficacy of a theory of sustainability: Did the outcomes sustain and did they contribute to sustainable impacts? The postprogram evaluation provides insights for redesign and/or the confidence to scale up.

The postprogram evaluation is especially important when assessing how activities affect an environmental resource or service. It often takes time for a resource or a service to be degraded or altered and it takes time for a resource to recover. External support for a program may end before environmental harm is detected or before recovery of a resource or service is obvious. Papua New Guinea is a country with many ongoing mangrove restoration programs, one of which illustrates the importance of a postprogram evaluation. Mangroves provide communities with a buffer from extreme weather events and their extensive root systems serve as nurseries to maintain fish and shrimp populations. In response to extreme degradation of mangroves, an intervention was designed to educate the communities about the services the mangroves provide. The intervention also worked with the community to replant and reduce harvesting. The program outcomes assessed in the final evaluation were participants' knowledge, and a target increase in the stock of mangroves. The expected impacts of the program were the restoration of fish and shrimp populations to improve food security and livelihoods, and the reestablishment of the buffer services of mature mangroves. However, no postprogram evaluation was conducted, so the opportunity to learn from this intervention was lost. For example, is understanding the services provided by mangroves enough for a community to change practices of harvesting? Will replanting and decreased harvesting continue after external resources end? At what point will the nurseries reach a sustainable yield? What protection do the recovering stands of mangrove provide during coastal weather events? This is just some of the information that could have been collected 3–5 years after the program that would have contributed to future design and effectiveness of mangrove restoration.

As discussed previously, a precondition for sustainability is the increased capabilities of participants/communities. Confirmation of enhanced capabilities will emerge after external support of the program has ended. Evidence may include the continuity and efficacy of governance structures to manage the program and of participants to adapt to changing conditions, demonstrate problem solving, and innovate. Evidence of emerging sustainable impacts—impacts that come about due to participant effort and resources—include continuous investment using community resources, such as asset transfers, training and mentoring, collective action, or maintenance of public infrastructure.

Postprogram evaluation will also reveal how the initial program may have stimulated change in other sectors. In this chapter's Malawi case study, the postprogram evaluation revealed how the initial dairy program catalyzed backward and forward linkages, making it more likely that the nascent program would thrive. These spillover effects aren't always obvious during planning and they may not be developed sufficiently at the final evaluation to be seen as an indirect benefit that supports sustainable change.

Evaluating programs that aspire to have sustainable impacts necessitates a framework that includes process, an understanding of the synergy between the intervention and environmental limits, and a commitment to postprogram assessment. The following case study of an asset transfer program in Malawi illustrates elements of program design that have enhanced capabilities but not accounted for environmental limits. This program provides an example of a postprogram evaluation that focused on process and the role that capabilities played in nurturing livelihood outcomes. This evaluation used social network analysis (SNA), a proxy for social capital, to measure relationships of reciprocity and trust and how these relationships become pathways for knowledge, innovation, and livelihood opportunities. SNA is also used to trace out the spill-

over effects of a program. This is important for sustainability because networks create a more diverse economic and social ecosystem, one that is more resilient to potential shocks.

Dairy Development Asset Transfer—Malawi

An older woman cupped my face in her hands and looked directly into my eyes. She said, "Look at me, I am a poor old woman, and I, me, put my granddaughter through college. I was able to do this because of my cows. I gave the morning milk to my family and those in need and I sold the evening milk to the association. In two years, I was able to send my granddaughter to college in Lilongwe."[2]

The goal of this program was to enhance the livelihoods among smallholder farmers, mostly women, whose income and access to resources put them at or below the poverty line. The project included an asset transfer in the form of a dairy cow[3] and training to enhance the capabilities of participant households. This training included care of the cow (nutrition, health, shelter), improvement in home gardens, and management skills. Households that received a pregnant cow agreed to pass on the first calf to another household that completed the training and was ready to receive it. The initial household also provided support and guidance in the care of the animal to the recipient household.

The theory of change for this community development intervention asserted that the transfer of the asset of livestock (physical capital) combined with a set of trainings (human capital) would enhance social capital and productive capacity among beneficiary households. This new productive capacity would then stimulate an increase in income. The model further assumed that participants, through collective action, would link to markets and sustain advances made in

---

[2]This story is drawn from the author's personal experience.

[3]The cows and transferred calves are the property of the women in the household.

income. The ultimate goal was for the intervention to catalyze a movement out of poverty for the participants and the larger community.

This program began in 2009 with the final evaluation completed in 2012 and the postprogram evaluation in 2015. The purpose of the postprogram evaluation was to determine if participant capabilities and other outcomes increased while maintaining or improving the environment.

The evidence of enhanced capabilities among the participants included ongoing community governance of the asset transfer, innovation and problem solving, and continuous investment using community resources. Participants in the program were initially asked by the implementing partner, in 2009, to establish a governing committee to manage the transfer of livestock and facilitate ongoing training of new participants. From the close of the program to the postprogram evaluation, this committee had rotated in new members as terms expired, and successfully monitored and enforced the process of passing on the livestock.

The participants and key informants provided evidence of their application and extension of the knowledge and skills to support dairy production. This indicates the effectiveness of participants' initial training and the community's ability to extend and adapt knowledge over time. The program also trained community volunteers as community animal health workers and equipped them to support participants to manage basic animal health issues. These two complementary activities—the extension of participants' knowledge and the embedding of a community-based paraprofessional—proved to be important to internalizing capabilities in the community. The postprogram evaluation revealed a decrease in livestock mortality and an increase in live births after the close of the program, indicating participants' ability to solve problems and innovate[4] (Heifer International, 2012).

Another indicator of sustainability was the community's continuous investment in program-related activities. The evidence from the postprogram evaluation included continuation of the asset transfer, active mentoring of new participants, and the use of collective funds to improve and enlarge the cooling facility.[5] During the 3 years after the program, the collection and chilling system increased throughput and maintained the community association's reputation for high-quality product established during the program. Typically, one of the most common challenges for newly formed associations is financial management. The postprogram evaluation found that bills were paid, a savings account maintained, and payments to farmers were prompt.

While the program demonstrated institutions and capabilities that contribute to sustainable impacts, a complementary phenomenon was the stimulating effects this program had on other sectors. One of the surprises from the postprogram evaluation was the growth of backward linkages brought about by the program's expansion of the dairy sector. This development became visible as we analyzed changes in social networks from the midterm to the postprogram evaluation. We expected the networks to become larger as more households in the communities were able to get a calf, the training, and support. But the postprogram evaluation found an increasing number of people in the networks who were not dairy producers but provided backward linkages: farmers providing inputs into dairy mash and feed supplements, and new hammer mills. The demand for transport for milk and inputs increased employment and contributed to these spillover effects.[6] We also observed spillover effects of knowledge. Eighty percent of nonprogram households in the network reported that they adopted an improved practice due to their relationship with a program participant. Nonprogram households also

---

[4]One example of innovation was the cross breeding of Zebus (an indigenous cow) with the program Jersey cows. Jersey are known for the high fat content of their milk, but Zebus consume less water and are more robust in a rugged environment. Although milk production with the cross-bred cows was lower, the incidence of disease declined significantly.

[5]The program supported participants in establishing a milk collection and chilling center by acquiring milk chilling equipment while the participants contributed locally available resources and labor to construct the collection and chilling house.

[6]Analysis from the postprogram evaluation showed that a 1% change in income of participants resulted in a .78% change in the income of nonparticipants in the network.

reported that they relied on the program-trained community animal health workers for information and vaccinations. These investments, which became significant after the close of the program, signaled the strength of the economic changes taking place in the livestock sector. These could be considered emerging outcomes, those that were unintended but occurred as a result of the program and the efforts and resources of project participants.

One of the clearly articulated outcomes of this program was the increase in income from advances in dairy productivity. As part of the postprogram evaluation, we examined the change in real net farm income for participants and compared it to a measure of living income (Fitzpatrick & Akgungor, 2019).[7] The calculation for the living income benchmark for this project was based on the Anker and Anker (2014) study of the living wage in the tea-growing area of Malawi. Because this project site was in the central region of Malawi, we made adjustments based on differential costs of food and housing. All prices were adjusted to a 2015 base, allowing for a linear measurement across time, similar to a poverty line.

The purpose of measuring changes in the farm income of households was twofold, seeking to determine (a) if positive change occurred and (b) if it was sustained after the end of the intervention. If participants were able to achieve a level of net income equal to or greater than the living income benchmark, and at least sustain this level of income after the close of the project, we can conclude that livelihoods have improved and have exhibited evidence of sustainability and resilience to unexpected events. Figure 1 illustrates changes in real farm income at baseline (2009), end of project (2012), and postprogram evaluation (2015).

One of the limitations of living income as a benchmark of improved livelihoods is that it does not account for how income is distributed among household members or over the year, which may have important implications for food security. Assessing household income over a multiyear period provides an estimate of the reliability of the income. It is therefore important that measuring income to a benchmark be done over time. Both of these issues are important for food security and household resilience.

We found that the change in net income for the program households was sufficient to afford participants a modest but decent living at the close of the project in 2012. When we measured net income in 2015, 3 years after the end of the project, net incomes had increased 60%, indicating a sustainable change in income. The increases that we saw in net income were likely to level off, at least in the short to medium run, as supply and demand conditions adjusted. What is important is that this infusion of capital allowed participants to increase incomes above the living income and, as such, increased the economic resiliency of this group.

Postprogram evaluation also allowed us to assess whether collective and individual capabilities increased. We did this by examining whether new institutions fulfilled their function, whether sufficient knowledge was transferred to maintain and increase the dairy population, and whether participants were able to pass on knowledge and support to new participants without the support of project personnel. Continuous investment also confirmed participants' confidence and commitment to program outcomes. Last, while analyzing the role of social capital in facilitating many of these outcomes, we discovered that the program catalyzed spillovers into other sectors, further increasing the likelihood that this new growth will be sustainable. This postprogram evaluation took place 3 years after the project's close, which provided enough time to assess the likelihood of sustainable outcomes and sustainable impact. Five to seven years would provide the time for testing innovations, revealing design flaws, and observing resilience. Waiting longer before postprogram evaluation would also provide more time to see how participants innovated in response to changing market conditions, climate, and the inevitable challenges of managing change in close communities.

[7] Living income is a concept adopted from the work of Anker and Anker (2014). Living income is different from a poverty line in that it encompasses the idea of a "decent living," one that measures the ability of households to meet all their basic needs including food, water, education, housing, healthcare, clothing, and a provision for unexpected events.

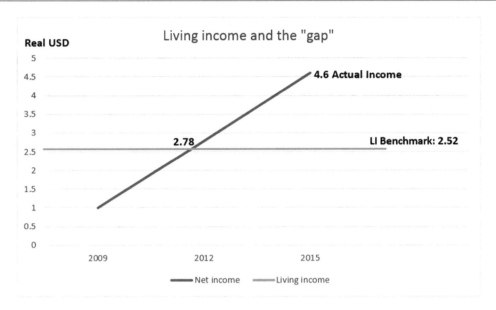

Fig. 1 Comparison of living income benchmark and average net real incomes, 2009–2015
Note. 2009 figure is the baseline at the start of the program, 2012 is at the end of the program, at 2015 is at post- program evaluation (daily income of a household in 2015 USD). Source: Fitzpatrick and Akgungor (2019; this source discusses the statistical analysis of living income 2009–2015)

Environmental Effects

As programs modify economic and social systems to improve well-being, sustainability requires that the design and implementation consider that we are operating in a closed system and that environmental resources and services are limited. To this end, the design and evaluations of a program should address two questions: Has the program been designed to maintain or improve environmental resources and services? Are there missed opportunities or unintended effects that were caused by program activities or that evolved after the program ended?

Recall that one objective of this program was to increase the income of participants and their communities through the development of a dairy sector. At the beginning of the program in 2009, 60 pregnant cows were delivered to the first round of participants. By 2015, when the postprogram evaluation was conducted, the communities had approximately 1200 dairy cows that were project related. This does not include the increase in dairy population that resulted indirectly from the program, namely the increases in Zebu-Jersey crosses or the male calves that were often sold to nonparticipants. But the design and evaluation of the program did not take into account the environmental impacts of such a large change in livestock population.

The environmental resources and services that a greatly enlarged dairy sector require include water for animal hydration and hygiene and increased water for producing dairy feed. The backward linkages that stimulated feed and fodder production put more marginal land into cultivation, potentially contributing to soil erosion and loss of fertility. Another concern was the pressure that the demand for feed and fodder put on the conversion of forests to farmland. These were not considerations when the program was designed and implemented. The environmental impacts of development interventions, especially those associated with improving livelihoods through the expansion of market activity, are rarely considered in the design stage or incorporated into a theory of change. A postprogram evaluation can estimate the environmental impacts because many do not become apparent until years after the project has ended. If an organization or government is committed to sustainable development and the primacy of the environment in enabling economic growth, an environmental impact assessment should inform the design of the program and a postprogram evaluation should surface the intended and unintended impacts on the environment. Identifying environmental problems can support work with communities to design remediation strategies.

A postprogram evaluation may also reveal missed opportunities to contribute to environmental health. For example, a major environmental and health problem in Malawi is the exposure to wood-based cooking fuels; about 98% of households depend on this source (Gercama & Bertrams, 2017). This has serious health effects and also contributes to widespread deforestation. In the long run, this deforestation creates food security risks as forests diminish, soil erodes, and regional droughts become more common. This program missed an opportunity to address this health and environmental problem. The increasing sources of dairy manure could have been used to make biogas for cooking and lighting. This would have created a demand for new products: biogas, biogas equipment, and dried organic fertilizer from the sludge. If attention to environmental impacts had been integrated into the design of this program, these opportunities would have been apparent and would have strengthened the sustainability of program impacts. A theory of sustainability embedded in the theory of change with explicit links from activities to outcomes and outcomes to impact could have revealed potential environmental concerns and incorporated plans to address concerns and opportunities into the program design.

Discussion

Sloppy, opportunistic, and superficial treatment of the concept of sustainability has contributed to the ineffectiveness of many development interventions and has, in many cases, led to environmental degradation. As development professionals and government planners, we have an obligation to promote programs with sustainable impacts to improve the social, economic, and environmental returns on investments, to be respectful of the time and resource commitments that participants make in the expectation of improved well-being, and to contribute to intergenerational equity.

If outcomes are to be sustainable, interventions need to deepen the capabilities of participants; they will be the ones to maintain the program, to make it dynamic—they are they innovators, the problem solvers. Otherwise, the cows are eaten!

Development professionals and governments should incorporate two interrelated elements of sustainability to the design and evaluation of programs. First is incorporating the primacy of environment into interventions, and second is formulating outcomes and impacts that result in streams of benefits that continue beyond the life of the program. Finally, to know whether sustainable impacts occurred, a commitment to assessing this work well after the programs have ended is essential.

To secure a stream of benefits after the close of a program, an intervention should deepen the communities' abilities to create and manage governance structures, solve problems, and innovate. Postprogram evaluation allows us to determine whether governance structures have fulfilled their functions and whether outcomes have continued. In the case presented, we observed the income of the participants increasing 3 years after the program ended. This indicated that participants have managed, solved problems, and created new opportunities. This postprogram evaluation also revealed an intervention that contributed to environmental degradation. Well-meaning interventions focused on improving human well-being often overlook negative impacts to environmental well-being or assume none will occur. We are not accustomed to putting the environment front and center and we are often ill equipped to project the impact of interventions on the environment in the long run. Environmental boundaries must be incorporated into initial program planning, and when our planning isn't sufficient, evaluation plays a role in integrating knowledge of the damage or potential damage back into the redesign process.

While the results of this study pertain to a specific project, they also provide preliminary evidence of the importance of catalyzing and engaging physical, human, and social capital in an intervention to influence the structure of a local community in a way that leads to a sustainable set of new opportunities for the most vulnerable. When this occurs, outcomes are more likely to be sustainable because the intervention triggered new skills and institutions that enhanced capability. We see this in the continued asset transfers, the deepening of social networks, the growth of backward linkages, and the participants reinvestment in program activities. These enhanced capabilities and innovations allow for endogenous change in the communities that is regenerative, signaling a capacity for sustainability.

References

Almond, R. E. A., Grooten, M., & Petersen, T. (Eds). (2020). *Living planet report 2020: Bending the curve of biodiversity loss*. WWF. https://www.worldwildlife.org/publications/living-planet-report-2020

Anker, R., & Anker, M. (2014). *Living wage for rural Malawi with a focus on the tea growing area*. Fair Trade International.

Boulding, K. (1966). The economics of the coming spaceship Earth. In H. Jarret (Ed.), *Environmental quality in a growing economy* (pp. 3–14). John Hopkins University Press.

Farley, H., & Smith, Z. (2014). *Sustainability: If it's everything, is it nothing?* Routledge.

Fitzpatrick, E., & Akgungor, S. (2019). Evaluating the asset transfer model in facilitating sustainable livelihood in rural Malawi. *Food Economy, 21*(1), 129–152. https://doi.org/10.3280/ECAG2019-001007.

Gercama, I., & Bertrams, N. (2017, October 26). Smoke and mirrors: Malawi's untold crisis. *New Internationalist*. https://newint.org/features/interactives/2017/10/26/malawi-health-crisis

Hartmann, D. (2014). *Economic complexity and human development*. Routledge.

Heifer International. (2012). *Final report for Malawi smallholder dairy development program*. Lilongwe, Malawi.

Lin, N. K. (2008). *Social capital, theory and research*. Transaction Publishers.

Mog, J. (2004). Struggling with sustainability. *World Development, 32*(12), 2139–2160. https://doi.org/10.1016/j.worlddev.2004.07.002.

Woolcock, M. (2000). *Social capital: Implications for development theory, research and policy*. World Bank Research Observer.

World Bank. (2018, September 19). *Decline of extreme global poverty continues but at a slowed rate* [Press release]. https://www.worldbank.org/en/news/press-release/2018/09/19/decline-of-global-extreme-poverty-continues-but-has-slowed-world-bank

World Bank. (2020, December 12). *Net official development assistance and official aid received* [Data set]. https://data.worldbank.org/indicator/DT.ODA.ALLD.CD

World Bank. (2021, January 5). *Poverty data* [Data set]. https://data.worldbank.org/topic/poverty

World Commission on the Environment and Development. (1987). *Our common future*. Oxford University Press.

Can We Assume Sustained Impact? Verifying the Sustainability of Climate Change Mitigation Results

Jindra Cekan and Susan Legro

Abstract

The purpose of this research was to explore how public donors and lenders evaluate the sustainability of environmental and other sectoral development interventions. Specifically, the aim is to examine if, how, and how well post project sustainability is evaluated in donor-funded climate change mitigation (CCM) projects, including the evaluability of these projects. We assessed the robustness of current evaluation practice of results after project exit, particularly the sustainability of outcomes and long-term impact. We explored methods that could reduce uncertainty of achieving results by using data from two pools of CCM projects funded by the Global Environment Facility (GEF).

Keywords

Climate change mitigation · Sustainability · Climate finance · Ex-post evaluation · Post project evaluation · Postcompletion evaluation · Sustained impact · Impact

Evaluating sustainable development involves looking at the durability and continuation of net benefits from the outcomes and impacts of global development project activities and investments in various sectors in the post project phase, i.e., from 2 to 20 years after donor funding ends.[1] Evaluating the sustainability of the environment is, according to the Organisation for Economic Co-operation and Development (OECD, 2015), at once a focus on natural systems of "biodiversity, climate change, desertification and environment" (p.1) that will need to consider the context in which these are affected by human systems of "linkages between poverty reduction, natural resource management, and development" (p. 3). This chapter focuses more narrowly on the continuation of net benefits

---

[1] We use the term "postproject" evaluations to distinguish these longer term evaluations from terminal evaluations, which typically occur within 3 months of the end of donor funding. While some donors (JICA, 2004; USAID, 2019) use the term "ex-post evaluation" to refer to evaluations distinct from the terminal/final evaluation and occurring 1 year or more after project closure, other donors use the terms "terminal evaluation" and "ex-post evaluation" synonymously. Other terms include postcompletion, postclosure, and long-term impact.

J. Cekan (✉)
Valuing Voices at Cekan Consulting, LLC, Prague, Czech Republic

S. Legro
Independent Consultant, Prague, Czech Republic
e-mail: susan@ecoltdgroup.com

© The Author(s) 2022
J. I. Uitto, G. Batra (eds.), *Transformational Change for People and the Planet*, Sustainable Development Goals Series, https://doi.org/10.1007/978-3-030-78853-7_8

from the outcomes and impacts of a pool of climate change mitigation (CCM) projects (see Table 1). The sustainability of CCM projects funded by the Global Environment Facility (GEF), as in a number of other bilateral and multilateral climate funds, rests on a theory of change that a combination of technical assistance and investments contribute to successfully durable market transformation, thus reducing or offsetting greenhouse gas (GHG) emissions.

CCM projects lend themselves to such analysis, as most establish ex-ante quantitative mitigation estimates and their terminal evaluations often contain a narrative description and ranking of estimated sustainability beyond the project's operational lifetime, including the achievement

Table 1 Changes in OECD DAC Criteria from 1991 to 2019

1991	2019
SUSTAINABILITY:	SUSTAINABILITY: WILL THE BENEFITS LAST?
Sustainability is concerned with *measuring whether the benefits of an activity are likely to continue after donor funding has been withdrawn.* Projects need to be *environmentally as well as financially sustainable.*	*The extent to which the net benefits of the intervention continue, or are likely to continue.* Note: Includes an examination of the *financial, economic, social, environmental, and institutional capacities of the systems needed to sustain net benefits over time.* Involves analyses of resilience, risks, and potential trade-offs.
IMPACT:	IMPACT:
The positive and negative *changes produced by a development intervention, directly or indirectly, intended or unintended.* This involves the main impacts and effects resulting from the activity on the local social, economic, environmental, and other development indicators.	The extent to which the intervention has generated or is expected to *generate significant* positive or negative, intended or unintended, *higher-level effects. . . .* It seeks to identify social, environmental, and economic effects of the intervention that are *longer-term or broader in scope.*

Source: OECD/DAC Network on Development Evaluation, (2019); italics are emphais added by Cekan

of project objectives. The need for effective means of measuring sustainability in mitigation projects is receiving increasing attention (GEF Independent Evaluation Office [IEO], 2019a) and is increasingly important, as Article 13 of the Paris Agreement mandates that countries with donor-funded CCM projects report on their actions to address climate change (United Nations, 2015). As several terminal evaluations in our dataset stated, better data are urgently needed to track continued sustainability of past investments and progress against emissions goals to limit global warming.

Measuring Impact and Sustainability

Although impactful projects promoting sustainable development are widely touted as being the aim and achievement of global development projects, these achievements are rarely measured beyond the end of the project activities. Bilateral and multilateral donors, with the exception of the Japan International Cooperation Agency (JICA) and the U.S. Agency for International Development (USAID),[2] have reexamined fewer than 1% of projects following a terminal evaluation, although examples exist of post project evaluations taking place as long as 15 years (USAID) and 20 years (Deutsche Gesellschaft fur Internationale Zusammenarbeit [GIZ]) later (Cekan, 2015). Without such fieldwork, sustainability estimates can only rely on assumptions, and positive results may in fact not be sustained as little as 2 years after closure. An illustrative set of eight post project global development evaluations analyzed for the Faster Forward Fund of Michael Scriven in 2017 showed a range of results: One project partially exceeded terminal evaluation results, two retained the sustainability assumed at inception, and the other five showed a decrease in results of 20%–100% as early as 2 years post-exit (Zivetz et al., 2017a).

---

[2] In a 2013 meta-evaluation, Hageboeck et al. found that only 8% of projects in the 2009–2012 USAID PPL/LER evaluation portfolio (26 of 315) were evaluated post-project following the termination of USAID funding.

Since the year 2000, the U.S. government and the European Union have spent more than $1.6 trillion on global development projects, but fewer than several hundred post project evaluations have been completed, so the extent to which outcomes and impacts are sustained is not known (Cekan, 2015). A review of most bilateral donors shows zero to two post project evaluations (Valuing Voices, 2020). A rare, four-country, post project study of 12 USAID food security projects also found a wide variability in expected trajectories, with most projects failing to sustain expected results beyond as little as 1 year (Rogers & Coates, 2015). The study's Tufts University team leaders noted that "evidence of project success at the time of exit (as assessed by impact indicators) did not necessarily imply sustained benefit over time." (Rogers & Coates, 2015, p. v.). Similarly, an Asian Development Bank (ADB) study of post project sustainability found that "some early evidence suggests that as many as 40% of all new activities are not sustained beyond the first few years after disbursement of external funding," and that review examined fewer than 14 of 491 projects in the field (ADB, 2010). The same study described how assumed positive trajectories post funding fail to sustain and noted a

> tendency of project holders to overestimate the ability or commitment of implementing partners— and particularly government partners—to sustain project activities after funding ends. Post project evaluations can shed light on what contributes to institutional commitment, capacity, and continuity in this regard. (ADB, 2010, p. 1)

Learning from post project findings can be important to improve project design and secure new funding. USAID recently conducted six post project evaluations of water/sanitation projects and learned about needed design changes from the findings, and JICA analysed the uptake of recommendations 7 years after closure (USAID, 2019; JICA, 2020a, 2020b). As USAID stated in their 2018 guidance,

> An end-of-project evaluation could address questions about how effective a sustainability plan seems to be, and early evidence concerning the likely continuation of project services and benefits after project funding ends. Only a post project evaluation, however, can provide empirical data about whether a project's services and benefits were sustained. (para. 9)

Rogers and Coates (2015) expanded the preconditions for sustainability beyond only funding, to include capacities, partnerships, and ownership. Cekan et al. (2016) expanded ex-post project methods from examining the sustainability of expected project outcomes and impacts post closure to also evaluating emerging outcomes, namely "what communities themselves valued enough to sustain with their own resources or created anew from what [our projects] catalysed" (para. 19). In the area of climate change mitigation, rigorous evaluation of operational sustainability in the years following project closure should inform learning for future design and target donor assistance on projects that are most likely to continue to generate significant emission reductions.

How Are Sustainability and Impact Defined?

The original 1991 OECD Development Assistance Committee (DAC) criteria for evaluating global development projects included sustainability, and the criteria were revised in 2019. The revisions related to the definition of sustainability and emphasize the continuation of benefits rather than just activities, and they include a wider systemic context beyond the financial and environmental resources needed to sustain those benefits, such as resilience, risk, and trade-offs, presumably for those sustaining the benefits. Similarly, the criteria for impact have shifted from simply positive/negative, intended/unintended changes to effects over the longer term (see Table 1).

In much of global development, including in GEF-funded projects, impact and sustainability are usually estimated only at project termination, "to determine the relevance and fulfilment of objectives, development efficiency, effectiveness, impact and [projected] sustainability" (OECD DAC, 1991, p. 5). In contrast, actual sustainability can only be evaluated 2–20 years after all project resources are withdrawn, through desk studies, fieldwork, or both. The new OECD definitions present an opportunity to improve the measurement of sustained impact across global development, particularly via post project evaluations. Evaluations need to reach beyond projected to actual measurement across much of "sustainable development" programming, including that of the GEF.

GEF evaluations in recent years have been guided by the organization's 2010 measurement and evaluation (M&E) policy, which requires that terminal evaluations "assess the likelihood of sustainability of outcomes at project termination and provide a rating" (GEF Independent Evaluation Office [IEO], p. 31). Sustainability is defined as "the likely ability of an intervention to continue to deliver benefits for an extended period of time after completion; projects need to be environmentally as well as financially and socially sustainable" (GEF IEO, 2010, p. 27).

In 2017, the GEF provided specific guidance to implementing agencies on how to capture sustainability in terminal evaluations of GEF-funded projects (GEF, 2017, para. 8 and Annex 2): "The overall sustainability of project outcomes will be rated on a four-point scale (Likely to Unlikely)":

- Likely (L) = There are little or no risks to sustainability;
- Moderately Likely (ML) = There are moderate risks to sustainability;
- Moderately Unlikely (MU) = There are significant risks to sustainability;
- Unlikely (U) = There are severe risks to sustainability; and
- Unable to Assess (UA) = Unable to assess the expected incidence and magnitude of risks to sustainability

Although this scale is a relatively common measure for estimating sustainability among donor agencies, it is not a measure that has been tested for reliability, i.e., whether multiple raters would provide the same estimate from the same data. It has also not been tested for construct validity, i.e., whether the scale is an effective pre-

dictive measure of post project sustainability. Validity issues include whether an estimate of risks to sustainability is a valid measure of the likelihood of post project sustainability, whether the narrative estimates of risk are ambiguous or double-barreled; and the efficacy of using a ranked, ordinal scale that treats sustainability as an either/or condition rather than a range (from no sustainability to 100% sustainability).

Throughout this chapter, we identify projects by their GEF identification numbers, with a complete table of projects provided in the appendix.

The Limits of Terminal Evaluations

Terminal evaluations and even impact evaluations that mostly compare effectiveness rather than long-term impact were referenced as sources for evaluating sustainability in the GEF's 2017 Annual Report on Sustainability (GEF IEO, 2019a). Although they can provide useful information on relevance, efficiency, and effectiveness, neither is a substitute for post project evaluation of the sustainability of outcomes and impacts, because projected sustainability may or may not occur. In a terminal evaluation of Mexican Sustainable Forest Management and Capacity Building (GEF ID 4149), evaluators made the case for ex-post project monitoring and evaluation of results:

> There is no follow-up that can measure the consolidation and long-term sustainability of these activities. . . . Without a proper evaluation system in place, nor registration, it is difficult to affirm that the rural development plans will be self-sustaining after the project ends, nor to what extent the communities are readily able to anticipate and adapt to change through clear decision-making processes, collaboration, and management of resources. . . . They must also demonstrate their sustainability as an essential point in development with social and economic welfare from natural resources, without compromising their future existence, stability, and functionality. (pp. 5–9)[3]

---

[3] Page numbers provided with GEF ID numbers only refer to project terminal evaluations; see Appendix.

Returning to a project area after closure also fosters learning about the quality of funding, design, implementation, monitoring, and evaluation and the ability of those tasked with sustaining results to do so. Learning can include how well conditions for sustainability were built in, tracked, and supported by major stakeholders. Assumptions made at design and final evaluation can then also be tested, along with theories of change (Sridharam & Nakaima, 2019). Finally, post project evaluations can verify the attributional claims made at the time of the terminal evaluation. As John Mayne explained in his 2001 paper:

> In trying to measure the performance of a program, we face two problems. We can often—although frequently not without some difficulty—measure whether or not these outcomes are actually occurring. The more difficult question is usually determining just what contribution the specific program in question made to the outcome. How much of the success (or failure) can we attribute to the program? What has been the contribution made by the program? What influence has it had? (p. 3)

In donor- and lender-funded CCM projects, emission reduction estimates represent an obvious impact measure. They are generally based on a combination of direct effects—i.e., reductions due to project-related investments in infrastructure—and indirect effects—i.e., reductions due to the replication of "market transformation" investments from other funding or an increase in climate-friendly practices due to improvements in the policy and regulatory framework (Duval, 2008; Legro, 2010). Both of these effects are generally estimated over the lifetime of the mitigation technology involved, which is nearly always much longer than the operational duration of a given project (see Table 2).

The increasing use of financial mechanisms such as concessional loans and guarantees as a component of donor-funded CCM projects, such as those funded by the Green Climate Fund (https://www.greenclimate.fund/), can also limit the ability of final evaluations to capture sustainability, because the bulk of subsequent invest-

Table 2 Typology of GHG Reductions Resulting from Typical Project Interventions

Type of GHG reductions	Project lifetime (quarterly annual monitoring)		Post project lifetime (post project evaluation)
Direct reductions	Reductions directly financed by donor-funded pilot project(s) or investment(s)	TERMINAL EVALUATION	Continuing reductions from project-financed investments (through the end of the technology lifetime; e.g., 20 years for buildings, 10 years for industrial equipment, etc.)
Indirect reductions	Reductions from policy uptake (e.g., reduced fossil fuel use from curtailment of subsidies, spillover effects from tax incentives, increased government support for renewable energy due to strategy development) (co-) funded by the donor		Continuing reductions from policy uptake (e.g., reduced fossil fuel use from curtailment of subsidies, spillover effects from tax incentives, increased government support for energy efficiency or renewable energy due to strategy development)
	Reductions from market transformation (changes in availability of financing, increased willingness of lenders, reduction in perceived risk) supported by pilot demonstrations and/or outreach and awareness raising (co-)funded by the donor		Continuing reductions from market transformation (changes in availability of financing, increased willingness of lenders, reduction in perceived risk) as a legacy of the pilot demonstrations and/or outreach and awareness raising funded by the donor-funded project
			New reductions from the continuation of the investment or financing mechanism established by the donor-funded project

ments in technologies that are assumed with revolving funds will not take place during the project lifetime. A 2012 paper by then-head of the GEF Independent Evaluation Office, Rob van den Berg, supported the need for post project evaluation and importantly included:

> Barriers targeted by GEF projects, and the results achieved by GEF projects in addressing market transformation barriers . . . facilitate in understanding better whether the ex-post changes being observed in the market could be linked to GEF projects and pathways through which outcomes and intermediate states . . . [and] the extent GEF-supported CCM activities are reducing GHGs in the atmosphere . . . because it helps in ascertaining whether the incremental GHG reduction and/or avoidance is commensurate with the agreed incremental costs supported by GEF. . . . It is imperative that the ex-ante and ex-post estimates of GHG reduction and avoidance benefits are realistic and have a scientific basis. (GEF IEO, 2012, p. 13)

This description of GHG-related impacts illustrates the difficulties associated with accurately drawing conclusions about sustainability from using a single scale to estimate "the *likely ability* [emphasis added] of an intervention to continue to deliver benefits for an extended period of time" (GEF IEO, 2010, p. 35) due to several factors. First, the GEF's 4-point scale is supposed to capture two different aspects of continuation: ongoing benefits from a project-related investment, and new benefits from the continuation of financing mechanisms. Without returning to evaluate the continued net benefits of the now-closed investment, such assumptions cannot be fully claimed. Second, the scale is supposed to capture benefits that can be estimated in a quantitative way (e.g., solar panels that offset the use of a certain amount of electricity from diesel generators); benefits that can be evaluated through policy or program evaluation (e.g., the introduction of a law on energy efficiency); and benefits that will require careful, qualitative study to determine impacts (e.g., training programs for energy auditors or awareness-raising for energy consumers, leading to knowledge and decision changes). Aggregating and weighing such an array of methods into one ranking is methodologically on shaky ground, especially without post

project measurements to confirm whether results happened at any time after project closure.

Methodology

The impetus for this research was a sustainability analysis conducted by the GEF IEO that was summarized in the 2017 GEF Annual Performance Report (GEF IEO, 2019a). The study stated: "The analysis found that outcomes of most of the GEF projects are sustained during the postcompletion period, and a higher percentage of projects achieve environmental stress reduction and broader adoption than at completion" (p. 17). Learning more about postcompletion outcomes and assessing how post project sustainability was evaluated was the aim of this work.

This chapter's research sample consists of two sets of GEF project evaluations. We chose projects funded by the GEF because of the large size of the total project pool. For example, the Green Climate Fund lacks a large pool of mitigation projects that would be suitable for post project evaluation. Our first tranche was selected from the pool of CCM projects cited in the sustainability analysis, which included a range of projects with the earliest start date of 1994 and the latest closing date of 2013 (GEF IEO, 2019a). These constituted $195.5 million dollars of investments. The pool of projects in the climate change focal area ($n = 17$), comprising one third of the GEF IEO sample, was then selected from the 53 projects listed in the report for further study. We then classified the selected projects by which ones had any mention of field-based post project verification according to an evaluability checklist (Zivetz et al., 2017a). This list highlights methodological considerations including: (a) data showing overall quality of the project at completion, including M&E documentation needed on original and post project data collection; (b) time postcompletion (at least 2 years); (c) site selection criteria; and (d) proof that project results were isolated from concurrent programming to ascertain contribution to sustained impacts (Zivetz et al., 2017a).

Next, we reviewed GEF documentation to identify any actual quantitative or qualitative measures of post project outcomes and impacts. These could include: (a) changes in actual energy efficiency improvements against final evaluation measures used, (b) sustained knowledge or dissemination of knowledge change fostered through trainings, (c) evidence of ownership, or (d) continued or increased dissemination of new technologies. Such verification of assumptions in the final documents typically explores why the assumptions were or were not met, and what effects changes in these assumptions would have on impacts, such as CO_2 emissions projections.

The second tranche consisted of projects in the climate change focal area that were included in the 2019 cohort of projects for which the GEF received terminal evaluations. As the GEF 2019 Annual Performance Report explained:

> Terminal evaluations for 193 projects, accounting for $ 616.6 million in GEF grants, were received and validated during 2018–2019 and these projects constitute the 2019 cohort. Projects approved in GEF-5 (33 percent), GEF-4 (40 percent) and GEF-3 (20 percent) account for a substantial share of the 2019 cohort. Although 10 GEF Agencies are represented in the 2019 cohort, most of these projects have been implemented by UNDP [United Nations Development Programme] (56 percent), with World Bank (15 percent) and UNEP [United Nations Environment Programme] (12 percent) also accounting for a significant share. (GEF IEO, 2020, p. 9)

We added the second tranche of projects to represent a more current view of project performance and evaluation practice.

The climate change focal area subset consisted of 38 completed GEF projects, which account for approximately $155.7 million in GEF grants (approximately 20% of the total cohort and 25% of the overall cohort budget). Projects included those approved in 1995–1998 (GEF-1; $n = 1$) and 2003–2006 (GEF-3; $n = 2$), but 68% were funded in 2006–2010 (GEF-4; $n = 26$), and 24% in 2010–2014 (GEF-5; $n = 9$), making them more recent as a group than the 2019 cohort as a whole. Six GEF agencies were represented: Inter-American Development Bank (IDB), International Fund for Agricultural Development

(IFAD), UNDP, UNEP, United Nations Industrial Development Organization (UNIDO), and the World Bank.

We eliminated three projects listed in the climate focal area subset from consideration in the second tranche because they had not been completed, leaving a pool of 35 projects. Ex-ante project documentation, such as CEO endorsement requests, and terminal evaluation reports were then reviewed for initial estimates of certain project indicators, such as GHG emission reductions, and ratings of estimated sustainability on the 4-point scale, including the narrative documentation that accompanied the ratings.

Findings

The question of whether post project sustainability was being measured was based on the first tranche of projects and on the sustainability analysis in which they were included. Most of the documents cited in the sustainability analysis were either terminal or impact evaluations focused on efficiency (GEF IEO, 2019a), and most of the documents and report analysis focused on estimated sustainability. Of the 53 "postcompletion verification reports," as they are referred to in the review (GEF IEO, 2019a, p. 62), we found only 4% to contain adequate information to support the analysis of sustainability. Our wider search for publicly available post project evaluations, which would have constituted an evidence base for sustained outcomes and environmental stress reduction and adoption cited in the GEF IEO 2019 analysis, did not identify any post project evaluations. We were unable to replicate the finding that "84% of these projects that were rated as sustainable at closure also had satisfactory postcompletion outcomes. . . . Most projects with satisfactory outcome ratings at completion continued to have satisfactory outcome ratings at postcompletion" (GEF IEO, 2019a, p. 3) or to compare the CCM subset of projects with this conclusion. The report stated that "the analysis of the 53 selected projects is based on 61 field verification reports. *For 81 percent of the projects, the field verification was con-*

ducted at least four years after implementation completion [emphasis added]." However, we found no publicly accessible documentation that could be used to confirm the approach to field verification for 8 of the 17 projects.

Similarly, the available documentation for the projects lacked the most typical post project hallmarks, such as methods of post project data collection, comparisons of changes from final to post project outcomes and impacts at least 2 years post closure, and tracing contribution of the project at the funded sites to the changes. Documentation focused on a rating of estimated sustainability with repeated references to only the terminal evaluations and closure reports. In summary, of the 17 projects selected for review in the first tranche, 14 had data consisting of terminal evaluations, and none was 2–20 years post closure. We did not find publicly available evidence to support *measurement* of post project sustainability other than statements that such evidence was gathered in a handful of cases. Of the pool of 17 projects, only two (both from India) made any reference to post project data regarding the sectors of activity in subsequent years. However, these two were terminal evaluations within a country portfolio review and could not be substantiated with publicly accessible data.

We then screened the first tranche of projects using the Valuing Voices evaluability checklist (Zivetz et al., 2017b):

• High-quality project data at least at terminal evaluation, with verifiable data at exit: Of 14 projects rated for sustainability, only six were rated likely to be sustained and outcome and impact data were scant.
• Clear ex-post methodology, sufficient samples: None of the evaluations available was a post project evaluation of sustainability or long-term impact. Although most projects fell within the evaluable 2–20 years post project (the projects had been closed 4–20 years), none had proof of return evaluation. There were no clear post project sampling frames, data collection processes including identification of beneficiaries/ informants, site selection, isolating legacy

effects of the institution or other concurrent projects, or analytic methods.
• Transparent benchmarks based on terminal, midterm, and/or baseline data on changes to outcomes or impacts: M&E documents show measurable targets and indicators, baseline vs. terminal evaluations with methods that are comparable to methods used in the post project period: For some of the 17 projects, project inception documents and terminal evaluations were available; in other cases, GEF evaluation reviews were available. Two had measurable environmental indicators that compared baseline to final, but none were after project closure.
• Substantiated contribution vs. attribution of impacts: Examples of substantiated contribution were not identified.

Evaluation reports revealed several instances for which we could not confirm attribution. For example, evaluation of the project Development of High Rate BioMethanation Processes as Means of Reducing Greenhouse Gas Emissions (GEF ID 370), which closed in 2005, referenced the following subsequent market information:

> As of Nov 2012, capacity installed from waste-to-energy projects running across the country for grid connected and captive power are 93.68MW and 110.74 MW respectively [versus 3.79KW from 8 sub-projects and 1-5 MW projects]. . . . The technologies demonstrated by the 16 sub-projects covered under the project have seen wide-scale replication throughout the country. . . . An installed capacity of 201.03MW within WTE [waste to energy] projects and the 50% of this is attributed to the GEF project. (GEF IEO, 2013, vol. 2, p. 64)

Claims of "the technical institutes strengthened as a result of the project were not fully effective at the time of project completion but are now actively engaged in the promotion of various biomethanation technologies" are unsubstantiated in publicly available information; as a result, the ex-post methods of contribution/attribution data are not clear. Another project in India, Optimizing Development of Small Hydel [hydroelectric] Resources in Hilly Areas (GEF ID 386), projected that later investments in the government's

5-year plans would happen, and the resulting hydropower production would be attributable to the original project (GEF IEO, 2013); again, this attributional analysis was not documented. Analysis of a third project in India, Coal Bed Methane Capture and Commercial Utilization (GEF ID 325), which closed in 2008, claimed results that could not be reproduced: "Notable progress has been made through replication of projects, knowledge sharing, and policy development" and "expertise was built" (GEF IEO, 2013, Vol. 2, p. 90). Further claims that the project contributed to "the total coal bed methane production in the country and has increased to 0.32 mmscmd [million metric standard cubic meters per day], which is expected to rise to 7.4 mmscmd by the end of 2014" is without proof. The evaluation reported estimates of indirect GHG emission reduction, based on postcompletion methane gas production estimates of 0.2 million m³ per day:

> 1.0 Million tons equivalent per year, considering *an adjustment factor of 0.5 as the GEF contribution* [emphasis added], the indirect GHG emission reduction due to the influence of the project is estimated to be 0.5 million tons of CO_2 equivalent per annum (2.5 million tons over the lifetime period of 5 years). (GEF IEO, 2013, Vol. 2, p. 91)

Yet without verification of coal bed methane capture and commercial utilization continuing, this impact cannot be claimed.

How Is Sustainability Being Captured?

Fifteen of the 17 CCM projects we reviewed in the first tranche were rated on a 4-point scale at terminal evaluation. Of those 15, 12 had overall ratings of either satisfactory or marginally satisfactory, and one highly satisfactory overall. Eleven of the sustainability ratings were either likely or marginally likely. Only two projects were rated marginally unlikely overall or for sustainability, and only one project received marginally unlikely in both categories (the Demand Side

Management Demonstration energy conservation project that ended in 1999 [GEF ID 64]). Although none of the documents mentioned outcome indicators, eight of the 17 rated estimated CO_2 direct and indirect impacts.

In the second pool of projects—the CCM subset of the 2019 cohort—63% of the projects were rated in the likely range for sustainability ($n = 22$; nine were rated likely and 13 marginally likely). This is slightly higher than the 2019 cohort as a whole, in which 59% were rated in the likely range. In turn, the 2019 annual performance report noted that "the difference between the GEF portfolio average and the 2019 cohort is not statistically significant for both outcome and sustainability rating" (GEF IEO, 2020, p. 9). It is slightly lower than the percentage of CCM projects receiving an overall rating of marginally likely or higher in the 2017 portfolio review (68%, $n = 265$; GEF IEO, 2017, p. 78).

In this second set of projects, only two received a rating of marginally unlikely and only one received a sustainability rating of unlikely. The remainder of the projects could not be classified using the 4-point rating scale, either because they had used an either/or estimate (one project), a 5-point scale (one project), or an estimate based on the assessment of risks to development outcome (two projects). Six projects or could not be assessed due to the absence of a publicly accessible terminal evaluation in the GEF and implementing agency archives.

How Effectively Is Sustainability Being Captured?

Throughout the first set of reports on which the sustainability was claimed, "84% of these projects that were rated as sustainable at closure also had satisfactory postcompletion outcomes, as compared with 55% percent of the unsustainable projects" (GEF IEO, 2019a, p. 29). The data did not support the claim, even during implementation.

- As a Brazilian project (GEF ID 2941) showed, sustainability is unlikely when project achievements are weak, and exit conditions and benchmarks need to be clear: The exit strategy provided by IDB Invest77 is essentially based on financial-operational considerations but does not provide answers to the initial questions how an EEGM [energy efficiency guarantee mechanism] should be shaped in Brazil, how relevant it is and for whom, and to whom the EEGM should be handed over (p. 25).
- In Russia, the terminal evaluation for an energy efficiency project (GEF ID 292) cited project design flaws that seemed to belie its sustainability rating of likely: "From a design-for-replication point of view the virtually 100% grant provided by the GEF for project activities is certainly questionable" (Global Environment Facility Evaluation Office [GEF EO], 2008, p. 20). Further, the assessment that "the project is attractive for replication, dissemination of results has been well implemented, and the results are *likely to be sustainable* [emphasis added] for the long-term, as federal and regional legislation support is introduced" (GEF EO, 2008, p. 39), makes a major assumption regarding changes in the policy environment. (In fact, federal legislation was introduced 2 years post project, and the extent of enforcement would require examination.)
- A Pacific regional project (GEF ID 1058) was rated as likely to be sustained, but its report notes that it "does not provide overall ratings for outcomes, risks to sustainability, and M&E" (p. 1).
- The Renewable Development Energy project in China (GEF ID 446) that closed in 2007 was evaluated in 2009 (not post project, but a delayed final evaluation). The report considered the project sustainable with a continued effort to support off-grid rural electrification, claiming, "the market is now self-sustaining, and thus additional support is not required" (p. 11). The project estimated avoided CO_2 emissions and cited 363% as achieved; however, calculations were based on 2006 emis-

sions values for thermal power sector and data from all wind farms in China, without a bottom-up estimate. The interpolation of this data lacks verification.

- Similar sampling issues emerge in a project in Mexico (GEF ID 643): "A significant number of farmers . . . of an estimated 2,312 farmers who previously had had no electricity" (p. 20) saw their productivity and incomes increase as a result of their adoption of productive investments (e.g., photovoltaic-energy water-pumping systems and improved farming practices). A rough preliminary estimate is extrapolated from an evaluation of "*three* [emphasis added] beneficiary farms, leading to the conclusion that in these cases average on-farm increases in income more than doubled (rising by139%)" (p. 21).

Baseline to terminal evaluation comparisons were rare, with the exception of photovoltaic energy projects in China and Mexico, and none were post project. Two were mid-term evaluations, which could not assess final outcomes much less sustainability. Ex-post project evaluations far more typically focus on the contributions that projects made, because only in rare cases can the attribution be isolated, especially for a project pool, where the focus is often on creating an enabling environment reliant on a range of actors. One such example is the Indian energy efficiency project approved in 1998 (GEF ID 404), in which

> the project resulted in a favorable environment for energy-efficiency measures and the sub-projects inspired many other players in similar industries to adopt the demonstrated technologies. Although quantitative data for energy saved by energy efficiency technologies in India is not available, it is evident that due to the change in policy and financial structure brought by this project, there is an increase in investment in energy efficiency technologies in the industries. (GEF IEO, 2013, Vol. 2., p. 95)

And while such GEF evaluators are asking for ex-post evaluation, in an earlier version of this book, *Evaluating Climate Change Action for Sustainable Development* (Uitto et al., 2017), the authors encouraged us to be "modest" in expectations of extensive ex-post evaluations and exploration of ex-post's confirmatory power seemingly has not occurred:

> The expectations have to be aligned with the size of the investment. The ex-post reconstruction of baselines and the assessment of quantitative results is an intensive and time-consuming process. If rigorous, climate change-related quantitative and qualitative data are not available in final reports or evaluations of the assessed projects, it is illusive to think that an assessment covering a portfolio of several hundred projects is able to fill that gap and to produce aggregated quantitative data, for example on mitigated GHG emissions. When producing data on proxies or qualitative assessments, the expectations must be realistic, not to say modest. (p. 89)

Project Evaluability

Following an analysis of the sustainability estimates in the first pool of projects, we screened project documentation and terminal evaluations for conditions that foster sustainability during planning, implementation, and exit. We also analyzed how well the projects reported on factors that could be measured in a post project evaluation and factors that would predispose projects to sustainability. These sustained impact conditions consisted of the following elements: (a) resources, (b) partnerships and local ownership, (c) capacity building, (d) emerging sustainability, (e) evaluation of risks and resilience, and (f) CO_2 emissions (impacts).

Although documentation in evaluations did not verify sustainability, many examples exist of data collection that could support post project analyses of sustainability and sustained impacts in the future. Most reports cited examples of resources that had been generated, partnerships that had been fostered for local ownership and sustainability, and capacities that had been built through training. Some terminal evaluations also captured emerging impacts due to local efforts to sustain or extend impacts of the project that had not been anticipated ex-ante.

The Decentralized Power Generation project (GEF ID 4749) in Lebanon provides a good example of a framework to collect information on elements of sustainability planning at terminal (see Table 3).

Table 3 Sustainability Planning from a Decentralized Power Generation Project in Lebanon (GEF ID 4749)

Resources	Are there financial risks that may jeopardize the sustainability of project outcomes?
	What is the likelihood of financial and economic resources not being available once GEF grant assistance ends?
Ownership	What is the risk, for instance, that the level of stakeholder ownership (including ownership by governments and other key stakeholders) will be insufficient to allow for the project outcomes/benefits to be sustained?
	Do the various key stakeholders see that it is in their interest that project benefits continue to flow?
	Is there sufficient public/stakeholder awareness in support of the project's long-term objectives?
Partnerships	Do the legal frameworks, policies, and governance structures and processes within which the project operates pose risks that may jeopardize sustainability of project benefits?
Benchmarks, risks, & resilience	Are requisite systems for accountability and transparency, and required technical know-how, in place?
	Are there ongoing activities that may pose an environmental threat to the sustainability of project outcomes?
	Are there social or political risks that may threaten the sustainability of project outcomes?

Source: 4749 Terminal Evaluation, p. 45. Note: Capacity Building and Emerging Sustainability were missing from project 4749

Tangible examples of the above categories at terminal evaluations include the following.

Resources

The most widespread assumption for sustainability was sufficient financial and in-kind resources, often reliant on continued national investments or new private international investments, which could be verified. National resources that could sustain results include terminal evaluation findings such as:

Funding for fuel cell and electric vehicle development by the Chinese Government had increased from Rmb 60 million (for the 1996-2000 period) to more than Rmb 800 million (for the 2001-2005 period). More recently, policymakers have now targeted hydrogen commercialization for the 2010-2020 period. (GEF ID 445, p. 17)

Another example is: "About 65 percent of [Indian] small Hydro electromechanical Equipment is sourced locally" (GEF ID 386; GEF IEO, 2013, Vol.2, p. 76). The terminal evaluation of a global IFC project stated that "Moser Baer is setting up 30 MW solar power plants with the success of the 5 MW project. Many private sector players have also emulated the success of the Moser Baer project by taking advantage of JNNSM scheme" (GEF ID 112, p. 3).

Local Ownership and Partnerships

The Russian Market Transformation for EE Buildings project (GEF ID 3593) showed in its recommendation to governmental stakeholders that their ownership would be essential for sustainability, describing "a suitable governmental institution to take over the ownership over the project web site along with the peer-to-peer network ensuring the sustainability of the tools [to] support the sustainability of the project results after the project completion" (p. xi). An Indian project (GEF ID 386) noted how partnerships could sustain outcomes:

> By 2001, 16 small hydro equipment manufacturers, including international joint ventures (compared to 10 inactive firms in 1991) were operational. . . . State government came up with policies with financial incentives and other promotional packages such as help in land acquisition, getting clearances, etc. These profitable demonstrated projects attracted private sector and NGOs to set up similar projects. (GEF IEO, 2013, Vol. 2, p. 74)

Capacity Building

The Renewable Energy for Agriculture project in Mexico (GEF ID 643) established the "percentage of direct beneficiaries surveyed who learned of the equipment through FIRCO's promotional activities" (86%), "number of replica renewable energy systems installed" (847 documented rep-

licas), and "total number of technicians and extensionists trained in renewable energy technologies" (p. 33). This came to 3022, or 121% of the original goal of 2500, which provides a good measure of how the project exceeded this objective.

Emerging Sustainability

Recent post project evaluations also address what emerged after the project that was unrelated to the existing theory of change. These emerging findings are rarely documented in terminal evaluations, but some projects in the first pool included information about unanticipated activities or outcomes at terminal evaluation, and these could be used for future post project fieldwork follow-up. As a consequence of the hydroelectric resource project, for example, the Indian Institute "developed and patented the designs for water mills" (GEF ID 386; GEF IEO, 2013, Vol. 2, p. 73). The terminal evaluation for another project stated that "following the UNDP-GEF project, the MNRE [Ministry of New and Renewable Energy] initiated its own programs on energy recovery from waste. Under these programs, the ministry has assisted 14 projects with subsidies of US$ 2.72 million" (GEF ID 370; GEF IEO, 2013, Vol. 2, p. 62).

Benchmarks, Risks, and Resilience

As the GEF's 2019 report itself noted, "The GEF could strengthen its approach to assessing sustainability further by explicitly addressing resilience" (GEF IEO, 2019a, p. 33). Not doing so is a risk, as our climate changes. Two evaluations noted "no information on environmental risks to project sustainability;" these were the Jamaican pilot on Removal of Barriers to Energy Efficiency and Energy Conservation (GEF ID 64; p. 68) and a Pacific regional project (GEF ID 1058). For likelihood of sustainability, the Jamaican project was rated moderately unlikely and the Pacific Islands project was rated likely but "does not provide overall ratings for outcomes, risks to sustainability, and M&E" other than asserting that

> the follow-up project, which has been approved by the GEF, will ensure that the recommendations entailed in the documents prepared as part of this

project are carried out. Thus, financial risks to the benefits coming out of the project are low. (p. 3)

Greenhouse Gas Emissions (Impacts)

In GEF projects, timeframe is an important issue, which makes post project field verification that much more important. As the GEF IEO stated in 2018, "Many environmental results take more than a decade to manifest. Also, many environmental results of GEF projects may be contingent on future actions by other actors." (GEF IEO, 2018, p. 34).

Uncertainty and Likelihood Estimates

Estimating the likelihood of sustainability of greenhouse gas emissions at terminal evaluation raises another challenge: the relatively high level of uncertainty concerning the achievement of project impacts related to GHG reduction. GHG reductions are the primary objective stated in the climate change focal area, and they appear as a higher level impact across projects regardless of the terminology used. For a global project on bus rapid transit and nonmotorized transport, the objective was to "reduce GHG emissions for transportation sector globally" (GEF ID 1917, p. 9). For a national project on building sector energy efficiency, the project goal was "the reduction in the annual growth rate of GHG emissions from the Malaysia buildings sector" (GEF ID 3598; Aldover & Tiong, 2017, p. i). For a land management project in Mexico, the project objective was to "mitigate climate change in the agricultural units selected . . . including the reduction of emissions by deforestation and the increase of carbon sequestration potential" (GEF ID 4149, p. 21). For a national project to phase out ozone-depleting substances, the project objective was to "reduce greenhouse gas emissions associated with industrial RAC (refrigeration and air conditioning) facilities in The Gambia" (GEF ID 5466, p. vii). Clearly, actual outcomes in GHG emissions need to be considered in any assessment of the likelihood of sustainability of outcomes.

Unlike projects in the carbon finance market, GEF projects estimate emissions for a project period that usually exceeds the duration of the GEF intervention. In most cases, ex-ante estimated GHG reductions in the post project period are larger than estimated GHG reductions during the project lifetime. In practice, this means that for projects for which the majority of emissions will occur after the terminal evaluation, evaluators are being asked to estimate the likelihood that benefits will not only continue, but will increase due to replication, market transformation, or changes in the technology or enabling environment. Table 4 provides several examples from the GEF 2019 cohort of how GHG reductions may be distributed over the project lifecycle.

The range in Table 4 shows the substantial variation in uncertainty when estimating the likelihood of long-term project impacts. For projects designed to achieve all of their emission reductions during their operational lifetimes, the achievement of GHG reductions can be verified as a part of the terminal evaluation. However, most projects assume that nearly all estimated GHG reductions will occur in the post project period, so uncertainty levels are much higher and estimates may be more difficult to compile. In other evaluations, evaluators may identify inconsistent GHG estimates (e.g., GEF ID 4157 and 5157), or recommend that the ex-ante estimates be downsized (e.g., GEF ID 3922, 4008, and 4160). These trends may also be difficult to capture in likelihood estimates.

Conclusions and Recommendations

While sustainability has been estimated in nearly all of the projects in the two pools we considered, it has not been measured. Assessing the relationship between projected sustainability and actual post project outcomes was not possible due to insufficient data. Further, findings from the first pool of climate change mitigation projects did not support the conclusion that "outcomes of most of the GEF projects are sustained during the post-completion period" (GEF IEO, 2019a, p. 17).

Table 4 Distribution of Estimated GHG Reductions Ex-Ante for Selected Projects in the CCM Subset of the GEF 2019 Cohort

GEF ID	Country	Sub-Sector	Ex-ante GHG reduction estimates		% of reductions achieved by the terminal evaluation
			During project lifetime (tCO$_2$e)	Total reductions (tCO$_2$e)	
2941	Brazil	EE Buildings	705,000	9,588,000	7
2951	China	EE Financing	5,400,000	111,500,000	5
3216	Russia	EE Standards / Labels	7,820,000	123,600,000	6
3555	India	EE Buildings	454,000	5,970,000	8
3593	Russia	EE Industry	0	3,800,000	0
3598	Malaysia	EE Buildings	2,002,000	18,166,000	11
3755	Vietnam	EE Lighting	2,302,000	5,268,000	44
3771	Philippines	EE Industry	560,000	560,000	100

Sources: 2941 Project Document, pp. 35–37; 2951 PAD/CEO Endorsement Request, p. 88; 3216 Project Document, pp. 80–90; 3555 Terminal Evaluation; 3593 Terminal Evaluation, p. 23; 3598 Terminal Evaluation, p. 24; 3755 GEF CEO Endorsement Request; 3771 Terminal Evaluation pp. 8–9

In the absence of sufficient information regarding project sustainability, determining post project GHG emission reductions is not possible, because these are dependent on the continuation of project benefits following project closure.

We also conclude that although the 4-point rating scale is a common tool for estimating the likelihood of sustainability, the measure itself has not been evaluated for reliability or validity. The scale is often used to summarize diverse trends in the midst of varying levels of uncertainty limits. The infrequency of the unlikely rating in terminal evaluations may result from this limitation—evaluators believe that some benefits (greater than 0%) will continue. However, the 4-point scale cannot convey an estimate of what percentage of benefits will continue. Furthermore, the use of market studies to assess sustainability is not effective in the absence of attributional analysis linking results to the projects that ostensibly caused change.

As a result, the current evaluator's toolkit still does not provide a robust means of estimating post project sustainability and is not suitable as a basis for postcompletion claims. That said, M&E practices in the CCM projects we studied supported the collection of information that documented conditions (e.g., resources, partnerships, capacities, etc.) in a way that projects could be evaluable, or suitable for post project evaluation. We recommend that donors provide financial and administrative support for project data repositories to retain data in-country at terminal evaluation for post project return and country-level learning, and include evaluability (control groups, sampling sizes, and sites selected by evaluability criteria) in the assessment of project design. We also recommend sampling immediately from the 56 CCM projects in the two sets of projects that have been closed at least 2 years.

Donors' allocation of sufficient resources for CCM project evaluations would allow verification of actual long-term, post project sustainability using the OECD DAC (2019) definition of "the continuation of benefits from a development intervention after major development assistance has been completed" (p. 12). It would also enable evaluators to consider enumerating project components that are sustained rather than using an either/or designation (sustained/not sustained). Evaluation terms of reference should clarify the methods used for contribution vs. attribution claims, and they should consider decoupling estimates of direct and indirect impacts, which are difficult to measure meaningfully in a single measure. For the GEF portfolio specifically, the development of a postcompletion verification approach could be expanded from the biodiversity focal area to the climate change focal area (GEF IEO, 2019b), and lessons could also be learned from the Adaptation Fund's (2019) commissioned work on post project evaluations.

Bilateral donors such as JICA have developed rating scales for post project evaluations that assess impact in a way that captures both direct and indirect outcomes (JICA, 2017).

Developing country parties to the Paris Agreement have committed to providing "a clear understanding of climate change action" in their countries under Article 13 of the agreement (United Nations, 2015), and donors have a clear imperative to press for continued improvement in reporting on CCM project impacts and using lessons learned to inform future support.

Appendix

Projects Discussed in Chapter

Unless indicated otherwise in the text, see project documentation at the GEF project database (https://www.thegef.org/projects) under the project ID number.

Name	GEF ID	Implementing Agency	Country
Demand Side Management Demonstration	64	The World Bank	Jamaica
Photovoltaic Market Transformation Initiative (IFC)	112	The World Bank	Global, India, Kenya, Morocco
Capacity Building to Reduce Key Barriers to Energy Efficiency in Russian Residential Building and Heat Supply	292	UNDP	Russian Federation
Coal Bed Methane Capture and Commercial Utilization	325	UNDP	India
Development of High Rate BioMethanation Processes as Means of Reducing Greenhouse Gas Emissions	370	UNDP	India
Optimizing Development of Small Hydel Resources in Hilly Areas	386	UNDP	India
Energy Efficiency	404	The World Bank	India
Barrier Removal for the Widespread Commercialization of Energy-Efficient CFC-Free Refrigerators in China	445	UNDP	China
Renewable Energy Development	446	The World Bank	China
Renewable Energy for Agriculture	643	The World Bank	Mexico
Demonstration of Fuel Cell Bus Commercialization in China (Phase II-Part I)	941	UNDP	China
Pacific Islands Renewable Energy Programme (PIREP)	1058	UNDP	Regional: Cook Islands, Fiji, Micronesia, Kiribati, Marshall Islands, Nauru, Niue, Papua New Guinea, Palau, Solomon Islands, Tonga, Tuvalu, Vanuatu, Samoa
Reducing Greenhouse Gas Emissions with Bus Rapid Transit	1917	UNEP	Global, Columbia, Tanzania
Market Transformation for Energy Efficiency in Buildings	2941	UNDP	Brazil

(continued)

Name	GEF ID	Implementing Agency	Country
RUS Market Transformation Programme on Energy Efficiency in GHG-Intensive Industries In Russia	3593	European Bank for Reconstruction and Development	Russian Federation
Buildings Sector Energy Efficiency Project (BSEEP)	3598	UNDP	Malaysia
SPWA-CC: Promoting Renewable Energy Based Mini Grids for Productive Uses in Rural Areas in The Gambia	3922	UNIDO	Gambia
Reducing GHG Emissions from Road Transport in Russia's Medium-sized Cities	4008	UNDP	Russian Federation
SFM Mitigating Climate Change through Sustainable Forest Management and Capacity Building in the Southern States of Mexico (States of Campeche, Chiapas and Oaxaca)	4149	IFAD	Mexico
Promotion of Biomass Pellet Production and Utilization in Georgia	4157	UNDP	Georgia
Technology Transfer and Market Development for Small-Hydropower in Tajikistan	4160	UNDP	Tajikistan
Small Decentralized Renewable Energy Power Generation	4749	UNDP	Lebanon
ESCO Moldova - Transforming the market for Urban Energy Efficiency in Moldova by Introducing Energy Service Companies (ESCO)	5157	UNDP	Moldova
Reducing Greenhouse Gases and ODS Emissions through Technology Transfer in the Industrial Refrigeration and Air Conditioning Sector	5466	UNIDO	Gambia

References

Adaptation Fund. (2019). Report of the Adaptation Fund Board, note by the chair of the Adaptation Fund Board – Addendum. AFB/B.34–35/3. Draft – 8 November 2019. https://www.adaptation-fund.org/document/report-of-the-adaptation-fund-board-note-by-the-chair-of-the-adaptation-fund-board-addendum/

Aldover, R. Z., & Tiong, T. C. (2017). UNDP/GEF project PIMS 3598: Building sector energy efficiency project (BSEEP): Terminal evaluation report. Global Environment Facility and United Nations Development Programme. https://erc.undp.org/evaluation/evaluations/detail/8919

Asian Development Bank. (2010). Post-completion sustainability of Asian Development Bank-assisted projects. https://www.adb.org/documents/post-completion-sustainability-asian-development-bank-assisted-projects

Cekan, J. (2015, March 13). When funders move on. Stanford Social Innovation Review. https://ssir.org/articles/entry/when_funders_move_on#

Cekan, J., Zivetz, L., & Rogers, P. (2016). Sustained and emerging impacts evaluation. Better Evaluation. https://www.betterevaluation.org/en/themes/SEIE

Duval, R. (2008). A taxonomy of instruments to reduce greenhouse gas emissions and their interactions. Organisation for Economic Co-operation and Development. https://doi.org/10.1787/236846121450.

Global Environment Facility. (2017). Guidelines for GEF agencies in conducting terminal evaluation for full-sized projects. https://www.gefieo.org/evaluations/guidelines-gef-agencies-conducting-terminal-evaluation-full-sized-projects

Global Environment Facility Evaluation Office. (2008). Evaluation of the catalytic role of the GEF. https://www.gefieo.org/sites/default/files/ieo/ieo-documents/gef-catalytic-role-qualitative-analysis-project-documents.pdf

Global Environment Facility Independent Evaluation Office. (2010). GEF monitoring and evaluation policy. https://www.gefieo.org/sites/default/files/ieo/evaluations/gef-me-policy-2010-eng.pdf

Global Environment Facility Independent Evaluation Office. (2012). Approach paper: Impact evaluation of the GEF support to CCM: Transforming markets in major emerging economies. https://www.gefieo.org/

sites/default/files/ieo/ieo-documents/ie-ccm-markets-emerging-economies.pdf

Global Environment Facility Independent Evaluation Office. (2013). *Country portfolio evaluation (CPE) India.* http://www.gefieo.org/evaluations/country-portfolio-evaluation-cpe-india

Global Environment Facility Independent Evaluation Office. (2017). *Climate change focal area study.* https://www.thegef.org/council-meeting-documents/climate-change-focal-area-study

Global Environment Facility Independent Evaluation Office. (2018). *Sixth overall performance study of the GEF: The GEF in the changing environmental finance landscape.* https://www.thegef.org/sites/default/files/council-meeting-documents/GEF.A6.07_OPS6_0.pdf

Global Environment Facility Independent Evaluation Office. (2019a). *Annual Performance Report 2017.* https://www.gefieo.org/evaluations/annual-performance-report-apr-2017

Global Environment Facility Independent Evaluation Office. (2019b). *A methodological approach for post-project completion.* https://www.gefieo.org/council-documents/methodological-approach-post-completion-verification

Global Environment Facility Independent Evaluation Office. (2020). *Annual performance report 2019.* https://www.gefieo.org/evaluations/annual-performance-report-apr-2019

Hageboeck, M., Frumkin, M., & Monschein S. (2013). *Meta-evaluation of quality and coverage of USAID evaluations.* USAID. https://www.usaid.gov/evaluation/meta-evaluation-quality-and-coverage

Japan International Cooperation Agency. (2004). Issues in ex-ante and ex-post evaluation. In *JICA Guideline for Project Evaluation: Practical Methods for Project Evaluation* (pp. 115–197). https://www.jica.go.jp/english/our_work/evaluation/tech_and_grant/guides/pdf/guideline01-01.pdf

Japan International Cooperation Agency. (2017). Ex-post evaluation results. In *JICA annual evaluation report 2017* (Part II, pp. 1–34). https://www.jica.go.jp/english/our_work/evaluation/reports/2017/c8h0vm0000d2h2gq-att/part2_2017_a4.pdf

Japan International Cooperation Agency. (2020a). *Ex-post evaluation (technical cooperation).* https://www.jica.go.jp/english/our_work/evaluation/tech_and_grant/project/ex_post/index.html

Japan International Cooperation Agency. (2020b). *Ex-post evaluation (ODA loan).* https://www.jica.go.jp/english/our_work/evaluation/oda_loan/post/index.html

Legro, S. (2010, June 9–10). *Evaluating energy savings and estimated greenhouse gas emissions in six projects in the CIS: A comparison between initial estimates and assessed performance* [paper presentation]. International Energy Program Evaluation Conference, Paris, France. https://energy-evaluation.org/wp-content/uploads/2019/06/2010-paris-027-susan-legro.pdf

Mayne, J. (2001). Assessing attribution through contribution analysis: Using performance measures sensibly.

The Canadian Journal of Program Evaluation, 16(1), 1–24.

OECD/DAC Network on Development Evaluation. (2019). *Better criteria for better evaluation: Revised evaluation criteria definitions and principles for use.* Organisation for Economic Co-operation and Development. http://www.oecd.org/dac/evaluation/revised-evaluation-criteria-dec-2019.pdf

Organisation for Economic Co-operation and Development. (2015). *OECD and post-2015 reflections. Element 4, Paper 1: Environmental Sustainability.* https://www.oecd.org/dac/environment-development/FINAL%20POST-2015%20global%20and%20local%20environmental%20sustainability.pdf

Organisation for Economic Co-operation and Development, Development Assistance Committee. (1991). *DAC criteria for evaluating development assistance.* https://www.oecd.org/dac/evaluation/2755284.pdf

Rogers, B. L., & Coates, J. (2015). *Sustaining development: A synthesis of results from a four-country study of sustainability and exit strategies among development food assistance projects.* FANTA III, Tufts University, & USAID. https://www.fantaproject.org/research/exit-strategies-ffp

Sridharam, S., & Nakaima, A. (2019). Till time (and poor planning) do us part: Programs as dynamic systems—Incorporating planning of sustainability into theories of change. *The Canadian Journal of Program Evaluation.* https://evaluationcanada.ca/system/files/cjpe-entries/33-3-pre005.pdf

Uitto, J., Puri, J., & van den Berg, R. (2017). *Evaluating climate change action for sustainable development.* Global Environment Facility Independent Evaluation Office. https://www.gefieo.org/sites/default/files/ieo/documents/files/cc-action-for-sustainable-development_0.pdf

United Nations. (2015, December 12). *Paris agreement.* https://unfccc.int/sites/default/files/english_paris_agreement.pdf

United States Agency for International Development. (2018). *Project evaluation overview.* https://www.usaid.gov/project-starter/program-cycle/project-design/project-evaluation-overview

United States Agency for International Development. (2019). *USAID's impact: Ex-post evaluation series.* https://www.globalwaters.org/resources/ExPostEvaluations

Valuing Voices. (2020). *Catalysts for ex-post learning.* http://valuingvoices.com/catalysts-2/

Zivetz, L., Cekan, J., & Robbins, K. (2017a). *Building the evidence base for post project evaluation: A report to the faster forward fund.* Valuing Voices. http://valuingvoices.com/wp-content/uploads/2013/11/The-case-for-post-project-evaluation-Valuing-Voices-Final-2017.pdf

Zivetz, L., Cekan, J., & Robbins, K. (2017b). *Checklists for sustainability.* Valuing Voices. http://valuingvoices.com/wp-content/uploads/2017/08/Valuing-Voices-Checklists.pdf

Part III

Evaluating Climate Change Mitigation and Adaptation

Introduction

A. Viggh (✉) e-mail: aviggh@thegef.org
Global Environment Facility Independent Evaluation Office, Washington, DC, USA

One of the most profound challenges of our time is climate change. In this part, authors discuss approaches in evaluating both climate change mitigation and adaptation interventions. Two chapters relate to important aspects of evaluating climate change mitigation of carbon finance and a community-level energy efficiency project. Concerning adaptation to climate change, three chapters discuss aspects of evaluation's implications for monitoring, evaluation, and learning (MEL) including innovative MEL opportunities; structures, processes, and resources supporting MEL; and the implications of evidence reviews with respect to MEL.

The first chapter on climate change mitigation, "Using a Realist Framework to Overcome Evaluation Challenges in the Uncertain Landscape of Carbon Finance," discusses a framework for using the realist evaluation method to overcome contextual uncertainties of carbon market finance. The framework was applied in the midterm evaluation of the 12-year Carbon Market Finance Programme (CMFP). The UK Department for International Development and the Department of Business, Energy, and Industrial Strategy published the CMFP business case under the UK's International Climate Finance in 2013. The CMFP aims to support sub-Saharan African least developed countries in obtaining financing through the carbon market. The collapse of the carbon market and its uncertain future led the evaluation team to consider a realist evaluation approach that seeks to explain how projects work or do not work, and under what circumstances. Murdoch, Keppler, Burlace, and Wörlen describe the methodology, benefits, and challenges of realist evaluation as an approach, argue that using realist evaluation is practical, and provide recommendations for a revised approach for future evaluations.

In "Evaluation's Role in Development Projects: Boosting Energy Efficiency in a Traditional Industry in Chad," Yakeu Djiam discusses the importance of evaluation in climate change mitigation projects that contribute to sustainable development. The chapter presents the findings of an evaluation of a project that aimed to improve energy efficiency of traditional industries and reduce demand for firewood by promoting energy-efficient cook stoves in micro- and small-scale food-processing industries in Chad. The project, financed by the Global Environment Facility (GEF) and implemented by the United Nations Industrial Development Organization (UNIDO), targeted the beer-brewing and meat-grilling sectors, which are large consumers of firewood. The introduction of locally developed energy-efficient stoves in these sectors faced several barriers that the project addressed. The chapter highlights evaluation findings related to project performance, project coordination and management, gender mainstreaming, and other cross-cutting issues, followed by a discussion of the value of evaluation in ensuring a project's long-term relevance and sustainability for beneficiaries, and achievement of transformational change.

Gregorowski and Bours, in "Enabling Systems Innovation in Climate Change Adaptation: Exploring the Role for MEL," start the discussion on adaptation to climate change by stating that established MEL approaches no longer reflect the complexity of current problems of climate change, environmental degradation, and global pandemics. The chapter presents the findings of a study of MEL approaches and technologies commissioned by the Adaptation Fund's Technical Evaluation Reference Group (AF-TERG). The study aimed to provide new insights for innovative MEL approaches to support climate change adaptation interventions. The authors describe the purpose and approach of the study, including its scan-search-appraise methodology based on a three-step hypothesis to explore a set of innovative MEL approaches and methods. The chapter concludes with seven recommendations and provides illustrative examples of innovative MEL approaches, processes, and technical interventions for the future of MEL, to enhance a systems innovation approach to climate change adaptation.

Addressing evaluability of climate change adaptation-focused interventions, MacPherson, Jersild, Bours, and Holo explore in "Assessing the Evaluability of Adaptation-Focused Interventions: Lessons from the Adaptation Fund" how evaluability assessments can help identify opportunities for strengthening evaluability and MEL of a project. To assess whether Adaptation Fund projects have monitoring and evaluation plans and budgets to support useful MEL, AF-TERG developed an assessment framework and applied it to the whole AF portfolio. The chapter outlines the history and purpose of evaluability assessments, the approach of the assessment, and the analysis of the information collected. The authors discuss the findings from applying the framework and how the process of evaluability assessments can strengthen both a project's evaluability and its MEL. A key conclusion of the assessment is that focusing on improvements of a project's overall MEL should lead to improved evaluability.

The final chapter of the climate change part, "Evaluating Transformational Adaptation in Smallholder Farming: Insights from an Evidence Review," draws on the findings of research funded by the International Fund for Agricultural Development (IFAD) Independent Office of Evaluation as part of a broader thematic evaluation of their support for smallholder farmers' adaptation to climate change. Silici, Knox, Rowe, and Nanthikesan discuss evaluating transformational adaptation in smallholder farming through the findings of an evidence synthesis of literature searches and review. The review was executed in line with well-established guidelines and provided a synthesis of evidence on smallholder adaptation to climate change from the past 10+ years. The authors summarize the key messages from the evidence synthesis and discuss their implications with respect to MEL under the framework of transformational change, with a focus on the challenges that MEL faces. The chapter ends with comments on the future role of evidence reviews in thematic evaluation beyond climate change adaptation and on the value such synthesis can add to MEL.

Using a Realist Framework to Overcome Evaluation Challenges in the Uncertain Landscape of Carbon Finance

Callum Murdoch, Lisa Keppler, Tillem Burlace, and Christine Wörlen

Abstract

In 2013, the United Kingdom Department for International Development and the Department of Business, Energy, and Industrial Strategy published a business case for the Carbon Market Finance Programme (CMFP). The core mandate: to build capacity and develop aids for least developed countries in sub-Saharan Africa to access finance via the carbon market. The chosen strategy involved signing emission reduction purchase agreements with private sector enterprises, using the United Nations Framework Convention for Climate Change's Clean Development Mechanism (CDM) to verify generation of tradeable certified emissions reductions. The World Bank's Carbon Initiative for Development (Ci-Dev) would implement the 12-year program. The team for the 2019 midterm evaluation found that program uncertainty—from sociopolitical challenges in pilot markets to global indecision on the future of Article 6 and carbon markets—would complicate assessing progress toward business case objectives. The collapse and failed recovery of the carbon market impacted underlying assumptions of the CMFP's theory of change, and uncertainty about CDM's future complicated evaluation of program sustainability. This chapter presents a practical approach to using realist evaluation to overcome the contextual uncertainties of the carbon market landscape, providing strengths and weaknesses of the approach applied and recommending a revised approach for future evaluations.

The Evaluation of the Carbon Market Finance Programme

In 2013, the UK Department for International Development (DFID) and the Department of Business, Energy, and Industrial Strategy (BEIS; then the Department of Energy and Climate Change) published a business case for the Carbon Market Finance Programme (CMFP) under the UK's International Climate Finance (ICF). The core mandate of the program was to build capacity and develop tools and methodologies to help least developed countries in sub-Saharan Africa access finance via the carbon market. The business case explored several options before settling on a strategy that involved signing emission reduction purchase agreements with private sector enterprises seeking to improve energy access

C. Murdoch (✉) · T. Burlace
NIRAS-LTS International, Edinburgh, UK
e-mail: callum-murdoch@ltsi.co.uk

L. Keppler · C. Wörlen
Arepo, Berlin, Germany

© The Author(s) 2022
J. I. Uitto, G. Batra (eds.), *Transformational Change for People and the Planet*, Sustainable
Development Goals Series, https://doi.org/10.1007/978-3-030-78853-7_9

in least developed countries, using the United Nations Framework Convention for Climate Change's (UNFCCC) Clean Development Mechanism (CDM) to verify the generation of tradeable certified emissions reductions (CERs). The business case team selected the World Bank's Carbon Initiative for Development, or Ci-Dev, for the implementation of the program over a 12-year period.

On paper, the CMFP is a relatively straightforward results-based finance (RBF) program using carbon credits as the underlying result. However, the market for certified emissions reductions— and indeed, nearly all emission trading schemes— changed drastically since 2013 and was in a state of such uncertainty as of 2019 that market insiders were reticent to discuss it.[1] Starting in late 2011, the price for carbon trading instruments, including certified emissions reductions, began to decline. At the release of the CMFP business case, the market was in the middle of what most (including those at the UNFCCC) would describe as a collapse, and the outlook for recovery was difficult to determine (CDM Policy Dialogue, 2012). A further development that contributed to the uncertainty was the establishment of the Paris Agreement at the 21st Conference of Parties (COP21) in 2015, especially Article 6.4, which stated:

> A mechanism to contribute to the mitigation of greenhouse gas emissions and support sustainable development is hereby established under the authority and guidance of the Conference of the Parties serving as the meeting of the Parties to this Agreement for use by Parties on a voluntary basis. (UNFCCC, 2015)

The implication of this Article, and indeed the rest of Article 6, was that a new carbon verification and trading instrument would be established by the Paris Agreement, likely replacing the CDM. What Article 6 did not do was clearly establish what that mechanism will look like or how it will operate. The quest for the elusive Paris Rulebook, which would provide the foundations on which this new mechanism will be built, has not yet yielded any results (as of 2020). With the review of the nationally determined contributions scheduled for 2020, it was hoped that COP25 in 2019 would generate clarity. Although some progress was made, the future of carbon markets remains as undefined and uncertain as it has since the signing of the Paris Agreement.

The UK is implementing the CMFP in this context, accompanied by the evaluation described in this chapter. The 11-year evaluation kicked off in late 2014 and will run until the conclusion of the program in 2025. In 2019, the evaluation team conducted a midterm evaluation to gauge the program's progress to date (LTS International, 2020). We quickly found that the uncertainty surrounding the program, from the local sociopolitical challenges of the markets where the projects are being piloted to the global indecision on the future of Article 6 and carbon markets, would make assessing the program's progress toward its stated business case objectives a challenging process. The collapse and failed recovery of the market struck at many of the underlying assumptions of the CMFP's theory of change. Moreover, uncertainty about the future of CDM complicated any evaluative judgement on program sustainability.

The team considered a number of approaches before deciding that the realist evaluation approach (Pawson & Tilley, 1997) would be best suited to evaluating the program given the uncertain landscape. Although systematic evaluation methods are most common, they focus on explaining whether or not an intervention led to a certain outcome. Instead, realist evaluation helps to open the black box of program theory—it tries to explain how and why projects work or do not work, for whom, and under what circumstances. It recognizes that the context in which individual projects are operating makes important differences to the projects' results. It also shows that no project intervention is likely to work every-

---

[1] Even though the team for the evaluation discussed in this chapter noted this risk and limitation throughout the preparation of the evaluation approach, we were nonetheless surprised how unwilling critical senior stakeholders were to share information and opinions when the topic of post-Paris carbon markets was raised.

where, under all circumstances, and for everyone. The application of realist evaluation, therefore, provides the opportunity to evaluate projects within their unique and changing local, national, and global contextual factors.

Another benefit of realist evaluation is that it does not prescribe a specific, regimented approach. A plethora of literature offers different applications for realist principles in evaluation and various philosophical discussions about the nature of realist thinking. This chapter fits within the former category and seeks to offer a field guide of sorts for applying realist evaluation to interventions for which the underlying theory has been affected by the uncertain contextual landscape in which it operates. We first describe the methodology applied in the CMFP midterm evaluation, then weigh the positives and negatives of this approach, finally presenting a revised methodology taking into account the learning emerging from the evaluation.

Overview of Methodology

To systematically address the described complexity of both the program itself and of its embedded environment, the team developed an evaluation framework drawing on realist evaluation principles. Complementing this were other evaluative analysis methods, each of which gives specific insights into the program's dynamic implementation and progress. Descriptive analysis focused on verifiable and quantitative data, including reporting against the program's logical framework, a value-for-money analysis, benchmark assessments, and, to a lesser extent, a qualitative comparative analysis. We used explanatory analysis for qualitative, interpretative data and where the evaluation required greater consideration of the contextual factors contributing to the program's progress. For this explanatory analysis, the team used a realist evaluation approach in two tranches: first, as a specific evaluation method to gather, code, and analyze data from a variety of sources, and second, as a synthesis framework against which other explanatory evaluation methods, such as a contribution analysis and an energy market barrier analysis based on the Theory of No Change (Wörlen et al., 2011), could be assessed in the context of the wider portfolio findings. Figure 1 shows the chosen realist evaluation framework.

Realist Evaluation as an Approach

The leading questions of a realist evaluation ask how, why, for whom, and under what circumstances the program works or does not work. Answering these questions requires identifying the underlying generative mechanisms and causal relationships of the program's dynamic through a continuous, multistage hypotheses development process. This retroductive process moves back and forth between inductive and deductive logic based on assumptions, continuous learning, and the expertise of its developers (Greenhalgh et al., 2017a). Inductive reasoning generates a new theory from collected data and multiple observations showing a logical pattern, whereas deductive reasoning starts with a theory and the formulation of hypotheses which are tested and verified by observations. The retroductive approach applied by realist evaluation draws from both.

Hypothesis Development

The first step in the process was the formulation of hypotheses in the form of intervention-context-mechanism-outcome statements, or ICMOs, which were developed with a top-down approach. Based on the program's theory of change and a review of program- and context-related literature, these statements were formulated in an abstract way to be valid to the whole program itself, and to the project portfolio. ICMO configurations are the core analytical elements of realist evaluation. As shown in Fig. 2, they bring together in one statement

- a program's intervention (I)
- the context (C) in which the intervention takes place and that influences whether an intervention activates a mechanism (M), which is the

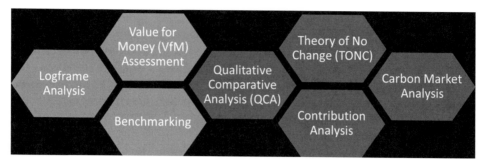

Fig. 1 Realist evaluation framework combining descriptive and explanatory evaluation methods

Fig. 2 Elements of an ICMO statement

response of the intervention target to the intervention

- the outcome (O), the desired end result of the other three components' interactions.

The intervention is the only factor under the direct control of the program; context, mechanism, and outcome are outside its direct control. The mechanism is the center of the realist explanation for how and why change occurs. It is a non-observable process, often described as changes in the reasoning and behavior of individuals or different levels of systems, that leads from the intervention to the outcome interconnected with contextual factors (Greenhalgh et al., 2017b).

Overall, our evaluation team developed four ICMO statements: two addressing the direct results of the program and two addressing the program's impact level, focusing on (a) barrier removal in energy markets and (b) the transformation of the carbon market, including replication of the program's approach (LTS International, 2020). An example ICMO statement (paraphrased from the CMFP evaluation) is provided in Box 1.

> **Box 1: Example ICMO Statement**
>
> By providing carbon-results-based financing and business development support funding (I), in a context with sufficient customer demand for the energy technologies, access to finance for the pilot enterprises, and a supportive policy framework (C), revenue and capacity for projects will be sufficient to overcome the operational challenges in providing rural energy access technologies (M), resulting in increased energy access and the generation and sale of certified emissions reductions (O).

To improve the explanatory value, each ICMO configuration consisted of sub-statements for each of the elements where the hypotheses involved compound statements (i.e., I1a, I1b, I1c, C1a, C1b, etc.; see Box 2). For example, the CMFP funding provided for carbon credit purchase often sought to trigger the same or similar mechanisms as funding provided for project readiness or capacity development. Thus, these two interventions were often considered in parallel. However, a particular piece of evidence collected might support one intervention more than the other and therefore would need to be independently assessed.

Box 2: Example ICMO Sub-statements

By providing carbon-results-based financing (I1a) and business development support funding (I1b), in a context with sufficient customer demand for the energy technologies (C1a), access to finance for the pilot enterprises (C1b) and a supportive policy framework (C1c)...

After splitting the ICMOs into sub-statements, the team identified critical components for each sub-statement that would prove the sub-statement's accuracy (see Box 3). The components listed were not exhaustive but served to guide what evidence would either confirm or disprove a sub-statement. This was to ensure consistency across users of the realist methodology and to enhance the deductive side of the analysis.

Box 3: Example ICMO Sub-statement Components

Sub-statement: "By providing carbon-results-based financing (I1a)".

Sample components:

1. An emissions reduction purchase agreement (ERPA) has been signed with the project entity.
2. The ERPA price is within a suitable range.
3. Finance has been transferred in exchange for carbon credits.

A fundamental element of realist evaluation is the involvement of core program or intervention stakeholders in the ICMO development process. In the CMFP evaluation, the ICMOs were refined through consultations with both the evaluation commissioners (BEIS) and the implementation agents (Ci-Dev) to ensure that they contained all of the relevant elements to depict the theory of change and its influences. The ICMOs passed through several iterations, often with minor phrasing or order changes.

Coding System

The evaluation team then tested the ICMO hypotheses by coding initial primary data (interviews) and secondary data (program documents, previous evaluation exercises, etc.) against the hypotheses to find out whether these theories were pertinent, productive, and appropriately designed. In an inductive process, we revised the hypotheses where the coded data gave indication about contexts, mechanisms, or outcomes that had not yet been considered in the ICMO statements, but where data showed relevance to the program's development. During the process, the team again consulted stakeholders on the interpretation of the available data to reach a reasonable judgement about the most useful findings. We repeated these steps at key stages of the evaluation process, leading to the retroductive nature of the approach.

For coding the ICMOs, we adapted the approach first adopted by the midterm evaluation of the UK Government's Climate Public Private Partnership Program (Climate Policy Initiative & LTS International, 2018). A matrix was designed that not only coded the evidence against the ICMO hypotheses, but also assessed the strength of each point of evidence and the overall evidence saturation. The intention of the coding matrix was to develop a quantifiable scoring process for the evidence collected that would allow the evaluation team to determine how accurate the initial hypotheses were. We devised a simple scoring system (see Table 1) that ranged from 3 (when a particular piece of evidence demonstrated high accuracy of the statement or sub-statement) to -3 (when evidence strongly disproved or contradicted the statement or sub-statement). The scoring also included a neutral value, X, which marked evidence as being relevant to the ICMO but not applicable to the current sub-statements. The team closely reviewed

Table 1 ICMO evidence scoring guide

Score	Description
3	Evidence strongly supports ICMO statement. Multiple or all components are met, or particularly strong evidence toward select components is provided.
1	Evidence partially supports ICMO statement. Some components have been met, or evidence supports the overall statement without actually meeting the component.
−1	Evidence partially contradicts ICMO statement. Evidence disproves or creates doubt that some components have been met, or evidence contradicts the overall statement without opposing specific components.
−3	Evidence strongly contradicts or disproves ICMO statement. Multiple or all components are countered or disproved, or particularly strong evidence negating select components is provided.
X	Evidence does not support or contradict the sub-statement but is relevant to the overall ICMO. For example, evidence provides a specific contextual factor that the statement hasn't captured, or an outcome that differs from those anticipated.

these evidence points and ensured the continuous verification of the relevance and revisions of the ICMO configurations. This scoring approach was designed to be simple, intuitive, and quantifiable while providing a traceable roadmap of how evidence was used to formulate specific evaluation findings.

To assess the strength of evidence, we categorized each piece of evidence according to Table 1, then used a modifier to weight the evidence (see Table 2). Verifiable evidence, either factual information or evidence from highly authoritative sources, received a two-times modifier to reflect the inherent strength of such evidence. Plausible evidence largely refers to data such as stakeholder interviews or discussions, qualitative or subjective secondary literature, or any other data source that would require further triangulated evidence to verify. As such, we applied no modifier to this type of evidence, on the understanding that evidence from one stakeholder would need to be cross-referenced and validated by evidence from other stakeholders or sources. If no data or argument supported the evidence point or if relevant contrary evidence was provided, the evidence was not coded but still used as information to improve the further analysis.

Table 2 Strength of evidence scoring scheme

Type	Description	Modifier
Verifiable evidence	Refers to data that are both plausible and possible to verify. Such evidence generally describes quantifiable measures that can be physically counted	×2
Plausible evidence	This includes evidence that may make a plausible claim but may draw heavily on assumptions from secondary literature. Alternatively, it may refer to evidence that is the plausible conclusion drawn by an expert stakeholder or observer. Presented evidence may justify this view but lack methodology against which the validity of the conclusion can be verified	×1
Minimal evidence	Some documents may simply claim an outcome but provide no information about the data or methodology used to evidence this claim. Alternatively, a claim may be supported by some evidence, but other contrary evidence is also provided. This evidence was not coded but used to signpost potential data and a need for further analysis	×0

Evidence Saturation

For each sub-statement, we calculated the convergence of all data to score the saturation of its content in relation to the components and how strongly it supported or contradicted the underlying ICMO statement. The convergence was calculated for positive and negative data points using the banding shown in Table 3 to determine

Table 3 Saturation rating

Evidence saturation level	Rating
> 75% convergence of relevant evidence supporting hypothesis	Green saturation level – treated as confirmed evidence
60–75% convergence of evidence supporting hypothesis	Amber saturation level – treated as partially confirmed but level of saturation and divergent views pointed out
< 60% convergence of evidence supporting hypothesis	Red saturation level – not treated as confirmed; discussed within the analysis' findings

Table 4 Categories of ICMO scoring results

Category	Description
Total statement score	Total score for each sub-statement, adjusted for strength of evidence
Total data points	Total number of evidence data points scored against each sub-statement
Average statement score	Average score for each sub-statement, adjusted for strength of evidence
Positive data convergence (%)	Saturation of data in relation to the components and its support of the specific sub-statement
Negative data convergence (%)	Nonsaturation of data in relation to the components and its contradiction to the specific sub-statement
Saturation of data (%)	Predominant data convergence for the specific sub-statement

how different statements would be discussed and analyzed depending on their overall data saturation. Where the majority of evidence for a hypothesis was scored positively, the saturation level would support claims that the hypothesis was accurate; where the evidence was mostly negative, the saturation level supported the opposite, indicating that the hypothesis did not hold true. For the high saturation threshold, more than 75% of the evidence needed to be scored either positive or negative. Low saturation of evidence implied that less than 60% of the evidence scored either positively or negatively.

Coding Results

Table 4 summarizes the score categories that resulted from the coding of evidence against the

ICMO statements in the matrix. Overall, we coded more than 800 individual data points against the ICMO statements, providing almost 2000 total scores. The matrix then generated average scores for each statement or sub-statement, adjusting for the strength of evidence modifiers, which were used to assess the statement's accuracy. The matrix also generated overall saturation scores, which increased confidence in the accuracy of the score achieved. On this basis, the evaluation team formulated the findings of the ICMO analysis.

Table 5 provides three example scores adapted from the evaluation to demonstrate coding results for hypotheses that are shown to be accurate, inaccurate, or divergent due to significant differences in project performances within the portfolio.

Overall, the results of the coding in the CMFP evaluation varied significantly, although this was to be expected. The scatter graph in Fig. 3 shows the average score of all sub-statements coded in relation to their data saturation. High saturation was achieved with around half the sub-statements coded, but very few sub-statements achieved an average score of greater than 1 or less than −1. This is indicative of high reliance on plausible data, often receiving a weaker score with no modifier, unlike verifiable evidence. The chart shows several outliers, highlighting where significantly strong evidence was found supporting the accuracy of the sub-statements.

In the following two charts, the data is split to show specifically the scoring for the interventions (Fig. 4) and the mechanism sub-statements (Fig. 5). For the interventions, all average scores were positive and, with the exception of one outlier, all achieved high saturation and relatively high accuracy scores. We found greater variance and less saturation with the mechanism scores, and on average they received a lower accuracy score with none breaking beyond an average score of 1 or −1. This is not especially surprising as interventions are more observable and verifiable than the mechanisms they hope to trigger, and thus receive higher scores.

These charts say little about the evaluation findings, but they do illustrate several important factors that we explore in the next sections. First,

Table 5 Example scoring results

Category	Hypothesis 1	Hypothesis 2	Hypothesis 3
Total statement score	98	−49	2
Total data points	32	63	44
Average statement score	3.06	−0.78	0.05
Positive data convergence (%)	94	24	54
Negative data convergence (%)	6	76	46
Saturation of data (%)	94	−76	54
Result	Very high likelihood hypothesis is accurate	High likelihood hypothesis is partially inaccurate	Highly divergent evidence indicative of differences at a project level

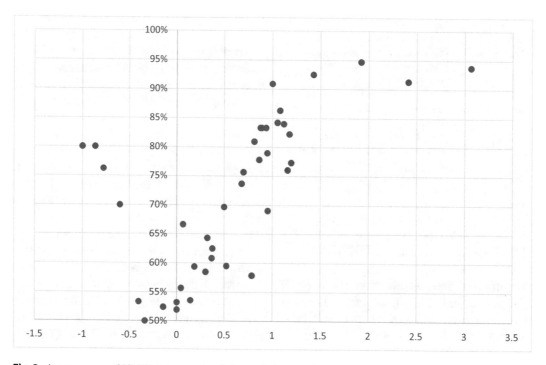

Fig. 3 Average score of ICMO statements in relation to their saturation

more than half the sub-statements received a convergence score of less than 75%, indicating moderate to significant evidence divergence. This was primarily driven by the heterogeneity of the program portfolio, which, due to its small size, could be significantly offset by a single outlying project or small project cluster. Second, few sub-statements scored beyond an average of 1 or −1 for accuracy. This is reflective of the context in which this coding took place, where uncertainty abounds and clear, verifiable evidence is limited. It is also indicative of the maturity of the projects (the latest commencing mid-2018), which limited the availability of strong evidence, especially for mechanism and outcome sub-statements.

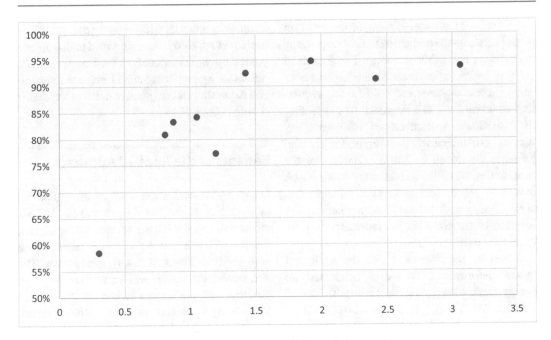

Fig. 4 Average score of sub-statements for interventions in relation to their saturation

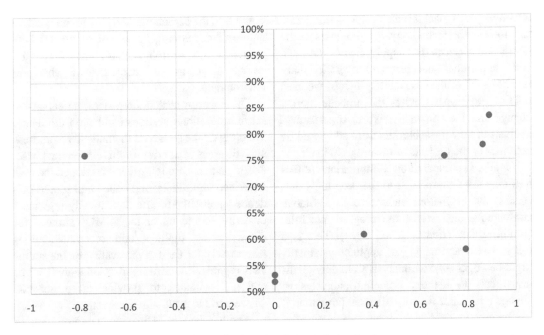

Fig. 5 Average score of sub-statements for mechanisms in relation to their saturation

Realist Evaluation as a Framework

The evaluation team also conducted other descriptive and explanatory analyses to increase the quality of the evaluation findings. To promote consistency, improve the overall robustness of the data scoring, and guide the application of these methods, all data that emerged from these different analyses undertaken were synthesized under the realist evaluation framework and coded accordingly.

To improve the understanding of the program portfolio, we conducted case studies for projects representing half of the portfolio. These covered different technologies, business models, states of energy access markets, and political and regulatory environments. The approach to conducting the case studies was developed using contribution analysis principles, drawing on the six-step guidance developed by Mayne (2008). For the case studies, we collected additional data through interviews with the projects' implementing actors, policy stakeholders, and market experts for each of the technologies represented in the case study portfolio.

Based on the interviews and the additional market information, the team conducted an energy market barrier analysis using Theory of No Change (TONC; Wörlen et al., 2011). TONC is a program-based evaluation approach that looks at the four main groups of stakeholders of the energy access market that can influence the effectiveness of market transformation programs: the users of the technology, the providers of the goods and services (the supply chain), the local and international financiers, and the policy makers. For each of the case study projects, we used a TONC to reveal barriers that impede market change and their intensity. We also performed analysis of how these barriers were addressed by activities of the projects or the program itself.

Finally, to compare how combinations of factors may have contributed to the program's outcomes, the evaluators undertook a qualitative comparative analysis of the program portfolio (Ragin, 2000; Thomas et al., 2014). This is a theory-based approach that applies systematic, logic-based, cross-case analysis to largely qualitative data to identify potential pathways of change (Baptist & Befani, 2015). In particular, it can be used to identify different combinations of conditions necessary to achieve a desired outcome. This is particularly useful in complex settings where contextual and intervention characteristics vary across cases and interdependencies exist between contextual and intervention conditions. The qualitative comparative analysis approach is remarkably compatible with realist thinking—a theory-based approach to

complexity analysis with limited generalizability (Befani et al., 2007)—and provided both a unique avenue by which to analyze evidence regarding causes of project success and evidence generation to parallel and triangulate much of the realist coding (Olsen, 2014).

Benefits of the Applied Approach

Using realist evaluation both as an evaluation method and as the basis for a mixed-methods evaluation framework has several benefits. First, the method offers the possibility of exploring complexity and context in a systematic way. The aim of the evaluation was to analyze program progress and the impact it had on the various local energy markets and the carbon market. Therefore, during the design stage, the evaluation team had to take into account global, national, and local contextual factors and the uncertain outlook of the carbon market. Due to the iterative nature of developing the realist framework, the approach was flexible enough after the initial design stage to be adapted to a changing environment.

During prior evaluation phases, an expansive set of evaluation questions had been formulated. We grouped the questions thematically to address specific areas of interest for this evaluation phase. Based on these groupings, we developed the four ICMO configurations and coded and scored evidence against them. The thematically organized evidence scored under the ICMOs allowed for extraction of findings and recommendations respectively for each of the evaluation questions. This organization also highlighted significant or outlying evidence to provide more nuanced answers to the evaluation questions.

Another benefit is that the configuration of ICMOs is a continuous, retroductive process and is therefore able to consider new insights acquired during the evaluation or major changes of the program's embedded environment. Like the evaluation framework itself, the ICMO configurations can be iteratively adjusted and used for subsequent evaluation phases. For example, if contextual factors have changed or new mechanisms and outcomes are identified, the statements can be modified without requiring a full reset of the evaluation framework. If additional areas of interests arise at a later phase of the evaluation process, evaluators can also develop new ICMO configurations. This ensures that the evaluation methodology is able to keep up with the shifting carbon landscape and developments in its implementation while maintaining rigor and consistency of approach.

A top-down, program-focused approach was used for the ICMO formulation because much time had been spent in the previous evaluation stages to develop and improve the program's theory of change. Developing ICMOs based on program-level information is more time efficient than formulating statements for each project and later abstracting them to make them valid for the whole project portfolio. Program-level statements also make the program theory more visible and give it a clear structure. The abstract structure of the ICMO configurations enabled us to incorporate a wide variety of existing data. To improve the ICMO configurations, the evaluation team refined them based on specific project findings when these findings were also likely to help explain outcomes of other projects. Using the ladder of abstraction, we formulated each specific finding as a general phrase to make it valid for other projects of the portfolio.

The created ICMO matrix allows for high evidence traceability. With the inclusion of the data in the matrix and the developed coding system, we could extract the specific data sources that led to a given finding and categorize the strength of evidence against each finding. We could also filter the evidence regarding individual projects or countries and analyze the ICMO statements separately at the country or project level. This can be advantageous when findings need to be formulated according to different clusters, based on technology, business model or country. For example, in a situation such as the results under Hypothesis 3 in Table 5, the evaluation team could filter data in the matrix to identify which projects were causing the divergence in evidence and investigate those projects further. Once the system is in place, the data is systematically scored according to its plausibility and importance for the evaluation. In contrast to other evaluation methods, this method minimizes the subjective assessment of individual evaluators: listing the components with signifiers for each sub-statement ensures consistency in the subjective coding process. Moreover, the elaborate framework and coding system can be used for subsequent evaluation phases and can be iteratively improved as the availability of data increases and the understanding of individuals working on the evaluation grows.

Challenges of the Applied Approach

As described above, the realist evaluation methodology has several advantages. However, the development and application of the method during this evaluation also revealed challenges and limitations.

First, the method described requires significant levels of effort. The establishment of the components supporting the accuracy of the sub-statements and the coding process itself were time consuming. Regarding the components, reaching consensus across the team and key stakeholders as to which components would act as signifiers took extensive consultation. The coding process, which is based on a line-by-line isolation approach, required personnel sufficiently familiar with the ICMOs and the evidence base. Had the midterm evaluation not been sufficiently resourced, the approach chosen likely would not have been as effective as it was. In addition to the resource requirements, although the scoring system was highly robust and provided clear, quantitative assessments of hypotheses' accuracy, it was limited in its ability to assess negative findings and did not offer a useful option for assessing which other interventions might have led to a given mechanism, and vice versa.

Second, the coding process was somewhat inflexible, requiring positive or negative scoring against predetermined hypothesis statements, which, if incorrectly set out or phrased, could lead the evaluation team down a narrow path in the wrong direction. Given the highly variable contexts in which the CMFP operates and the diverse range of outcomes expected at the project level, this approach to coding did not support effective and efficient capture of unstated or unforeseen contextual factors and program outcomes. Although the coding process did include an investigation marker score, the X, to highlight where evidence indicated unpredicted contextual factors or outcomes, in practice this was challenging to implement given the limited opportunity for qualitative description.

Third, the program consisted of 12 projects and developing robust ICMOs for this number proved to be challenging. Working from the bottom up, developing 12 sets of ICMOs—one for each project—would have been time consuming and significantly increased the evaluation data requirements and the ICMO-related consultations. However, the top-down approach, developing ICMOs at the portfolio level and testing them at the project level, was somewhat hindered by the heterogeneity of the projects, which resulted both in ICMO statements that were too general to effectively capture the nuance of the different projects and in high divergence of evidence. This also meant the resulting findings were not always generalizable, and, despite efforts to produce synthesized scores for the wider portfolio, extensive analysis and discussion was required in the evaluation report to explicitly draw out where projects landed on a particular findings curve. Identifying the right balance between specific project-level results versus more abstract portfolio-level results—the right rung on the ladder of abstraction—in such a portfolio is a recognized challenge of realist evaluation and well exampled in this case (Punton et al., 2020).

Finally, although the realist approach addressed many of the challenges created by the uncertain landscape, it could not resolve several fundamental issues. The fact that post-Paris Agreement stakeholder engagement was limited, for example, was not inherently improved by the realist approach beyond highlighting where saturation was low and more evidence needed. The nature of the market also resulted in higher availability of negative evidence, particularly in relation to the transformation of the carbon market, which, if not read and analyzed correctly, could lead to incorrect assumptions about the program results.

Improving the Methodology

Drawing on the lessons learned during the development and application of the realist evaluation approach, this section considers an alternative approach to developing ICMOs and applying realist evaluation in an uncertain landscape with a small but heterogeneous selection of evidence studies.

Bottom-Up Formulation of ICMO Statements

The formulation of the ICMOs themselves may benefit from a bottom-up approach, rather than top-down. Although each of the 12 projects involved had the same basic strategy—using results-based financing and supporting grants to implement commercial business models for energy access technology to generate certified emissions reductions—a core objective of the overall program was to test new business models in different markets, using different technologies. As such, the project models varied substantially, from traditional commercial cookstove sales, to biomass fuel utilities, to public aggregator-led solar home system distribution. Each of these models relies on different intervention strategies, seeks to trigger different mechanisms of change, and must contend with different contextual factors. In cases of such heterogeneity, developing ICMO hypotheses for each project (or cluster of projects, in the case of larger portfolios) is likely to yield more nuanced theories that can be more readily tested at the project level. Had the team developed 12 sets of ICMOs, or even eight sets reflecting the eight project countries, the task at the portfolio level would have been finding the correct level of abstraction at which to synthesize the hypotheses—too high and the aggregated theories risk becoming disconnected to the projects, too low and they will lack generalizability for the wider portfolio. Finding the right balance would allow for the formulation of portfolio-level ICMOs, which are well suited to assessing the overall success or impact of the portfolio and which can be effectively tested via project studies. However, this approach is not without trade-offs. For example, stakeholder engagement, feedback, and consultation are critical in the effective formulation of ICMOs. Increasing the overall number of ICMOs would increase the engagement requirements, particularly because each project, with a few exceptions, involved an entirely different set of implementors, funders, partners, and other key stakeholders.

Increasing Traceability of Causality by Tailoring the Coding to the Mechanism

ICMO statements are theoretically portable, meaning that a mechanism proven to operate as expected in one situation could feasibly be repeated elsewhere. However, to actually be portable, ICMOs must maintain a balance between having sufficient generalizability to be transferred and enabling appropriate analysis of how and why a given mechanism functioned or was triggered, to ensure any transference or replication is suitably tailored to new contexts (Pawson & Tilley, 1997).

Another option with ICMOs would be to take a different approach to the coding process. In the CMFP evaluation, the team used a coding system that sought to prove the accuracy of each hypothesis and sub-statement. As noted above, this system provided quantitative scoring on whether the overall hypothesis was correct but limited evidence on causality, particularly where the evidence differed from the original theory. In part, this was because each statement type in the ICMOs was treated the same, with equal weighting. An alternative approach would be to place the emphasis in coding on the actual mechanisms of change—the M in ICMO and the critical consideration for the ICMO portability. This approach would still use the concept of signifiers, but would only provide them for the mechanisms. This would reduce the level of effort required to agree on effective evidence thresholds while

allowing for deeper analysis of the causal mechanisms. During the coding process, these signifiers could then be directly tied to the accuracy rating for the whole statement.

In the example in Box 4, the intervention of providing carbon-RBF would ensure "revenue and capacity for projects will be sufficient to overcome the operational challenges in providing rural energy access technologies." Breaking this into signifiers, one might say that evidence of a project-supported company using revenue from the carbon-RBF to recruit new staff or to invest in distribution infrastructure would be good indicators that the mechanism was operating as theorized. Evidence of each of these signifiers would warrant a strong accuracy rating in coding. Thus, the mechanism hypothesis statements could be tested through traditional deductive reasoning.

> **Box 4: Example ICMO Statement**
> By providing carbon-results-based financing and business development support funding (I), in a context where there is sufficient customer demand for the energy technologies, access to finance for the pilot enterprises, and a supportive policy framework (C), revenue and capacity for projects will be sufficient to overcome the operational challenges in providing rural energy access technologies (M), resulting in increased energy access and the generation and sale of certified emissions reductions (O).

In this approach, complete ICMO hypotheses should still be constructed, but coding them as a complete unit is not necessary. With the primary focus on providing evidence and proving the mechanism, a more flexible approach applied to the interventions might benefit the overall analysis, particularly given the shifting landscape of the program. Each intervention could be given a tag (I1, I2, I3, etc.) to be used to link the relevant interventions to the mechanism being coded against. With the mechanism example in the previous paragraph, the ICMO was developed with two interventions: first, the commitment of carbon-RBF (I1); and second, the business development support provided by the program implementor (I2). Using the alternative approach, evidence could be binarily marked positively where it indicates the presence of one or both of these interventions, and scored based on the strength by which the evidence links them to the mechanism. A further consideration is that possible evidence might indicate that the mechanism was due to another intervention (such as the use of the program implementor's influence in the market, an intervention tested by our evaluation team) or an intervention that had not been captured by the original ICMO hypotheses, playing perhaps an even greater role than the originally linked intervention. The original coding approach would not have adequately captured this linkage. In the revised approach, such interventions could be tagged and scored for strength of linkage to the mechanism in question. At the conclusion of the coding process, the evaluation team could then assemble a more complete picture of which interventions contributed to which mechanisms and by how much, based on the actual evidence gathered.

Using this more flexible system allows evaluators to set out their hypotheses at the start of the process and assess how accurate they are, and also allows for effective evidence gathering on causal linkages that had not been drawn at the outset. It supports a more inductive approach to developing causal pathways without the need for the continuous stakeholder consultation that is required for reformulation of the hypotheses.

Increasing Variability of Contextual Factors

In analyzing the contextual factors, a similar inductive approach may be better suited to the uncertainties of the current carbon market. Although certain key contextual factors such as appetite for carbon trading or capacity of implementing organizations were evidently important from the outset, the evaluation team found a variety of surprising and unexpected contextual issues that often proved more critical to the success or

failure of each intervention than those identified and coded. Thus, staying open to updating and revising the ICMOs in the face of new evidence that does not fit within the existing framework is important, because narrowing the view of the analysis to specific factors for each mechanism can limit the nuance of the evaluation findings.

Evaluators can employ a revised approach drawing on the evidence-based ICMO assembly method that some favor. The pre-identified contextual factors could be grouped thematically to allow quick coding of the relevant factors demonstrated by specific evidence extracts. This also cuts down on repetition among the contextual factors. In coding evidence, qualitative description linked to a tagged context grouping is likely the most effective option, providing additional detail or analysis on the contextual factors identified by a specific piece of intervention or mechanism evidence. This qualitative coding would also support the capture of other contextual factors not previously identified by the evaluators, thus allowing for both a deductive and inductive approach to coding and developing understanding of the critical contextual factors involved.

A similar approach could be appropriate for outcomes to ensure qualitative capture of unexpected outcomes or outcomes with a stronger link to a given mechanism than envisioned, although generally the bond between mechanism and outcome is more consistent.

Summary of the Modified Methodology

Table 6 presents sample coding that evaluators might apply to the above-described approach. In this approach, only the existence of the mechanism and the links between the mechanism and the interventions or outcomes are quantitatively scored.

The original scoring system would still be applicable to this revised approach for the mechanisms, with each piece of data being scored twice: once for the strength of the evidence (Table 7); and once for the content of the evidence in relation to the signifiers (Table 1). The strength of evidence score applies a multiplier to the content

Table 6 Modified ICMO coding system

Statement	Score	Description
Intervention	Assigned intervention	Yes/no based on intervention established in ICMO
	Strength of linkage	Positive or negative score based on strength of connection to mechanism demonstrated by evidence
	Alternative intervention	Does the evidence indicate that other identified interventions led to mechanism?
	Strength of linkage	Positive or negative score based on strength of connection to mechanism of the alternative intervention
	Other	Does the evidence indicate that non-identified/external interventions contributed to mechanism?
Context	Contextual factors	Which identified contextual factors does the evidence support?
	Other	What other contextual factors are identified by the evidence?
Mechanism	Score	Score the accuracy of the mechanism statement per evidence scoring guide, based on evidence signifiers
Outcome	Assigned outcome	Yes/no based on outcome(s) established by ICMO
	Strength of linkage	Positive or negative score based on strength of connection to the mechanism demonstrated by evidence
	Alternative outcomes	Does the evidence indicate that other identified outcomes are caused by the mechanism?
	Other	Does the evidence indicate that unexpected outcomes are linked to the mechanism?

Table 7 Modified ICMO evidence scoring guide

Score	Definition
3	Evidence strongly supports mechanism statement. Multiple or all signifiers are met.
1	Evidence partially supports mechanism statement. One or more signifiers are met.
−1	Evidence partially contradicts or disproves mechanism statement. Evidence disproves or creates doubt in one or more signifiers.
−3	Evidence strongly contradicts or disproves mechanism statement. Multiple or all signifiers are countered or disproved.

score, recognizing that verifiable and authoritative sources provide more convincing evidence than plausible, subjective sources (see Table 7).

The modified coding system will generate several scores including the overall data score, the total data points, the average data score, and the data saturation in the form of positive and negative data convergence (a score reflecting what percentage of the total data points were pos-itive or negative) and the total evidence convergence. The revised approach to coding also allows for the inductive generation of alternative hypotheses, allowing users to efficiently reflect on the data gathered to reformulate and revise ICMO statements. Further, it provides a reference matrix linking evidence to findings during substantive evaluations.

Conclusion

The application of realist evaluation is appropriate when considering interventions operating in highly unstable or unpredictable landscapes. The method offers a sufficient balance of evaluation rigor and adaptability, allowing for retroductive analysis that can evolve over time. The use of ICMO statements allows for nuanced hypotheses to be tested (provided the right level of abstraction is achieved), which incorporates critical contextual factors at their core. The approach offers the opportunity to find out not only what has happened, but how it happened and, to a lesser extent, why. In the CMFP evaluation, this contextualized understanding was important to generating effective, balanced findings that fairly accounted for the market uncertainty in evaluating the outcomes and impact of the program interventions. The approach developed by the evaluation team for the CMFP was very effective in providing traceable, quantified findings. It also provided a robust framework against which other evaluation exercises could be designed, implemented, and scored.

However, as with all evaluation methodologies, realist evaluation is not without its limitations, some of which were apparent in this evaluation. The significant resource requirements, risks of overly narrow lines of analysis, and challenge in ensuring generalizability are all important lessons that should be considered when setting out to conduct a realist evaluation. Based on these learnings, we have presented a revised approach to conducting realist evaluation. This approach seeks to increase the flexibility of realist evaluation, ensure more nuanced analysis of the causal linkages between interventions and mechanisms, and open the approach further to unforeseen contextual factors or program outcomes.

This approach is unlikely to be suitable for all evaluators seeking to apply realist approaches, nor would it be appropriate for all evaluations. Nevertheless, the following key lessons are useful insights for any evaluator embarking on a realist evaluation:

- **Find the right level of abstraction for your ICMO statements**: When dealing with a portfolio of projects or interventions of any size, setting the right balance between portfolio- and project-level hypotheses is vitally important. Start from the bottom where possible, cluster projects by intervention type if needed, and remember to think about generalizability and portability of the hypotheses.
- **Engage stakeholders regularly but appropriately**: Stakeholder input to the ICMO development process is one of the realist evaluation pillars and consultations should be held at all key development stages, including iterations after data collection. However, striking a balance is important between sufficient engagement and the resource implications, not to mention the evaluator's biggest concern: burdening the commissioner. Agreeing on the process for consultations early in the ICMO development and testing adjustments before consultation may help to strike this balance.
- **Do not underestimate the resource requirements**: Developing, coding, and analyzing ICMOs using either of the methods described above is a time-consuming process. It requires not only individual subjective assessments for every piece of data collected, but also extensive stakeholder consultation and rigor in data gathering. Even if the described coding approach is not adopted, the development of ICMO statements, especially for portfolios of projects, is a difficult task that requires sufficient resources and time for feedback and iterations.
- **Remain open to emerging concepts (and do not be overly deductive)**: Developing hypotheses and scoring against them can result in tunnel vision, blinkering evaluators to emerging concepts and data. The retroductive approach described in this chapter, which allows for regular feedback loops between the evidence and the hypotheses, is a beneficial way of thinking about and analyzing the data that allows for multiple iterations of development and the incorporation of emerging ideas and trends.

References

Baptist, C., & Befani, B. (2015). *Qualitative comparative analysis – A rigorous qualitative method for assessing impact*. Coffey.

Befani, B., Ledermann, S., & Sager, F. (2007). Realistic evaluation and QCA: Conceptual parallels and an empirical application. *Evaluation, 13*(7), 171–192. https://doi.org/10.1177/1356389007075222.

Climate Policy Initiative & LTS International. (2018). *Climate public private partnership (CP3) monitoring and evaluation mid-term evaluation*. UK Foreign, Commonwealth & Development Office. http://iati. dfid.gov.uk/iati_documents/50221453.pdf

Department for International Development and Department of Energy & Climate Change. (2013). *International Climate Fund business case and intervention summary: Carbon Market Finance Programme.* https://www.gov.uk/government/publications/international-climate-fund-business-case-and-intervention-summary-carbon-market-finance-programme.

Greenhalgh, T., Pawson, R., Wong, G., Westhorp, G., Greenhalgh, J., Manzano, A., & Jagosh, J. (2017a). *Retroduction in realist evaluation*. RAMESES II Project, National Institute of Health Research, Health Services and Delivery Research Programme. http://www.ramesesproject.org/media/RAMESES_II_Retroduction.pdf

Greenhalgh, T., Pawson, R., Wong, G., Westhorp, G., Greenhalgh, J., Manzano, A., & Jagosh, J. (2017b). *What is a mechanism? What is a programme mechanism?* RAMESES II Project, National Institute of Health Research, Health Services and Delivery Research Programme. http://www.ramesesproject.org/media/RAMESES_II_What_is_a_mechanism.pdf

LTS International. (2020). *Evaluation of the carbon market finance programme*. UK Department for Business, Energy and Industrial Strategy. https://devtracker. fcdo.gov.uk/projects/GB-GOV-13-ICF-0025-CiDev/documents

Mayne, J. (2008). Contribution analysis: An approach to exploring cause and effect. *ILAC Brief 16*, 4. https://hdl.handle.net/10568/70124

Olsen, W. (2014). The usefulness of QCA under realist assumptions. *Sociological Methodology, 44*(1), 101–107. https://doi.org/10.1177/0081175014542080.

Pawson, R., & Tilley, N. (1997). *Realistic evaluation*. Sage.

Punton, M., Vogel, I., Leavy, J., Michaelis, C., & Boydell, E. (2020) *Reality bites: Making realist evaluation useful in the real world*. CDI Practice Paper 22. IDS. https://opendocs.ids.ac.uk/opendocs/handle/20.500.12413/15147

Ragin, C. C. (2000). *Fuzzy-set social science*. University of Chicago.

High-Level Panel on the CDM Policy Dialogue. (2012). *Climate change, carbon markets and the CDM: A call to action*. CDM Policy Dialogue. http://www.cdmpolicydialogue.org/report/rpt110912.pdf

Thomas, J., O'Mara-Eves, A., & Brunton, G. (2014). Using qualitative comparative analysis (QCA) in systematic reviews of complex interventions: A worked example. *Systematic Reviews, 3*(67). https://doi.org/10.1186/2046-4053-3-67.

United Nations Framework Convention on Climate Change. (2015). Article 6. *Paris Agreement*. https://unfccc.int/files/meetings/paris_nov_2015/application/pdf/paris_agreement_english_.pdf

Wörlen, C., Rieseberg, S., & Lorenz, R. (2011). *The theory of no change*. Arepo Consult. https://arepoconsult.com/wp-content/uploads/2020/02/2016_Woerlen-et-al_The-Theory-of-No-Change_International-Energy-Policies-Conference.pdf

Evaluation's Role in Development Projects: Boosting Energy Efficiency in a Traditional Industry in Chad

Serge Eric Yakeu Djiam

Abstract

This chapter illustrates the critical importance of evaluation in development projects. It explores the relevance, processes, and specifics of a project to introduce energy-efficient cook stoves in two traditional industries in Chad. Although Chad benefits from great solar potential given its location and being a Sahelian country, biomass accounted for 94% of the primary energy supply in 2008, and only 2.2% of Chadian households have access to electricity. The beer brewing and meat grilling sectors in particular use enormous quantities of limited and expensive firewood. Locally developed energy-efficient stoves for the two targeted sectors were available, but those technologies had not been commercialized and disseminated into the Chadian market. The project aimed to overcome issues of technology, financing, dissemination, resistance to change, and awareness to introduce and establish use of energy-efficient stoves in micro-scale food processing to achieve environmental and economic benefits, discussing the effectiveness of models introduced and adopted by project beneficiaries with related training. This chapter considers issues related to the project's financing and sustainability and concludes with lessons provided by the evaluation, including engagement with targeted beneficiaries, awareness of local context, and consideration of size and scale for a demonstration project that can be scaled up in future programs.

Introduction

Frequent and intense storms, widespread and destructive fires, shrinking water supplies, desertification, and changes to ocean environments increasingly evidence climate change. The impacts of climate change—which include slow or poor crop production, higher food and fuel prices, drought and famine, higher inflation, and slower economic growth—are especially severe for the poor, who often lack the resources to cope and adapt. As the Global Environment Facility (GEF, n.d.) explained,

> Taking action on climate change means adopting and implementing ambitious programs to limit emissions of greenhouse gases to levels compatible with the well-being of the ecosphere, while supporting communities around the world to adapt to the unavoidable impacts of the climatic changes that are already being observed. It also means embracing the potential of the green economy—a more sustainable way of life that balances economic, social, and environmental priorities. (para. 2)

S. E. Yakeu Djiam (✉)
Independent Evaluator, Yaounde, Cameroon

© The Author(s) 2022
J. I. Uitto, G. Batra (eds.), *Transformational Change for People and the Planet*, Sustainable Development Goals Series, https://doi.org/10.1007/978-3-030-78853-7_10

One such ambitious program that aimed to balance economic, social, and environmental priorities was a project to promote energy-efficient cook stoves in micro- and small-scale food-processing industries in Chad. Given its location and as a Sahelian country, Chad has a great solar potential of about 4.5 billion MWh/year and thus ranked 20 worldwide for its solar potential in 2008 (Price & Margolis, 2010). Despite this, biomass accounted for 94% of the primary energy supply in 2008, and only 2.2% of Chadian households had access to electricity; of these, only 1% were outside of the capital city of N'Djamena.[1] About 79% of the energy supply in urban and 90% in peri-urban and rural areas in Chad derive from ligneous sources. Burning firewood for fuel produces greenhouse gases directly, which has negative health effects on human health, and reduces the capacity of forests to act as carbon sinks. Moreover, the imbalance between the firewood supply and demand accelerates desertification and poses concerns for rural, peri-urban, and urban development. To curb high deforestation rates, the Government of Chad passed an act in 2009 that prohibited cutting green wood (Republic of Chad, 2017; United Nations Development Programme [UNDP], 2017).

However, an outcome of the act has been to triple the price of wood in the market, which has had a negative effect on micro and small entrepreneurs in Chad, especially beer brewers and meat grillers in N'Djamena. A 2010 study on firewood consumption estimated that the beer brewing and meat grilling sectors in N'Djamena alone consume around 14,000 tons of firewood per year, with more than 3300 cabarets (the equivalent of bars) and 2300 meat-grilling stands (Vaccari et al., 2012).

The traditional breweries produce a local beer called *bili bili,* a sorghum-based alcoholic drink that requires intensive cooking for 10–16 h per batch. The breweries are cottage industries that are exclusively operated by women entrepreneurs in the back yards of their houses. Each brewery employs four to 15 women and produces two to three 150- to 200-l batches of beer per week. Each batch of beer requires at least 54 kg of wood. The stoves used are outdated and highly inefficient, further increasing the amount of wood needed per batch.

The *tchélé,* in which men smoke or grill various types of meat, represents another sector that consumes a lot of firewood. Run by men, the tchélé is either a simple street booth or a small shop that includes an extension with benches where patrons can consume the meat on the spot. A tchélé employs between two and five people, depending on its size and on whether they also do butchery-related work.

To take advantage of the opportunity these traditional industries presented to improve energy efficiency and reduce demand for firewood, the Global Environment Facility (GEF) financed a project implemented by the United Nations Industrial Development Organization (UNIDO) in 2014, titled "Promoting Energy-Efficient Cook Stoves in Micro- and Small-Scale Food Processing Industries." The project drew on lessons learned from a similar intervention in Burkina Faso (UNIDO Office of Independent Evaluation, 2015). Project components included improving the design of cook stoves to achieve optimum fuel efficiency, creating sustainable financial programs to finance acquisition of the energy-efficient cook stoves, and improving the business performance of micro and small entrepreneurs. This project also included a monitoring and evaluation component with the express intent of facilitating smooth and successful project implementation and sound impact. This chapter explores the role of evaluation in a development project, with discussion of the project background; the evaluation findings including project performance, coordination and management, gender mainstreaming, cross-cutting issues, and conclusions; and a summary of the importance of evaluation in such a project.

---

[1] The Clean Energy information portal of the Renewable Energy and Energy Efficiency Partnership, reegle.org, is no longer active. For data available on the portal, contact info@reeep.org.

Project Background

High firewood consumption exerts a negative impact on the environment and on the livelihoods of micro and small entrepreneurs in Chad. Traditional food processing, such as brewing beer and grilling meat using large commercial cook stoves with low energy efficiency due to incomplete combustion of firewood, requires long cooking times and high consumption of firewood. The use of energy-efficient cook stoves could reduce firewood consumption by at least 50% and considerably decrease cooking times. As a corollary of the shift to energy-efficient cook stoves, entrepreneurs would also be able to increase their income and profit margins by decreasing their fuel costs and optimizing the production process. The shift would further contribute to decreasing greenhouse gas emissions and reducing deforestation rates and negative impacts on health.

Prior to this project, two local cook stove manufacturers had developed prototypes for energy-efficient stoves with support of the Association pour le Développement de Micro-Crédit and the Association of Appropriate Technology for the Protection of Environment (ATAPED). However, these prototypes had not yet been commercialized and disseminated into the Chadian market because the manufacturers faced constraints related to limited financial resources, high costs of raw materials, absence of relevant technical training and support, and weak market demand. These barriers prevented the local manufacturers from producing and selling their energy-efficient stoves at a market scale. The following were the main barriers to the introduction of energy-efficient stoves in the micro-scale food-processing sectors in Chad that the project aimed to overcome:

- *Technology*: Improved cook stove technology should not only be appropriate to the needs of each sector/usage, but also affordable, easy to use, durable, widely available, and socially acceptable or desirable.
- *Financing*: Producing and selling improved cook stoves would require significant invest-

ments to adapt the technology to users' needs and further improve the energy efficiency and performance of the stoves for market.
- *Dissemination*: Shifting from traditional, energy inefficient stoves to improved models would require adoption at a larger scale (demand vs. supply sides).
- *Resistance*: The introduction and dissemination of new energy-efficient technologies often faces reluctance to change; thus, a strategy was needed to overcome resistance and enhance acceptance among users and consumers.
- *Awareness*: Lack of awareness about the benefits of a new technology can stall buy-in and uptake. Benefits of potential new energy-efficient technologies are reduced fuel consumption and costs, and improved production processes, health conditions, and livelihoods.

The GEF-funded project sought to address some of these key barriers to the introduction of energy-efficient stoves among beer brewers and meat grillers in Chad. Therefore, the primary targets of the project were beer-brewing and meat-grilling micro enterprises, cook stove suppliers and local project developers, and experts from relevant policymaking and implementing institutions. The project aimed to create a market for energy-efficient stoves in three ways:

- Promoting energy-efficient stoves that consume 50–80% less firewood for processing food.
- Developing clusters[2] within the beer brewing and meat grilling sectors to support demand for improved cook stoves, generate collective gains, and empower female and male entrepreneurs.
- Facilitating access to finance to aid in acquiring improved cook stoves by implementing a

---

[2]Such clusters, or *tontines* as they are called in Chad, are self-help groups of five to ten members who are encouraged to develop savings behaviors and pool resources to purchase the new stoves. These are groups based on trust and mutual support that can more easily access microfinance because institutions are more likely to lend to them than to individuals.

credit and savings program and linking to voluntary carbon markets to generate additional revenues for end users switching to improved cook stoves.

Four intervention zones were identified as the focus areas for project activities based on two criteria: (a) high concentration of the selected types of enterprises and (b) proximity to N'Djamena, which is one of the main intervention zones, employing more than 40,000 beer brewers and 2300 tchélé workers. Table 1 lists the project's stakeholders and their roles in the project.

The initial project budget was $2,600,000 in both cash and in-kind support over 2 years (2015–2017), with a final deadline in 2018 to accommodate pending activities and outputs. A significant budget shortfall from the national government of Chad and the National Agency for Domestic and Energy and Environment Development (AEDE) reduced the project budget by more than half. This was due to increased engagement of the national government under the Special Fund for Environment (FSE) in supporting civil society organizations (CSOs) with small-grant interventions related to environmental preservation.[3]

Evaluation Findings

To enhance future projects, identify lessons learned, and provide recommendations, the evaluation team implemented evaluation activities (see the Appendix for the evaluation methodology) and prepared a final report. The team met with a sample of 18 key informants, four groups/cooperatives of blacksmiths, 14 groups/cooperatives of meat grillers, and 25 groups/cooperatives of beer brewers. The evaluation report was organized around project performance with targeted evaluation criteria and discussed project coordination and management, gender mainstreaming, and cross-cutting themes. It concluded with recommendations and lessons learned.

[3]Seventy-six microprojects address themes such as desertification, adaptation to climate change, biodiversity conservation, soil restoration, and capacity building.

Table 1 Project stakeholders (UNIDO, 2013)

Stakeholders	Roles in the project
Cofinanciers	**Global Environment Facility/ FEM:** Support part of the financial resource
	FSE: Support partial allocation of financial resource
	Ministry of Environment and Fishery (counterpart of the project): Support in the promotion of the project and energy-efficient stoves, especially in outreach and logistics
	Shell Foundation: 1. Support the dissemination of clean cook stove solutions 2. Support the development and scale-up of models to disseminate the use of clean cook stoves 3. Share knowledge and experience gained through projects implemented in other countries and regions
	Envirofit: 1. Develop well-engineered technology solutions to improve the energy efficiency of institutional stoves 2. Support and train local technicians on the assembly of the stoves 3. Support the development of related projects within the carbon market
Implementing agency	**UNIDO:** 1. Bears ultimate responsibility for implementing the project 2. Delivers planned outputs and expected outcomes 3. Leads the general management and monitoring of the project and reporting on project performance to the GEF

(continued)

Table 1 (continued)

Stakeholders	Roles in the project
Executing partner	**AEDE:** 1. Host location and provide close collaboration with the project management unit 2. Support various aspects of the project via its expertise in energy efficiency and stoves 3. Implement the project locally **Financial Development Microfinance Institution (FINADEV):** 1. Support financial training for beneficiaries 2. Support and manage the financial mechanism with loans to beneficiaries
Partner government agency	**Ministry of Women Empowerment, Social Action and National Solidarity:** 1. Leverage synergies between their activities and the project 2. Work with members of the steering committee in giving feedback and advice for the efficient implementation and sustainability of the project
Beneficiaries (more or less structured)	**Associations/cooperatives and individuals (blacksmiths, beer brewers, and meat grillers):** Support cluster development of micro-enterprises and generate collective gains
Other partners	**VERICHAD** (company): 1. Support the development of cook stove prototypes 2. Support training for blacksmiths **ATAPED:** 1. Support the production of cook stoves for beneficiaries 2. Provide follow-up training to other blacksmiths' cooperatives

Source: Adapted from UNIDO (2013) by the evaluation team

Project Performance

The project performance component of the evaluation targeted four criteria: project relevance, effectiveness, efficiency, and sustainability.

Relevance

The evaluation team found the project design to be relevant and aligned to several Chadian national policies. First, the project fit into the goals of the National Development Plan 2017–2021 (Republic of Chad, 2017) and the national poverty reduction plan (Republic of Chad, 2010) with respect to its five strategic axes:

1. Strengthening the resilience of agricultural, forestry, and fisheries systems
2. Promoting actions to mitigate greenhouse gas emissions
3. Improving sustainable access to diversified energy sources
4. Preventing risks and managing extreme weather events
5. Strengthening the capacity of institutions and actors in the fight against climate change and enhancing instruments and capacities for mobilizing climate finance

Moreover, as Chad is a Sahelian country that is exposed to advancing desertification, climate change, and environmental degradation, the project also supported the energy sector's policy framework (Ministry of Economy and Planning, Republic of Chad, 2008). The framework was established to promote technical and economic support for optimum energy development to reduce national dependency on wood fuel, which represents about 96.5% of energy consumption, thereby supporting the climate change strategy (Republic of Chad, 2017). Several of its priority actions aim to promote alternative energy sources and energy efficiency, such as extending the prohibition on cutting wood for fuel, using butane gas, and developing efficient domestic energy—actions that need to be strengthened to find substitutes for wood and charcoal, which are generally used to cook food and bake bricks for building houses.

Key informant assessments indicated that the project was unique in focusing on productive activities such as cook stoves for brewing beer and grilling meat. The project met one of the key priorities of both UNIDO and GEF strategic policy for renewable energy and energy efficiency (UNIDO, 2009b), which aims to help pro-poor actors in the Global South enhance their access to modern services and increase the viability of their small-scale industries by augmenting the availability of renewable energy for productive uses (UNIDO, 2009a). Thus, GEF resources allocated to support the dissemination of energy-efficient cook stoves in the traditional food-processing industry met one of the government's priorities to fight desertification and climate change (UNDP, 2017).

The project was relevant despite its limited scale; it targeted 15–30% of existing beer brewers and meat grillers in the selected project zones, a figure limited by the reduced financial resources of the project. The project linked to six of the Sustainable Development Goals (SDGs): 1 – poverty reduction; 3 – good health and well-being; 5 – gender equality; 7 – affordable energy and clean energy; 9 – industry, innovation, and infrastructure; and 13 – climate action. The project was also in line with Chad's Poverty Reduction Strategy Paper (Republic of Chad, 2010), reflecting the national development policy that prioritizes the food-processing sector while focusing on protecting the environment and promoting the private sector.

Effectiveness

Evaluation of the project's effectiveness included whether the project had achieved its expected outputs and outcomes. As a pilot program, the evaluation team considered the objective of improving the adoption of cook stoves to be about 70% achieved given the targeted number of beer brewers and meat grillers in the selected clusters. Three models of cook stoves were tested with these two cluster sectors. Beneficiaries rejected the first as too expensive and poorly adapted to local needs. The second model had limited capacity and was therefore also rejected, especially by women beer brewers. The third model, which the beneficiaries ultimately

adopted, provides some protection from fire and harms, is cost efficient, and consumes about 35% less firewood than previous cook stoves.

The second project output was to create a sustainable financial mechanism for beer brewers and meat grillers to purchase the new cook stoves. Although awareness meetings were conducted in the cluster zones, no loan had been granted to any cooperative at the time of the evaluation. This was due to delay on the part of UNIDO in signing the contract and providing the guarantee fund to FINADEV, attributable in part to UNIDO's not having a physical presence in Chad. Despite the lack of an agreement with the Union for Credit and Loan (UCEC) in Mandélia, 25 cooperatives opened and were operating accounts without loans. The project succeeded in linking cluster groups to the formal financial system by providing all members with an individual savings account at the FINADEV and at the UCEC. However, the project budget shortfall compromised several activities planned for Output 2, including lack of training for project developers on project identification and development.

The third and final output was to improve the business performance of micro and small enterprises through clustering. Gathered in cooperatives, beneficiaries were trained on several themes such as enterprise management, cooperative governance and financial literacy, and financial mechanisms using microfinance tools and marketing techniques. The objective was for cluster members to improve their business performance and profits by at least 40%. However, evidence indicated that the project focused primarily on cook stoves for energy efficiency to the exclusion of other training. Only about ten cluster associations were organized in all of the cluster zones, but all of these associations were formalized with official cooperative authorization. The positive feature of this output was the linkage of cooperative units to energy-efficient cook stove distributors.

The project completed about 75% of its objectives under the expected budget shortfall; the financial constraints did not permit completion of the overall planned activities. Evaluators did observe several unplanned changes among the meat grillers of N'Djamena and the beer brewers of Mandélia—the clusters with the greatest level of adoption of the energy-efficient stoves.

in environment and energy such as the Ministry of Environment, the Ministry of Oil and Petroleum of Chad, and AEDE.

Efficiency

The evaluation team found the project to be very cost-efficient. Despite the budget constraint, the project management unit was able to ensure a good value-for-money management process based mostly on signed memorandums of agreement or financial conventions, with payment based on delivery of technical reports and site visits when necessary, and/or on evidence from previous financial segment and reporting. Completion of project activities/outputs was high, considering that the project budget was 40% of the amount originally planned. Given this working environment, the evaluation team was impressed with the good quality of expertise provided.

The achieved project outputs were completed in a timely manner despite the lengthy administrative process for compliance with the UNIDO legal system and procurement obligations, exacerbated by the organization's lack of a physical presence in Chad. The evaluation did not find any deviation between the disbursements and project expenditures, and the level of achieved outputs affirmed satisfactory project cost-effectiveness.

Although the evaluation determined that the project was cost effective and project outputs were timely, monitoring of the production of improved cook stoves and distribution was less effective. The evaluation team found inconsistency in the new cook stoves distributed to cluster cooperatives even from the same production line, either in terms of size or the quality of iron used, which compromised the functionality and affected adoption of the new stoves. This situation had potential to significantly affect adoption of the technology and its dissemination. A final concern was that the evaluation found no established partnership between the project and other donors' projects; however, synergies were developed internally with sectoral ministries working

Sustainability

The evaluation team found thematic training and awareness meetings to be assets for the sustainability of the project. Moreover, training provided to the existing cooperative of blacksmiths was also helpful. Other assets could support the sustainability of the achieved outputs such as creating a community-based forest and providing support for existing traditional credit mechanisms, namely tontines, as informal opportunities for private ownership of cook stoves. However, the lack of knowledge among beneficiaries about the credit system approach and accountability and the delay in establishing the financial mechanism were negative factors that would likely affect replication of project results. Another factor was illiteracy on the part of most beneficiaries.

Implementing the planned exit strategy could sustain this pilot project's outcomes in the middle term, including an improved financial program coupled with a peer-to-peer learning strategy to support synergies among cluster groups. Training on the development and operationalization of future cluster teams and a coaching approach by cluster development officers would serve as key drivers for this skills-transfer strategy. At the time of this writing, the team expected the post-project sustainability plan of large-scale dissemination of energy-efficient cook stoves to have a significant impact on greenhouse gas reduction and improvement of beneficiaries' living conditions. These measures, coupled with the knowledge transfer, could support synergies within individual and cooperative unions at all levels. Cluster members are expected to fund the platform through a fee of $0.50 paid to the cluster cooperative on each energy-efficient cook stove sold. The collection of this symbolic fee will be encouraged to improve the platform continually and to meet user needs while protecting the environment. However, the evaluation did not find any practical measure related to establishing the fee, such as a signed agreement.

The evaluation team found that sustainability of the energy-efficient cook stoves was somewhat influenced by the operational costs for raw material. An in-depth assessment of whether new cook stoves built using local materials could be more efficient than the energy-efficient cook stoves made with iron could provide a useful further measure.

The project's visibility was relatively weak. For instance, limited attention was paid to the involvement of media, use of flyers, and development of publicity spots and/or brochures, which could have improved buy-in from beer brewers and meat grillers not targeted by the project. Another shortcoming was that the project was not able to schedule peer-to-peer learning missions with other countries' beneficiaries for knowledge sharing, such as in Burkina Faso, Sierra Leone, or Kenya, where similar projects have been tested.

The evaluation considered the project's original aim of linking the voluntary carbon markets to generate additional revenues for businesses that switched to the energy-efficient cook stoves. Evaluators found that because the stoves for the beer-brewing and meat-grilling sector consumed less firewood than bakers and brick construction, sectors that have previously benefitted from voluntary carbon markets, the demand was not great enough to pursue this approach.

Project Coordination and Management

The project adopted in-depth consultation with stakeholders. The evaluation team found that gathering key actors in a steering committee was an asset to managing the project activities and to achieving the project outputs given the ongoing budget shortfall. The project's approach incorporated a signed financial convention with service providers such as Verichad and FINADEV, and the overall procurement and provision of funds were managed by UNIDO with the local assis-

tance of AEDE. Final energy-efficient cook stoves were distributed to the five cluster groups later than originally planned due to delays in signing these contracts. The varied availability of the technical assistant from UNIDO headquarters also led to irregular scheduling of meetings of the steering committee; regular meetings to monitor and report on project activities would likely have reduced or eliminated delays.

Under the leadership of the project assistant, cluster agents ensured permanent monitoring of field activities by collecting progress information, including feedback from partners and beneficiaries. This information resulted in ten timely progress reports on the project. The technical assistant from UNIDO carried out five field missions, each of which provided opportunity for performance review and generating corrective measures or action plans. The evaluation team appreciated the adaptive exit strategy, which was based on performance review of achieved project outputs to support the financial program.

Gender Mainstreaming

Project design and implementation phases ensured a gender balance of women and men in project activities, including the project's management and participants. Strategically, the project was aligned with the National Gender Policy (NGP), which aims to serve as a guiding instrument for integrating the concerns and specific needs of women and men, especially for pro-poor targets (United Nations High Commissioner for Refugees [UNHCR], 2012). The project targets one of the most vulnerable small-scale industrial business groups in Chad, and the evaluation team found that the baseline study did include gender analysis with planned gender-related project indicators. About 65% of project beneficiaries were women and 35% were men. On the project management team, the project manager and the technical assistant were women, with three women among the eight people in the project management unit (38%), and two among the eight (25%) on the steering committee.

The evaluation found no negative factors that might affect the gender mainstreaming within the course of the project. The precariousness and inaccessibility of energy sources affect women more, but this factor decreases dramatically the longer women spend on production- and capacity-building activities. The training modules provided to women entrepreneurs helped to empower and integrate them into the energy-efficient stove value chain, especially in the business management, marketing, saving, and credit mechanism. The project's comparative advantage has been having a balanced gender equity focus on female and male business groups.

Cross-Cutting Issues

By considering both the context of and cultural practices around the beer-brewing and meat-grilling businesses in Chad, the project covered cross-cutting thematic issues. The participatory approach, adopted with in-depth consultative meetings, was a key asset in this regard.

First, all cluster agents were recruited locally and each was assigned to their zone of greatest concern given their long-term knowledge and working experience in the project zones. This approach enabled the project to work well with local and traditional authorities and with respect to local practices and concerns. The choice of business model was aligned with traditional food-processing practices with traditional stakeholders. For decades, non-Muslim women have managed beer brewing industrial processing, while Muslim men have largely dominated meat grilling industrial processing. Chad is primarily a Muslim country, with more than 75% of the total population identifying as Muslim, thus requiring that all animals be slaughtered as required by Muslim practice either for household consumption or for sale in a tchélé.

Finally, the project management, implementation, and evaluation fieldwork followed cultural and contextual practices. For instance, the last 3 days of the evaluation fieldwork coincided with the beginning of Ramadan. Therefore, meetings were scheduled within nonprayer times.

Conclusions

The evaluation team found that the project linked to the SDGs and aligned with Chad's Poverty Reduction and Growth Strategy (Republic of Chad, 2010), reflecting the national development policy that prioritizes the food-processing sector while focusing on environmental protection and promotion of the private sector. In terms of performance and results, the project design was relevant and aligned with several Chadian national policies, such as the national development plan (Republic of Chad, 2017); the policy framework of the energy sector, which considers measures as gathered by the energy sector master plan (Minister of Economy and Planning, Republic of Chad, 2008); the NGP, which considers gender mainstreaming to provide gender equity-focused interventions for both men and women (UNHCR, 2012); and the national strategy to combat climate change and support environmental preservation in Chad (UNDP, 2017). The financial support has been helpful to the beer-brewing and meat-grilling sectors.

The project was effective in achieving 75% of its goals. It included individual business processor actors in cooperatives and provided timely training on themes such as enterprise management, cooperative governance and financial literacy. Individual entrepreneurs appreciated the financial mechanism available, but establishing a sustainable financial mechanism remains an ongoing challenge.

Project management adopted in-depth consultation with stakeholders and engaged participation from female and male food processors. This equitable development of the gender dimension succeeded with consideration of cross-cutting themes such as working within the local context and cultural practices of the targeted zones.

The project has been cost-efficient, providing a good value for money, and cost-effective despite the budget shortfall of about 45%. However, the project failed to initiate partnership with other donor projects. The ongoing exit strategy could sustain the project outputs for a middle-term duration; this strategy focuses on an improved financial mechanism coupled with peer-to-peer learning to support synergies among cluster groups.

The evaluation team provided several recommendations to support continued progress on the project's objectives and for UNIDO's potential future collaboration in Chad. For the project management team, evaluators recommended continuing to monitor and coordinate efforts and increase attention to awareness and close collaboration with cluster groups, including the blacksmith cooperative to complete the cycle of the value chain. Second, the management team should continue to pursue the mobilization of resources committed by AEDE FSE to support the project.

The evaluators recommended that UNIDO consider signing tripartite agreements with AEDE and local contractors and promoting nonfinancial collaboration with other microfinance institutions, such as UCEC in Mandelia, to cover other clusters. A second recommendation was developing an evidence-based operational exit strategy and enhancing visibility of project outputs. Specifically, UNIDO should activate the financial program as soon as possible by coupling it with peer-to-peer learning to increase synergies among cluster groups. An audience-tailored approach to communication would further promote the project and improve awareness. Another recommended step for UNIDO was assessing other prototypes of energy-efficient cook stoves built using local material such as mot or bricks. Finally, the evaluators suggested that UNIDO consider sustaining the energy-efficient technology in future projects by including a field technical advisor and promoting partnership by aligning and enforcing the exit strategy with a framework on capacity development with other donors' projects.

Four overall lessons shared by the evaluation team offered insight for future projects.

1. In a sustainable development project dealing with energy-efficient cook stoves, consulting in depth with targeted beneficiaries and testing various prototypes are key steps toward the adoption of a technology by cluster groups.
2. In a cultural and contextual development environment with high female illiteracy, like Chad, having a woman cluster agent can generate buy-in from female participants.
3. Considering the size of this project and the huge amount of groundwork required at the community level, an important step is ensuring a manageable number of targeted clusters and beneficiaries to demonstrate success to scale up in future programs.
4. Considering local context is important when selecting energy-efficient technology to ensure suitable infrastructure and affordability for targeted beneficiaries.

Value of Evaluation in Development Projects

The energy-efficient cook stove project in Chad illustrates the critical importance of evaluation in development projects. Clearly, one important aspect is the ability to demonstrate effectiveness and efficiency in using donor funds, but perhaps more important to the long-term success of a project is ensuring its relevance and sustainability for the project participants and communities. In this case, the project was found to be relevant to several goals related to both the environment and to private sector business development. Evaluation identifies areas for improvement to enhance performance of a project while it is underway and for future similar efforts, and also recognizes aspects that work well. With this specific project, an identified difficulty was the delayed or reduced funding. Features of the project that were helpful and instructive for future undertakings were extensive stakeholder engagement and attention to the local context and customs.

As for sustainability, the evaluation hinted at a mutually reinforcing ecosystem in which beer brewers and meat grillers gain the financial benefits of reduced fuel cost and come together in clusters to improve their business skills and for mutual support for financing further purchases of energy-efficient equipment. This, in turn, supports the growth and development of a local industry for producing energy-efficient cook stoves.

Evaluation has an essential role in movement toward transformational change, as this project demonstrates. By identifying specifics of a project's successes and shortcomings and pinpointing the project aspects that directly affect its outcomes, evaluation highlights opportunities for scaling up an effort from a pilot to a broader initiative, which can deliver more successful outcomes for people and environmental measures. Also aiding in transformation is the cumulative effect of evaluation from one project to another. Just as this cook stove project in Chad built upon success identified in evaluation of similar projects in other nations, future efforts to develop environmentally beneficial opportunities for small entrepreneurs can build on this project's evaluation.

Appendix: Methodology

Sampling Methods and Data Collection

The evaluation followed United Nations Evaluation Group (UNEG) norms for standards for evaluation (UNEG, 2016) and employed the criteria of the Development Assistance Committee of the Organisation for Economic Co-operation and Development (OECD DAC) Network on Development Evaluation (2019). The purposive sampling method proposed by De Vaus (2001) was adopted to consider direct beneficiaries to visit and select actors among small businesses and traditional food-processing industries and sites. Respondents included men and women involved in either beer brewing or meat grilling and marketing. The client purposively proposed the five project locations (clusters) and beneficiary groups, with enough duplication to allow for substitutions if needed. Respondent selection was guided by (a) time constraints on fieldwork, (b) types of business activities, (c) the distance between N'Djamena and the three clusters outside of N'Djamena and related time constraints to allow at least two interviews and two focus groups in each loca-

tion, (d) the accessibility of the targeted location due to the ongoing security context of the country, and (e) the gender distribution within beneficiaries' cooperatives to include women and men. Other actors such as partners were selected based on their level of participation in project implementation.

The evaluation team used a mix of qualitative and quantitative analysis methods, including data triangulation and retroaction, to arrive at the evaluation findings that fully responded to the evaluation questions and make recommendations.

Desk Review

This process included identification and review of relevant technical reports and background documents such as those from the United Nations Industrial Development Organization (UNIDO, 2009a, b, 2013) and UNIDO OIE (2013), progress reports, baseline report, marketing report, workshop and training reports, and other documents collected during field interviews and field visits. Desk review helped the team understand the project logic model and implementation features, the operational contexts, and challenges.

Qualitative Methods

The evaluation team gathered data from key informant interviews (using semistructured guides, face-to-face or via Skype/phone calls) with a validated range of stakeholders engaged in the project implementation, and focus groups (with semistructured discussion guides) that considered gender for female beer-brewing and male meat-grilling entrepreneurs. These data revealed the activity's performance toward its primary objectives with tracking evidence.

Quantitative Methods

The evaluation used quantitative data such as socioeconomic information from focus group discussions with a purposive sample of direct project beneficiaries. The team also drew data from existing statistics and comparison figures from performance indicators and progress reports.

Site Visits

Field observation made use of a structured checklist of selected physical investments supported by the project to document how their function and benefit to the beer-brewing and meat-grilling beneficiaries.

Data Analysis Methods and Reporting

All analysis considered gender and socioeconomic characteristics of respondents. Correlation was used to assess relationships between and within indicators from the identified outcomes.

The team developed data analysis from its findings, analyzing qualitative data collected via the semistructured individual and group interviews using thematic and content analysis and based on a specific analytical framework. Content analysis helped the team go beyond descriptions of changes in practices and attitude to identify the most salient characteristics of these changes and map the main strengths and weaknesses in the decision-making process. The team designed rubrics to analyze the data and capture emerging themes, based on pattern analysis (convergent/divergent) and country distribution.

The evaluation analyzed the collected quantitative data and statistics with more focus on descriptive statistics such as frequency distribution using Excel. The team also validated qualitative responses with quantitative information from the project document, performance indicators, progress reports, community site visits, and any available findings. The evaluation employed triangulation (among sources, methods, and field information) to confirm or disconfirm findings from primary and secondary data sources.

Appendix References

De Vaus, D. (2001). *Research design in social research.* Sage.

OECD DAC Network on Development Evaluation. (2019). *Better criteria for better evaluation.* https://www.oecd.org/dac/evaluation/daccriteriaforevaluatingdevelopmentassistance.htm

United Nations Evaluation Group. (2016). *Norms and standards for evaluation.* www.unevaluation.org/document/download/2601

United Nations Industrial Development Organization. (2009a). UNIDO and energy efficiency: A low-carbon path for industry. https://www.unido.org/sites/default/files/2009-10/UNIDO_and_energy_efficiency_0.pdf

United Nations Industrial Development Organization. (2009b). UNIDO and renewable energy: Greening the industrial agenda. https://open.unido.org/api/documents/4747543/download/UNIDO%20and%20Renewable%20Energy%20-%20Greening%20the%20Industrial%20Agenda

United Nations Industrial Development Organization. (2013). Request for MSP approval for GEF Trust Fund: Promoting energy efficient cook stoves in micro and small-scale food processing industries. https://open.unido.org/api/documents/3079335/download/CEO%20Endorsement%20(GEF%20Project%20Document%20120617)

UNIDO Office for Independent Evaluation. (2013). Independent terminal evaluation: Burkina Faso: Promoting *energy efficiency technologies in beer brewery in Burkina Faso.* United Nations Industrial Development Organization. https://www.unido.org/sites/default/files/2015-08/GFBKF12001_BeerBrewingSec_TE-2014_EvalRep-F_150813_0.pdf

References

Global Environment Facility. (n.d.). *Climate change.* Retrieved December 28, 2020, from https://www.thegef.org/topics/climate-change

Ministry of Economy and Planning, Republic of Chad. (2008). *Master plan of the energy sector in Chad,* Annex 1, Section 1: Terms of Reference.

Price, S., & Margolis, R. (2010). *2008 solar technologies market report.* U.S. Department of Energy. https://www.nrel.gov/docs/fy10osti/46025.pdf

Republic of Chad. (2010). *Poverty reduction and growth strategy paper: NPRS2: 2008–2011.* https://www.imf.org/en/Publications/CR/Issues/2016/12/31/Chad-Poverty-Reduction-Strategy-Paper-24073

Republic of Chad. (2017). *Chad: Plan National de Développement (PND) 2017–2021.* https://www.refworld.org/docid/5b34ac4f4.html

United Nations Development Programme. (2017). *Chad national adaptation plan.* https://www.adaptation-undp.org/projects/chad-national-adaptation-plan

United Nations High Commissioner for Refugees. (2012). *SGBV strategy 2012–2016: Chad.* https://www.unhcr.org/56b1fd9f9.pdf

United Nations Industrial Development Organization. (2009a). *UNIDO and energy efficiency: A low-carbon path for industry.* https://www.unido.org/sites/default/files/2009-10/UNIDO_and_energy_efficiency_0.pdf

United Nations Industrial Development Organization. (2009b). UNIDO and renewable energy: Greening the industrial agenda. https://open.unido.org/api/docu-ments/4747543/download/UNIDO%20and%20Renewable%20Energy%20-%20Greening%20the%20Industrial%20Agenda

United Nations Industrial Development Organization. (2013). *Request for MSP approval for GEF Trust Fund: Promoting energy efficient cook stoves in micro and small-scale food processing industries.* https://open.unido.org/api/documents/3079335/download/CEO%20Endorsement%20(GEF%20Project%20Document%20120617)

UNIDO Office for Independent Evaluation. (2013). *Independent terminal evaluation: Burkina Faso: Promoting energy efficiency technologies in beer brewery in Burkina Faso.* United Nations Industrial Development Organization. https://www.unido.org/sites/default/files/2015-08/GFBKF12001_BeerBrewingSec_TE-2014_EvalRep-F_150813_0.pdf

Vaccari, M., Vitali, F., & Mazzú, A. (2012). Improved cookstove as an appropriate technology for the Logone Valley (Chad – Cameroon): Analysis of fuel and cost savings. *Renewable Energy, 47,* 45–54. https://doi.org/10.1016/j.renene.2012.04.008.

Enabling Systems Innovation in Climate Change Adaptation: Exploring the Role for MEL

Robbie Gregorowski and Dennis Bours

Innovation distinguishes between a leader and a follower.

Steve Jobs

What is now proved was once only imagined.

William Blake

Abstract

Traditional monitoring, evaluation, and learning (MEL) approaches, methods, and tools no longer reflect the dynamic complexity of the severe (or "super-wickcd") problems that define the Anthropocene: climate change, environmental degradation, and global pandemics. In late 2019, the Adaptation Fund's Technical Evaluation Reference Group (AF-TERG) commissioned a study to identify and assess innovative MEL approaches, methods, and technologies to better support and enable climate change adaptation (CCA) and to inform the Fund's own approach to MEL. This chapter presents key findings from the study, with seven recommendations to support a systems innovation approach to CCA:

1. Promote and lead with a CCA systems innovation approach, engaging with key concepts of complex systems, super-wicked problems, the Anthropocene, and socioecological systems.
2. Engage better with participation, inclusivity, and voice in MEL.
3. Overcome risk aversion in CCA and CCA MEL through field testing new, innovative, and often more risky MEL approaches.
4. Demonstrate and promote using MEL to support and integrate adaptive management.
5. Work across socioecological systems and scales.
6. Advance MEL approaches to better support systematic evidence and learning for scaling and replicability.
7. Adapt or develop MEL approaches, methods, and tools tailored to CCA systems innovation.

R. Gregorowski
Sophoi Ltd, Wivelsfield Green, Haywards Heath, UK
e-mail: robbie@sophoi.co.uk

D. Bours (✉)
Technical Evaluation Reference Group of the
Adaptation Fund (AF-TERG), Washington, DC, USA
e-mail: dbours@adaptation-fund.org

Introduction

The Adaptation Fund was established by the Parties to the Kyoto Protocol (CMP) of the United Nations Framework Convention for Climate Change (UNFCCC) to finance concrete

J. I. Uitto, G. Batra (eds.), *Transformational Change for People and the Planet*, Sustainable Development Goals Series, https://doi.org/10.1007/978-3-030-78853-7_11

adaptation projects and programs in developing countries that are particularly vulnerable to the adverse effects of climate change. At the Katowice Climate Conference in December 2018, the Parties to the Paris Agreement (CMA) decided that the Adaptation Fund shall also serve the Paris Agreement.

In late 2019, the Adaptation Fund's Technical Evaluation Reference Group (AF-TERG) commissioned a study to identify and assess innovative monitoring, evaluation, and learning (MEL) approaches, methods and technologies to better support and enable climate change adaptation (CCA). The study aimed to contribute to new knowledge on innovative MEL opportunities to both support and enable CCA and contribute to and inform the fund's own approach to MEL.

This chapter presents key findings from the study to a wider CCA MEL audience and concludes with a series of recommendations on future directions for MEL commissioners and practitioners working in the context of CCA and systems transformation.

Study Purpose and Approach

The study took a broad and open-ended approach to identifying innovative MEL for CCA, applying a scan-search-appraise method to look within and beyond the CCA sector to identify potentially innovative and useful MEL practices for CCA. The method is essentially a structured funneling and sieving process, refining from a broad field or landscape down to set of focused priorities or conclusions. The scan phase was a relatively rapid and high-level assessment of the whole innovative MEL field or landscape, comprising an open-ended online literature review and five open-ended key informant interviews. The search phase took the overview provided by the scan phase and refined it in the context of adaptation, using a more systematic document review process and 10 semi-structured key informant interviews.

The purpose of the study was to identify innovative MEL practices within and beyond the climate adaptation space (scan and search) of potential value and use to the Adaptation Fund (appraise). The scope of the study was explicitly

broad, to look beyond MEL methods, tools, and technologies to include wider and emerging MEL-relevant principles, approaches, and processes. The foundational concepts of innovation, adaptation, and MEL are defined as follows:

- Innovation – There is no single definition of innovation. The study adopted a simple yet comprehensive definition of innovation that resonates with innovative MEL as a concept and can be applied to both technological and process innovation: "Innovation is the renewing, advancing, or changing the way things are done" (Everett et al., 2011, p. 6).
- Adaptation – "The process of adjustment to actual or expected climate and its effects. In human systems, adaptation seeks to moderate or avoid harm or exploit beneficial opportunities. In some natural systems, human intervention may facilitate adjustment to expected climate and its effects" (Intergovernmental Panel on Climate Change, 2014, p. 1758).
- Monitoring, evaluation, and learning (MEL) – The emphasis of the definition in the context of the study was to break MEL down into its three separate but overlapping parts—monitoring, evaluation, and learning—as part of a virtuous project cycle informing project course correction, design, delivery, and learning in an ongoing process.
  - Monitoring – In the context of MEL, monitoring is a continuous assessment that aims at providing stakeholders with early, detailed information on the progress of an intervention. In the context of the Adaptation Fund, monitoring should support near to real-time learning as part of a wider approach to flexible and adaptive management.
  - Evaluation – Building on the definition from the Organisation for Economic Co-operation and Development (Development Assistance Committee Working Party on Aid Evaluation, 2002), evaluation refers to the process of determining the worth or significance of an activity, policy, program, or institution—an assessment, as systematic and objective as possible, of a planned, ongoing, or completed development inter-

vention. Useful and robust evaluation should inform both accountability and learning, depending on the emphasis of the evaluation questions.

- Learning – In essence, learning is about understanding what works, in what contexts, for whom, and why. Learning should support direct and rapid course correction of an intervention and generate evidence and knowledge on the scalability and/or transferability of interventions across contexts. In the context of the Adaptation Fund, learning should be linked to the building of capacities, particularly adaptive capacity, of all stakeholders—beneficiaries, implementers, managers, and wider interest audiences. It should also reflect how these stakeholders learn through double and triple-loop learning processes.

The study's scan-search-appraise methodology began from the following three-step hypothesis:

1. Climate change is a "super-wicked problem,"[1] shaped by the complex and dynamic interactions of social, economic, and environmental factors that define the Anthropocene era.
2. Successful CCA requires innovative and transformative ways of doing development, notably through a systems innovation approach.
3. MEL theory and practice needs to adapt and evolve to better support a systems innovation approach in CCA.

Complexity, Systems Innovation, and CCA

Social innovation and innovation for development in the context of today's global challenges are the focus of this section of the chapter. Innovation is increasingly seen as central to addressing the interlinked global challenges of poverty, inequality, and climate change. The latest thinking on social innovation and innovation for development is shaping how CCA is defined, understood, and addressed.

The concept of systems innovation is critical for successful CCA. In response to the finding that systems innovation approaches and business-as-usual MEL approaches are increasingly disconnected, we suggest that systems innovation and complexity offer the MEL community,[2] particularly those involved in CCA, an opportunity to evolve and advance MEL mindsets, approaches, and methods.

CCA, Complex Systems, and Innovation: Evolution to the Present Day

The Transformative Innovation Policy Consortium (TIPC) has produced a simple and elegant three-frame model (Fig. 1) that summarizes the major phases or frames of innovation theory and policy and places them in historical context. (Schot & Steinmueller, 2018; TIPC, 2018).

Frame 1 is identified as beginning with a post-World War II institutionalization of government support for science and research and development (R&D) with the presumption that this would contribute to growth and address market

---

[1]Wicked problems are difficult to clearly define, with understanding of the problem constantly evolving. They have many interdependencies, are often multicausal and socially complex, and exist in complex systems that exhibit unpredictable, emergent behavior. They usually have no right or wrong response, although responses might be worse or better; these problems cross governance boundaries, involve changing behavior, and are characterized by chronic policy failure (Rittel & Webber, 1973; Australian Public Service Commission, 2007). In the case of climate change as a super-wicked problem, time is running out, those who cause the problem also seek to provide a solution, the central authority needed to address the issues is weak or nonexistent, and irrational discounting occurs that pushes responses into the future (Levin et al., 2012).

[2]The authors recognize that a single, homogeneous MEL community does not exist. However, a consistent finding of the study is that mainstream or established MEL approaches and methods applied by the majority of MEL practitioners do not engage with the principles of systems innovation to address super-wicked problems. Only a small number of innovative MEL practitioners are already working on exploring MEL approaches that support systems innovation, with virtually none working in the context of CCA.

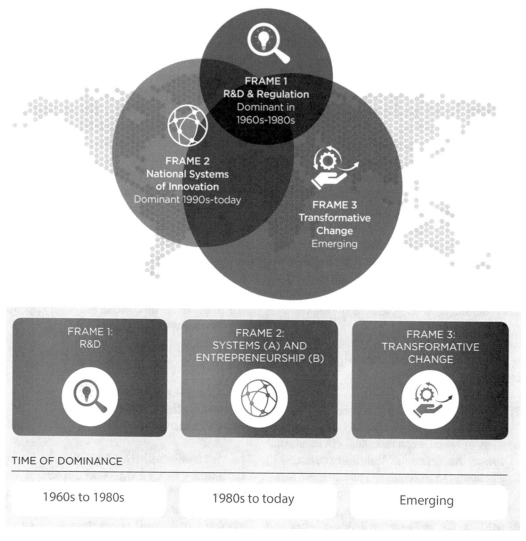

Fig. 1 Three frames of innovation

failure in private provision of new knowledge. Frame 1 builds on the thinking of some of the originators of innovation theory, notably Joseph Schumpeter (1883–1950) and his thinking that innovation and technological change of a nation comes from the interaction between government, scientists (inventors), and industry actors. Frame 1 produces a simple linear model of innovation focused on enhancing national economic growth through the overcoming of market failures in science and technology research. It identifies the discovery process (invention) in which technology is the application of scientific knowledge as the most important element of innovation.

Frame 2 represents the evolution of innovation theory and policy up to the present day and incorporates the perspectives of influential innovation thinkers such as Mariana Mazzucato[3] and Daniel Kahneman,[4] among others. While still focusing on innovation to support economic growth (rather than wider social and environmental needs), this

---

[3]Founder/director of the University College London Institute for Innovation and Public Purpose.

[4]Professor of psychology and public affairs emeritus at the Woodrow Wilson School, the Eugene Higgins Professor of Psychology emeritus at Princeton University, and a fellow of the Center for Rationality at the Hebrew University in Jerusalem.

frame emphasizes a more complex and dynamic relationship between key innovation stakeholders—actors, networks, and institutions—and stresses feedback loops between invention, innovation, and use. Innovation theory and policy draw on and have been influenced by the fields of behavioral economics, positive psychology, and public policy. Under this frame, Mazzucato and others recently have emphasized the critical role of entrepreneurs, and the relationship between the state/national systems and entrepreneurial actors.

Frame 3 is emerging and focuses on mobilizing the power of innovation to address a wide range of societal challenges, including inequality, unemployment, and climate change. It calls on "social innovation" to provide transformative change in the face of "grand challenges" (Schot & Steinmueller, 2018) and super-wicked problems. It notes that these challenges and problems extend across multiple scales that transcend national, sectoral, technological, and disciplinary boundaries. Solving these global environmental and societal problems requires the engagement of a much broader set of actors. Innovation theory (and emerging policy) in this area combines long-established academic thought in the areas of new institutional economics, public policy, common pool resources, and socioecological systems—drawing on the work of political economists such as Elinor Ostrom[5]—and combines this with new and emerging theory and practice in the areas of complex and adaptive systems, transformational change, and experimentation and active sensing. This third and latest frame of innovation builds on and embraces a small number of broader emerging concepts that relate to better conceptualizing and understanding the world. We present the key concepts in a short glossary, below.

The Anthropocene: The nature of humans' impact on the global biophysical system has become so dominant that scientists have proposed that the last 216 years of the existing Holocene period should become recognized as a new geological epoch, termed the Anthropocene. The concept of the Anthropocene has been suggested as a new geological era marked by global threats and challenges, the greatest of these being climate change, which are defined by the dynamic interactions between human and environmental systems (Olsson et al., 2017).

Super-wicked problems: Climate change is the greatest single threat facing the planet, threatening both natural and human systems. Addressing this urgent and intensifying threat is a complex, dynamic, and frequently contested challenge. Recognition is increasing that super-wicked problems (Balint et al., 2011; Levin et al., 2012), such as climate change, global pandemics, and rising inequality, require a fundamentally different approach from previous eras in history. Levin et al. (2012) define super-wicked problems as having four core characteristics: (a) time is running out, (b) those who cause the problem also seek to provide a solution, (c) the central authority needed to address the issue is weak or nonexistent, and (d) irrational discounting occurs that pushes responses into the future.

Complex adaptive systems: The notion of complexity as the property of a system is not new. But more recently, researchers have advanced the concept of complex adaptive systems (CAS) as a way of better understanding the global challenges characterized by complex interactions between human and environmental systems. David Snowden's Cynefin Framework (Snowden & Boone, 2007) was one of the first and is still the most elegant explanation of systems sense making. Framing climate change within a CAS has profound implications for how the challenge is understood and addressed (Preiser, 2018), and for how MEL of CCA is approached and delivered to support positive change.

---

[5](1933–2012) Distinguished professor, the Arthur F. Bentley Professor of Political Science, and co-director of the Workshop in Political Theory and Policy Analysis at Indiana University; also research professor and the founding director of the Center for the Study of Institutional Diversity at Arizona State University. Awarded the Nobel Memorial Prize in Economic Sciences in 2009.

Resilient and transformational change: Resilience is the capacity of a system to reorganize after a disruption (climatic shock or stress) without losing the essential functions of that system. The capacity for a social system (e.g., individuals, organizations, neighborhoods, communities, whole societies) to absorb disturbance and adapt where necessary, while undergoing significant change, is the defining characteristic of someone or something that is resilient. Although resilience is about maintaining the essential functions of a system, transformation is commonly interpreted as radical change requiring innovation and testing of new approaches. For climate change, this entails the generation of new knowledge and a markedly different way of doing things in order to address a threat of this scale (Climate Investment Funds, 2019). Both concepts—resilience and transformational change—have received heightened attention in response to the COVID-19 pandemic due to the increased recognition of economic and social systems shifting from those based primarily on efficiency to those defined by economic, social, and environmental resilience.

The third frame of the TIPC model and the concepts above are united and integrated in the field of social innovation, which draws on a tradition of broader innovation theory and policy and refines the innovation concept in recognition of complex and systemic social and environmental problems such as poverty, climate change, unemployment, discrimination, and biodiversity loss. According to the Center for Social Innovation (2020):

> Social innovation is the process of developing and deploying effective solutions to challenging and often systemic social and environmental issues in support of social progress. Social innovation is not the prerogative or privilege of any organizational form or legal structure. Solutions often require the active collaboration of constituents across government, business, and the non-profit world.

Frances Westley,[6] a global thought leader on social innovation, described social innovation as:

> any initiative (product, process, program, project, or platform) that challenges and, over time, contributes to changing the defining routines, resource and authority flows, or beliefs of the broader social system in which it is introduced. Successful social innovations have durability, scale and transformative impact.[7]

The concept of social innovation has several tenets that are particularly relevant to innovation for development and have profound implications for MEL. Resilient and transformational change is one of the tenets already mentioned as part of the third frame of innovation. Two other principles are:

Scales and their actors: Westley and others engaged in social innovation scaling talk in terms of three forms: (a) *scaling out*, which is based on market-based technological innovation scaling; (b) *scaling-up*, which involves scaling from individual social entrepreneurs to institutional and system entrepreneurs to take an innovation and find resources for it to reach a tipping point so that the institutional context of that innovation shifts to a new state (e.g., impact investing); and (c) *scaling deep*, which relates to deep, long-term, and profound shifts in culture, attitudes, and behaviors toward or triggered by an innovation (e.g., public support for action on climate change).

Solving complex social and environmental problems and challenges: The field of social innovation places particular emphasis on two elements: (a) CAS have certain unique and defining features, and (b) social innovation for sustainability and climate change requires a deeper focus on human–environmental interactions.

In the context of social innovation to help solve challenges such as climate change, the

---

complex systems category is useful because it suggests that solutions are discovered by developing a safe environment for experimentation. This experimentation allows us to discover important information that leads to the creation of new, emergent solutions. Evidence and knowledge are emergent through a probe-sense-respond process. Framing thinking on social innovation around the concept of the Anthropocene has led to an emerging belief (among scholars and practitioners) that "social innovation for sustainability lacks a deeper focus on human–environmental interactions and the related feedbacks, which will be necessary to understand and achieve large-scale change and transformations to global sustainability" (Olsson et al., 2017, p. 1) and "the social, environmental, and economic pillars often associated with sustainable development and 'triple bottom line' thinking have often led to trade-off decisions that either neglect the social-ecological, or strongly favor the economic" (p. 5).

Systems Innovation—The CCA Future

"Innovation-as-usual" – typically siloed and focused on "supplying" the market with technology-led solutions – is not delivering a 1.5-degree world. We need a new model of innovation to tackle climate change. . . one that is designed to generate options in the face of uncertainty and diversity, and to test for integrated and exponential solutions to address the complex, multi-faceted nature of the changes we need to make. . . . Using systems innovation as a key tool, our aim is to catalyze change in whole cities, regions, industries, and value chains by 2035. . . . Systems change not climate change.
Climate-KIC (2019)

So how best to address the super-wicked problem of climate change? A consensus is emerging that the answer lies in the concept of systems innovation. Systems innovation takes the concepts introduced in this chapter thus far, builds on the premise of social innovation, and is then applied in the context of urgent needs for systems-level transformation.

The EIT Climate Knowledge and Innovation Community (Climate-KIC) is a European Union-funded climate innovation initiative and a leading proponent of systems innovation. Climate-KIC aims to identify and support innovation that helps society mitigate and adapt to climate change. They explicitly recognize that climate change is a complex problem and that individual innovations, projects, and organizations are unlikely to meet the challenge. Rather, their approach applies a portfolio logic: They construct "portfolios of engagement" on a particular issue or challenge, based on the understanding that some elements of the portfolio will succeed and some will naturally fail. Systems innovation in this context is driven by what they call *levers of change*. "Systems innovation is not limited to technological improvements. It acts on a wide array of change levers all at once, testing for possibility, connecting different approaches to learn from one another, looking for integrations, mash-ups, and exponential effects" (Climate-KIC, 2019).

Climate-KIC then aims to identify "early signals of potential systemic change" to identify which innovations to further support and ultimately scale. This approach is very similar to the original probe-sense-response approach proposed by Snowden. In the context of climate change, this means an early and continuous focus on learning to support adaptive management and identify replicable and scalable opportunities.

A second concept central to Climate-KIC are the *levels of change*. Complexity and systems dynamics are addressed by working at many levels, from district and city level to countries, regions, sectors, and value chains.

The evolution of the field of innovation, from a post-World War II focus on the institutionalization of government support for science and R&D to the present-day focus on social and systems innovation, has been complemented by and runs parallel to continuing advances in understanding and enabling international development. Resonating with the growing interest in and support for systems innovation approaches are a set of *innovation for development* principles more explicitly focused on understanding and addressing the root causes of poverty and inequality. The

International Development Innovation Alliance (IDIA) has produced eight development innovation principles in practice (IDIA, 2019):

Principle 1. Promote inclusive innovation
Principle 2. Invest in locally driven solutions
Principle 3. Take intelligent risks
Principle 4. Use evidence to drive decision making
Principle 5. Learn quickly and iterate
Principle 6. Facilitate collaboration and co-creation
Principle 7. Identify scalable solutions
Principle 8. Integrate proven innovations

These eight principles can provide a blueprint for successful CCA when they are combined with a systems innovation approach, such as the Climate-KIC approach that proceeds from a portfolio logic, constructing portfolios of engagement on CCA as a particular challenge. In the next section, we explore the implications in terms of principles and approach for the MEL community—or at least for those practitioners who mainly focus on mainstream or established MEL approaches and methods.

MEL's Role in Enabling Systems Innovation for CCA

This third part focuses on the key implications, ideas, and lessons for MEL practitioners working in CCA, drawing on the latest thinking and practice from systems innovation and innovation for development. These implications, ideas, and lessons are framed from the outset by the earlier mentioned three-step hypothesis.

We also provide in this section illustrative examples of innovative MEL approaches, processes, and technological interventions according to the major themes that have emerged from MEL to support a systems innovation approach to CCA. These examples have the potential to support enhanced MEL in CCA either through development in related fields or by being piloted in a current CCA context. The AF-TERG MEL study indicated that these cases have the poten-

tial to advance CCA MEL to better support a systems innovation approach to addressing climate change. It also suggested that they would enable the MEL community more broadly to better engage with, support, and advance a systems innovation approach to super-wicked problems.

Seven Directions of Change for the CCA MEL Community

1. **Promote and lead with a CCA systems innovation approach including better engaging with the key concepts of complex systems, super-wicked problems, the Anthropocene, and socioecological systems.**

New terms and concepts are emerging in development discourse as fields such as systems innovation explore and more deeply define and understand global challenges and problems. With particular relevance to CCA, most prominent among these are the concepts of complex and adaptive systems, super-wicked problems, socioecological systems, the Anthropocene, and transformational change. Although the systems innovation community is pressing ahead in defining and exploring these concepts, they are not receiving consistent supported from the MEL community in terms of the critical MEL dimensions of these concepts—defining characteristics, indicator frameworks, and measurement approaches.

Taking transformational change as an example, we see little knowledge and consensus on defining, measuring, and assessing transformational change. The approach for CCA MEL practitioners should be to establish clear frameworks, pathways, and indicators for transformational change; establish an evidence base on which contexts and interventions are genuinely transformational (as opposed to other more incremental and iterative change processes); and develop relatively consistent indicator frameworks and measurement approaches to better assess and learn from transformational results.

2. **Engage better with participation, inclusivity, and voice in MEL.**

The issues of genuine participation, inclusion, and voice have rightly taken on increased prominence in MEL theory and practice in recent years. This is particularly the case in CCA MEL where locally led adaptation/action (LLA)[8] has gained recognition because people and communities on the frontlines of climate change are often the most active and innovative in developing adaptation solutions. More broadly, evidence has shown that for development to be effective and sustainable, people who are vulnerable must be empowered and their voices heard, and gender equality and the empowerment of women and girls is crucial to development progress, particularly in the context of CCA.

This commitment has been reaffirmed across the development space, perhaps most recently and prominently in the Sustainable Development Goals (SDGs), which commit to ensure "no one will be left behind" and to "endeavour to reach the furthest behind first" as overarching principles (United Nations General Assembly, 2015)[9] Organizations such as the United Nations Development Programme (UNDP) are advancing evidence and education about the underlying factors that cause people to be left behind. In a recent paper, they outlined five critical factors: discrimination, geography, governance, socioeconomic status, and shocks and fragility (UNDP, 2018). Again, these critical factors resonate with the big issues being engaged with and explored by innovation for development communities.

The combination of three factors, prevalent in wider development policy and practice, have enabled advances in locally and/or citizen-led MEL. These factors have not yet been fully explored in the context of CCA:

1. Demands for accountability – Increasingly energetic civil societies with a growing demand for greater transparency and public accountability
2. More mature civil societies – With civil society increasingly willing and capable to participate in MEL processes
3. The boom of information and communications technology (ICT) – Particularly in this context, the spread of internet and mobile phone technology

Out of these factors have emerged a number of citizen-led/citizen-generated data platforms, a prominent example being the Slum/Shack Dwellers International (SDI) Know Your City (KYC) initiative (SDI, 2018), which unites organized slum dwellers and local governments in partnerships anchored by community-led slum profiling, enumeration, and mapping. ICT-enabled participant reporting is not new to CCA programming but is still not widely included in the MEL components of programs. One likely reason is that it needs to be built into the design of the program itself from the onset rather than applied later as a MEL tool. The technology and tools are now proven, and cost and failure risks associated with ICT-enabled MEL have been reduced significantly over the last few years.

As such, MEL methods employing ICT should no longer be considered risky or unproven and citizen/participant reporting can provide data across the MEL cycle. ICT-enabled participant reporting as part of project monitoring can shift participants from passive beneficiaries of program activities to more active and empowered participants, reporting real-time reactions (positive and negative) to interaction with the program. On the accountability side, ICT-enabled participant reporting can generate data on program coverage and satisfaction. In terms of learning, open-ended questions such as "Provide an illustration of what you do differently as a result of the program" allow participants to demonstrate what a program may have enabled, often in video or photo form. Subjective reporting and learning are key areas that require further exploration, such as by asking partici-

---

[8] Also see https://gca.org/global-commission-on-adaptation/action-tracks/locally-led-action and https://www.wri.org/our-work/project/global-commission-adaptation/action-tracks/locally-led

[9] In addition to the overarching principle of "no one will be left behind," SGD 5 aims to achieve gender equality and empower all women and girls, and SDG 10 aims to reduce inequality within and among countries.

pants about the extent to which they feel more able to cope and adapt to the threats posed by climate change as a result of the program. This goes to the heart of many CCA programs that aim to build adaptive capacity in target groups, removing the need to develop proxy indicators of adaptation and then generate data against them. The added advantage is that this type of reporting can be done before, during, and after a climatic shock or stress, and overlaid with shock/stress intensity data.

Linked to this but at a more systemic level is the issue of the decolonization of MEL. Launched in 2018, the South to South Evaluation Initiative (S2SE; African Evaluation Association, 2017) aims to elevate the substantial but underutilized indigenous knowledge, theory, and capacities of the Global South, and aims to reverse the asymmetries in decision making, resources, and knowledge in the global evaluation ecosystem. CCA MEL is a particular case in point: Global South perspectives, knowledge, and capacity go to the heart of not only MEL but also of appropriate and effective CCA. Far too often, evaluations are designed and led by those designated international experts and only supported by national experts. This immediately places southern experts, who tend to hold much deeper and more relevant insights into local contexts, challenges, opportunities, and practices, into a subordinate, lower value role and position. It perpetuates north-south power dynamics and ultimately results in lower quality, less insightful, and less useful evaluations.

3. Overcome risk aversion in CCA and CCA MEL through field testing new, innovative, and often more risky MEL approaches.

The architecture, systems, and norms that are all pervasive in international development and CCA encourage risk aversion and the maintenance of traditional/established practices. This is the case in both CCA policy and programming and CCA MEL to support it. Despite increased recognition from the fields of complex systems

and adaptive management that simple, boilerplate solutions are rarely appropriate for complex problems, an aversion remains to identifying and testing new and innovative solutions.

The fear of failure leads CCA agencies to take limited risks in their policy, programming, and MEL. This results in the perpetuation of established but often inappropriate approaches, methods, and tools. These agencies are motivated and incentivized more by accountability-driven evaluation processes than by trying less-tested and more innovative evaluative approaches, processes, and tools that may deliver rich, insightful, and useful new knowledge, but come with a higher risk of failure until they are established.

Hence, the MEL-system norm—fixed-term results reporting systems paired with traditional accountability-focused, postproject, mixed-methods evaluations—remains the staple of most large bilateral and multilateral agencies working on CCA. In this context of risk aversion, it is deemed better to have a safely delivered end-of-project evaluation that produces a long, unengaging (often unread) evaluation report with little or no real insight than to risk engaging a more innovative evaluation approach or method but one that carries a higher risk of failing.

The challenge is not to find new and innovative MEL approaches, methods, and tools for CCA; the challenge is finding the opportunity and resources to field-test them. Ways exist to reduce the risk of failure, such as combining established MEL methods with the piloting of new and innovative data collection and data analysis methods and tools. Much as the process of bricolage has gained popularity in systems innovation, so could the same process in MEL for CCA (Patton, 2020). For example, a project that aims to strengthen and then track household climate resilience through face-to-face household surveys could also pilot more subjective (how resilient household residents feel) and real-time results reporting through mobile phone-based participant reporting in the face of a climatic shock or stress.

4. **Demonstrate and promote the use of MEL to support and integrate adaptive management.**

Recent years have seen an enhanced focus on adaptive management as the key concept at the heart of doing development differently. This is based on an increased recognition that development interventions are delivered in dynamic, unpredictable, and often contested contexts and systems; that in these contexts interventions need to be innovative; and that how best to deliver results in these contexts is uncertain. Therefore, operating effectively and efficiently in these contexts requires projects, programs, and institutions to be "adaptive." This means

- tailoring MEL systems, particularly monitoring systems, to generate robust evidence on program management and delivery;
- a focus on lesson learning in close to real time to support course correction;
- explicit focus on learning from unintended consequences and failures as well as from successes; and
- portfolio learning and sense making among coalitions of similar stakeholders, both within and outside of the program context, supporting evidence and learning on both scalability and transferability (Pasanen & Barnett, 2019; Wild & Ramalingam, 2018).

Adaptive management focuses on intentionally building in opportunities for structured and collective reflection, ongoing and real-time learning, course correction, and decision making in order to improve effectiveness. This means that adaptive programs require, at project inception, intentional MEL design in which learning and course correction are integrated into the program from the start.

The concept of adaptive management should resonate particularly with CCA projects, programs, and organizations given that they share the same concept at heart. While it seems obvious that organizations promoting climate change adaptive programming should themselves be adaptive in their own designs, actions, and behaviors, evidence that CCA practitioners are leading the way is limited.

5. **Work across socioecological systems and scales.**

A long-recognized, and largely unaddressed, challenge of the CCA MEL community is to develop and apply MEL approaches, methods, and capacities that integrate the social, economic, and environmental dimensions across systems and scales. At present, most MEL approaches tend to focus on individual domains—the human or the environmental/ecological domain—and not the interactions between them.

Earlier in this chapter, we introduced the concepts of super-wicked problems, complex adaptive systems (CAS), and the Anthropocene as critical, interrelated concepts that have a rising prominence in the context of CCA and with which the MEL community is beginning to engage. The fundamental issue underlying these concepts is the relationships among the social, economic, and environmental dimensions taking place in complex contexts and systems. In past development discourse, policy and practice has tended to focus on these elements in isolation and to prioritize the social and economic over the environmental.

Few CCA MEL frameworks systematically encourage a portfolio of indicators across all three domains—social, ecological, and economic. One potential solution would be to explicitly embrace a socioecological systems (SES) approach in CCA policy and program design. The SES approach describes the four essential dimensions (the natural system, livelihoods and people, institutions and governance, and external drivers) that provide the basis on which to situate and understand CCA results within a wider system. These also provide the basis for an MEL system based on the selection of a balanced set of CCA indicators under each dimension.

6. **Advance MEL approaches to better support systematic evidence and learning for scaling and replicability.**

Given that the effects of climate change are being felt globally—across contexts, locations, and scales—learning what works, in what contexts, and why is particularly relevant for the CCA MEL community. Along with locally led learning, the CCA community has rightly put considerable emphasis on evidence and learning to support scaling and replicability. However, these two key concepts are not yet systematically integrated into MEL frameworks either as indicators or as key evaluation and learning questions or criteria.

Systems innovation approaches place particular emphasis on identifying, scaling, and replicating successful interventions, whether these are products, technologies, or processes. IDIA (2017) has devised a high-level process for scaling innovation in a development context, and this process is particularly important in the context of CCA. CCA MEL frameworks—from strategy and design (theories of change) through monitoring frameworks (indicators) to evaluation and learning (key evaluation criteria and learning questions)—should more deeply engage with the concepts of scaling and replicability as defined in a systems innovation approach.

7. **Adopt or develop MEL approaches, methods, and tools tailored to CCA systems innovation.**

This final suggested direction engages more directly with the most appropriate MEL approaches, methods, and tools. Although several MEL approaches and frameworks aim to engage with scale, context, and system dynamics, the two most prominent are developmental evaluation (Patton, 2010) and its evolution into Blue Marble Evaluation[10] (Patton, 2020). Both approaches recognize some of the key issues raised in this chapter and Blue Marble Evaluation explicitly engages with systems innovation con-

cepts. Blue Marble Evaluation is founded around four overarching principles and 12 operating principles. A simple starting point for any CCA MEL approach, system or method (whether for a project, program, organization, or institution) would be to assess which and how many of the Blue Marble principles it is coherent with or supports. Commissioners of CCA MEL services have a large role to play in encouraging Blue Marble (systems innovation) principles in the MEL terms of references they draft and the MEL services they fund. As made clear throughout the chapter, this is a new and emerging area that is gaining momentum. Its implications have yet to be explored and defined in the context of CCA.

The AF-TERG study has focused on innovative MEL approaches that attempt to reframe the role played by the MEL community, suggesting that the way MEL is approached and delivered needs to fundamentally shift. The rationale is that a fundamental shift in approach is required from the MEL community and practitioners, rather than a more granular focus on adopting and integrating the latest MEL technological innovations. Consequently, this chapter has given little attention to innovative MEL technologies, despite the subject of technology for MEL being the focus of considerable attention through networks and events such as MERL Tech. Technological innovation in MEL tends to be related to data under three overlapping themes: (a) data collection and capture technologies, (b) data analysis technologies including artificial intelligence and machine learning, and (c) data presentation and visualization technologies. The data collection and capture technologies can be broken down into three core areas: (a) big data, which includes satellite imaging and remote sensing; (b) information and communication technology (ICT); and (c) the internet of things.

Despite the slow uptake, technological innovations provide solutions to core challenges in MEL: reaching isolated groups; monitoring behavior change; collecting qualitative and subjective data; compiling, integrating, and interpreting multiple datasets; enhancing quality control; post evaluation verification; and finding robust samples for comparison. Examples of use

---

[10] https://bluemarbleeval.org/

in data collection include decentralized data gathering through self-reporting, online data harvesting, and the use of real-time data (Raftree, 2016, 2020). In terms of analysis, Bamberger and Mabry (2019) outlined several data analysis techniques with direct applications for program MEL. Bruce et al., (2020) added machine learning and artificial intelligence for text analytics as technological innovations to the toolbox of the MEL practitioner.

Finally, in terms of data visualization, processed data or MEL evidence needs to succinctly communicate complex problems and solutions. Visualizing and packaging data in a meaningful way across stakeholders is a challenge. The rapid pace of change in data technology will eventually and inevitably shape, drive, and inform MEL, especially from learning and data visualization perspectives. This may mean that instead of a traditional narrative report, MEL products and outputs could (and should) increasingly become digital, interactive, and more widely available.

What matters is not how technological innovations collect, analyze, or present data in isolation, but how they are integrated into innovative MEL approaches and methods to advance the delivery of CCA programming and understanding of CCA more broadly. This is the bricolage Michael Quinn Patton refers to in Blue Marble Evaluation. For this to happen, the MEL community needs to better engage with, exchange with, and understand the data scientists and technologists, and vice versa.

Conclusion

In this chapter, we have argued that a systems innovation approach is required to address climate change adaptation, and that this in turn requires a fundamentally different and new approach from the CCA MEL community and practitioners. This new systems innovation approach in MEL would move the discipline in part away from the established static, uni-linear, project-program-country-region-bound MEL approaches, methods, and tools that have been standard practice for years. These traditional approaches, methods, and tools tend to be based on simple cause-effect results chains, tested ex-post through narrowly defined evaluation questions, and with established power relationships between evaluation commissioners, evaluation practitioners, participants, and intended audiences. They no longer reflect the dynamic complexity of super-wicked problems that define the Anthropocene: climate change, environmental degradation, and global pandemics.

References

African Evaluation Association. (2017). *South to south evaluation (S2SE): A call to action*. https://afrea.org/wp-content/uploads/2018/05/S2S_Brochure.pdf

Australian Public Service Commission. (2007). *Tackling wicked problems: A public policy perspective*. Australian Government. https://www.apsc.gov.au/tackling-wicked-problems-public-policy-perspective

Balint, P., Stewart, R., Desai, A., & Walters, L. (2011). *Wicked environmental problems: Managing uncertainty and conflict*. Springer.

Bamberger, M., & Mabry, L. (2019). *Real world evaluation: Working under budget, data and political constraints* (3rd ed.). Sage.

Bruce, K., Gandhi, V., & Vandelanotte, J. (2020). *Emerging technologies and approaches in monitoring, evaluation, research, and learning for international development programs*. MERL Tech. http://merltech.org/wp-content/uploads/2020/07/4_MERL_Emerging-Tech_FINAL_7.19.2020.pdf

Center for Social Innovation. (2020). *Defining social innovation*. Stanford Graduate School for Business. https://www.gsb.stanford.edu/faculty-research/centers-initiatives/csi/defining-social-innovation

Climate Investment Funds. (2019). *The CIF Transformational Change Learning Partnership: Pioneering joint learning to catalyze low-carbon, climate-resilient development*. Climate Investment Funds Transformational Change Learning Partnership. https://www.climateinvestmentfunds.org/knowledge-documents/cif-transformational-change-learning-partnership-pioneering-joint-learning

Development Assistance Committee Working Party on Aid Evaluation. (2002). *Glossary of key terms in evaluation and results based management*. Development Assistance Committee of the Organisation for Economic Co-operation and Development. http://www.oecd.org/development/evaluation/2754804.pdf

EIT Climate Knowledge and Innovation Community. (2019). *Work with us to achieve net zero, in time. Funding systems change is today's most important innovation*. https://www.climate-kic.org/

wp-content/uploads/2019/10/191029_EIT_Climate-KIC_FundingNetZero_Double.pdf

Everett, B., Barnett, C., & Verma, R. (2011). *Evidence review – Environmental innovation prizes for development.* DEW Point Enquiry No. A0405. DEW Point, the DFID Resource Centre for Environment, Water and Sanitation. https://assets.publishing.service.gov.uk/media/57a08abded915d622c00089b/61061-A0405EvidenceReviewEnvironmentalInnovationPrizesforDevelopmentFINAL.pdf

Intergovernmental Panel on Climate Change. (2014). *Climate change 2014: Impacts, adaptation, and vulnerability.* IPCC Working Group II Contribution to AR5. https://www.ipcc.ch/report/ar5/wg2/

International Development Innovation Alliance. (2017). *Insights on scaling innovation.* Author. https://www.idiainnovation.org/s/Insights-on-Scaling-Innovation.pdf

International Development Innovation Alliance. (2019). *Development innovation principles in practice - insights and examples to bridge theory and action.* Author. https://www.idiainnovation.org/s/8-Principles-of-Innovation_FNL.pdf

Levin, K., Cashore, B., Bernstein, S., & Auld, G. (2012). Overcoming the tragedy of super-wicked problems: Constraining our future selves to ameliorate global climate change. *Policy Sciences, 45,* 123–152. https://doi.org/10.1007/s11077-012-9151-0

Olsson, P., Moore, M.-L., Westley, F., & McCarthy, D. (2017). The concept of the Anthropocene as a game-changer: A new context for social innovation and transformations to sustainability. *Ecology and Society, 22*(2), 31. https://doi.org/10.5751/ES-09310-220231

Pasanen, T., & Barnett, I. (2019). *Supporting adaptive management: monitoring and evaluation tools and approaches* (Working Paper 569). Overseas Development Institute. https://www.odi.org/sites/odi.org.uk/files/resource-documents/odi-ml-adaptivemanagement-wp569-jan20.pdf

Patton, M. Q. (2010). *Developmental evaluation: Applying complexity concepts to enhance innovation and use.* Guilford.

Patton, M. Q. (2020). *Blue Marble evaluation: Premises and principles.* Guilford.

Preiser, R. (2018). *Key features of complex adaptive systems and practical implications for guiding action.*

GRAID by the Stockholm Resilience Centre. https://graid.earth/briefs/key-features-of-complex-adaptive-systems-and-practical-implications-for-guiding-action/

Raftree, L. (2016). *ICTs in evaluation practice.* https://lindaraftree.com/2016/05/31/icts-in-evaluation-practice/

Raftree, L. (2020). *MERL Tech state of the field: The evolution of MERL Tech.* MERL Tech. http://merltech.org/wp-content/uploads/2020/07/1_MERL_Tech-State-of-the-Field_FINAL_7.16.2020.pdf

Rittel, H. W. J., & Webber, M. M. (1973). Dilemmas in a general theory of planning. *Policy Sciences, 4*(2), 155–169. https://doi.org/10.1007/BF01405730

Schot, J., & Steinmueller, E. (2018). Three frames for innovation policy: R&D, systems of innovation, and transformative change. *Research Policy, 47*(9), 1554–1567. https://doi.org/10.1016/j.respol.2018.08.011

Slum/Shack Dwellers International. (2018). *Know your city: Slum dwellers count.* Author. https://knowyourcity.info/wp-content/uploads/2018/02/SDI_StateofSlums_LOW_FINAL.pdf

Snowden, D., & Boone, M. (2007, November). A leader's framework for decision making. *Harvard Business Review.* https://hbr.org/2007/11/a-leaders-framework-for-decision-making

Transformative Innovation Policy Consortium. (2018). *Three frames for innovation.* University of Sussex. http://www.tipconsortium.net/wp-content/uploads/2018/04/4173_TIPC_3frames.pdf

United Nations Development Programme. (2018). *What does it mean to leave no one behind?* A UNDP discussion paper and framework for implementation. Author. https://www.undp.org/content/dam/undp/library/Sustainable%20Development/2030%20Agenda/Discussion_Paper_LNOB_EN_lres.pdf

United Nations General Assembly. (2015). *Transforming our world: The 2030 Agenda for Sustainable Development,* A/RES/70/1. Author. https://sustainabledevelopment.un.org/post2015/transformingourworld

Wild, L., & Ramalingam, B. (2018). *Building a global learning alliance on adaptive management.* Overseas Development Institute. https://www.odi.org/sites/odi.org.uk/files/resource-documents/12327.pdf

Assessing the Evaluability of Adaptation-Focused Interventions: Lessons from the Adaptation Fund

Ronnie MacPherson, Amy Jersild, Dennis Bours, and Caroline Holo

Abstract

Evaluability assessments (EAs) have differing definitions, focus on various aspects of evaluation, and have been implemented inconsistently in the last several decades. Climate change adaptation (CCA) programming presents particular challenges for evaluation given shifting baselines, variable time horizons, adaptation as a moving target, and uncertainty inherent to climate change and its extreme and varied effects. The Adaptation Fund Technical Evaluation Reference Group (AF-TERG) developed a framework to assess the extent to which the Fund's portfolio of projects has in place structures, processes, and resources capable of supporting credible and useful monitoring, evaluation, and learning (MEL). The framework was applied on the entire project portfolio to determine the level of evaluability and make recommendations for improvement. This chapter explores the assessment's findings on designing programs and projects to help minimize the essential challenges in the field. It discusses how the process of EA can help identify opportunities for strengthening both evaluability and a project's MEL more broadly. A key conclusion was that the strength and quality of a project's *overall* approach to MEL is a major determinant of a project's evaluability. Although the framework was used retroactively, EAs could also be used prospectively as quality assurance tools at the pre-implementation stage.

R. MacPherson
Greenstate, Newtonmore, UK
e-mail: ronnie@greenstate.net

A. Jersild
Evaluation consultant and doctoral candidate in the Interdisciplinary Program in Evaluation (IDPE) at Western Michigan University (WMU), Kalamazoo, MI, USA
e-mail: amy.c.jersild@wmich.edu

D. Bours (✉) · C. Holo
Technical Evaluation Reference Group of the Adaptation Fund (AF-TERG), Washington, DC, USA
e-mail: dbours@adaptation-fund.org;
cholo1@adaptation-fund.org

Introduction

Background to Adaptation Fund

The Adaptation Fund was established by the Parties to the Kyoto Protocol (CMP) of the United Nations Framework Convention for Climate Change (UNFCCC) to finance concrete climate change adaptation (CCA) projects and programs in developing countries that are particularly vulnerable to the adverse effects of climate change. At the Katowice Climate Conference in December 2018, the Parties to the Paris Agreement (CMA) decided that the Adaptation Fund (AF) shall also serve the Paris Agreement.

J. I. Uitto, G. Batra (eds.), *Transformational Change for People and the Planet*, Sustainable Development Goals Series, https://doi.org/10.1007/978-3-030-78853-7_12

Since 2010, the Fund has committed $720 million in grants to more than 100 projects in developing countries, with projects working in a diversity of sectors including agriculture, disaster risk reduction, coastal management, food security, and urban development. The Fund provides grants to implementing entities that lead the development, implementation and monitoring of work on the ground, usually in partnership with other organizations. Implementing entities can be multilateral (e.g., UN and multilateral agencies), regional (regional development banks), or national (government ministries, national research institutions).

As with other comparable institutions, the Adaptation Fund uses evaluation as a tool for understanding project results, strengthening accountability, learning, and continuous improvement. An evaluation framework (AF, 2012) sets out the Fund's approach, defining objectives, requirements, roles, and processes that should be applied when evaluating Adaptation Fund supported projects. Central to this approach, implementing entities are required to commission independent final evaluations (AF, 2011a, b) for any Adaptation Fund-supported project, with independent midterm evaluations if a project is more than 4 years in length.

The Fund's evaluation function was initially outsourced. In 2019, the Fund internalized evaluation with the establishment of the Adaptation Fund Technical Evaluation Reference Group (AF-TERG). An early step by the AF-TERG was to commission a series of preliminary studies to inform and support the development of a multi-year work program. One of these studies was an evaluability assessment (EA) of the Adaptation Fund's portfolio of projects.

History and Purpose of Evaluability Assessment

Michael Scriven (1991) defines evaluability analogous to requiring serviceability in a new car. It may be thought of as "the first commandment in accountability," notes Scriven (p. 138). The technical use of the term originated with Joseph

Wholey (1979) and his colleagues at the Urban Institute in the 1970s as a response to the delays and low value found in summative evaluations of U.S. government programs. EAs were a means by which to examine a program's structure to determine whether it could lend itself to generating useful results from an outcome evaluation. They were also viewed as a preformative evaluation activity that was part of a cost-effective strategy in determining readiness for evaluation and enhancing use. For Wholey, EAs were the first of four tools in a "sequential purchase of information," including rapid feedback evaluation, performance monitoring, and impact evaluation (Wholey, 1979; Wholey et al., 2010, p. 82).

The Development Assistance Committee of the Organisation for Economic Co-operation and Development (OECD DAC, 2002) defined evaluability as "the extent to which an activity or a program can be evaluated in a reliable and credible fashion" (p. 21) with a focus on methods and design, in contrast to Wholey's stronger focus on utility from a cost perspective. The OECD DAC further described EAs as an "early review of a proposed activity in order to ascertain whether its objectives are adequately defined and its results verifiable" (p. 21). Scriven (1991) noted the possible confusion that EAs may pose with regard to taking the place of serious summative evaluation, or to support the greater tendency to rely on objectives-based evaluation (a pseudo-evaluative approach) when the time comes to evaluate (Stufflebeam & Coryn, 2014).

Wholey (1979) developed an eight-step approach to implementing EAs:

1. Define the program to be evaluated.
2. Collect information on the intended program.
3. Develop a program model.
4. Analyze the extent to which stakeholders have identified measurable goals, objectives, and activities.
5. Collect information on program reality.
6. Synthesize findings to determine the plausibility of program goals.
7. Identify options for evaluation and management.

8. Present conclusions and recommendations to management.

Over the years, Wholey's approach was modified and others have further elaborated and emphasized certain aspects while reducing the number of steps. Smith (1989), for instance, identified stakeholder awareness as a particularly vital part of EAs, noting the importance of perceptions about what a program is to accomplish and whatever defined needs there may be for evaluative information on a program, whereas Rutman (1980) focused on methods and the feasibility of achieving an evaluation's purpose. Trevisan and Walser (2015) simplified Wholey's approach into a model of four iterative components:

1. focusing the EA,
2. developing an initial program theory,
3. gathering feedback on program theory, and
4. using the EA.

Each component features a checklist with questions reflecting the Program Evaluation Standards (Yarbrough et al., 2011).

The application of EAs has been intermittent. Although Wholey's work initiated a decade with a flurry of EAs in the 1970s–1980s, a revival in their use, particularly at the international level, did not occur until the late 1990s. Few publicly available examples of EAs exist. As Scriven (1991) warned about the confusion EAs may present in relation to other evaluative activities, reviews of the available EAs reflect concern about inconsistent implementation and use, revealing a lack of clarity about EA as a unique concept (Davies, 2013; Davies & Payne, 2015; Trevisan, 2007). Confusion with needs assessments, formative evaluations, and process evaluations were found, as was mission creep, with EAs extending into other evaluative functions based on commissioners' interests and budgets. Davies and Payne (2015) identified a need for clearly bounded expectations of the outputs of an EA and for linking the contents of a checklist for implementing EAs to relevant wider theory and evidence.

In developing our EA approach to CCA programming at the portfolio level, we found few concrete examples of previous EAs, with the exception of the Green Climate Fund Independent Evaluation Unit's 2019 Summary of the Evaluability of Green Climate Fund Proposals (Fiala et al., 2019). Given the challenges noted in the literature on maintaining clear objectives for EAs, and a need to have clear links between the dimensions of our checklist and relevant theory and evidence, we were purposeful in developing our EA framework, discussed below.

Evaluation of Climate Change Adaptation

Climate change poses dire consequences for the world, and given the relatively short timeframe to reverse trends, evaluating CCA interventions is essential to understand how best to adapt. The challenges in evaluating CCA interventions are well known and prove to be particularly complex. These include, for example, assessing attribution, creating baselines, and monitoring over longer time horizons (Bours et al., 2014; Fisher et al., 2015; Uitto et al., 2017).

CCA programming poses particular challenges for evaluation due to adaptation performance stretching far beyond the project life cycle. As a result, the impacts of such programs are difficult to measure because their outcomes may manifest much later. Climate change patterns and the prediction of weather patterns and extremes also pose a level of uncertainty for both programming and evaluation. The level of uncertainty increases when moving from global to regional climate models, to regional scenarios, and then to local impacts on human and natural systems (Wilby & Dessai, 2010). Given this unpredictability, collecting baseline data against which progress can be tracked is difficult. With climate change interventions spanning sectors and areas, another challenge is thinking at a systems level and across multiple stakeholder groups.

Finally, like most complex problems, CCA presents a challenge for causal inferences between intervention and outcome. Given the cross-sector nature of CCA, with multiple influences from both the social and natural worlds, developing a coherent evidence base on which to make causal inferences is difficult (Bours et al., 2015). In spite of these challenges, MEL for CCA projects plays a central role for identifying how best to reduce vulnerability and build resilience to climate change (Bours et al., 2014). With a growing need for accountability and learning, having MEL systems in place that generate evidence that is fed back into adaptation practice is important. In this context, the role of evaluability is critical to ensuring that a project can be evaluated and has the foundations necessary for carrying out evaluations that will offer important lessons for the future.

Study Objectives

Given the need for clearer definition on EAs and the challenges CCA programming presents for evaluation, this chapter describes the Fund's evaluability framework and the process of developing and applying it in assessing the evaluability of the projects part of the Fund's portfolio. We reflect on areas of learning that have implications for both the evaluation and CCA fields.

Assessment Approach

Framework Development

The AF-TERG embarked on an EA in 2019 to examine all 100 projects approved by the Fund's board at the start of the assessment's inception in November 2019, making up the Fund's project portfolio. These projects were diverse, spanning a range of types of grantee organizations working within varying contexts, from small island states in the Caribbean and Pacific to locations in Central Asia, Africa, and South America. The projects also had a diverse set of implementing entities and stakeholders, including grassroots organizations, civil society organizations (CSOs),

government bodies, regional organizations, and multilateral stakeholders.

We adopted the OECD DAC (2002) definition of evaluability, "the extent to which an activity or a program can be evaluated in a reliable and credible fashion," and developed two objectives to guide the development and implementation of our EA:

1. Assess the extent to which the Adaptation Fund's projects have in place structures, processes, and resources capable of supporting credible and useful monitoring, evaluation and learning.
2. Based on the assessment's findings, provide advice on how to improve the evaluability of the Adaptation Fund's projects and portfolio.

The bounded outputs for the assessment included determining the extent of evaluability of the overall portfolio and identifying ways to improve evaluability. These findings were discussed internally to develop strategies to address both policy and operations, and to inform future evaluative work. Our EA design was guided by Scriven's (2007) discussion on checklists, an approach that implies a comprehensive approach to understanding the phenomenon under study. Practical considerations for implementation, including review of the MEL standards already applied by the Fund and standards not necessarily applied but identified as being of critical importance to credible MEL and/or evaluability, were based on the work of Davies (2013) and Davies and Payne (2015). The design was conceptualized in two phases, assessment and verification. The first phase, which we detail below, was concluded by July 2020, with the second phase of field verification planned for 2021.

Through a process of literature review, brainstorming, and multiple consultations, we constructed our assessment framework with seven categories, each associated to a key component of MEL: (a) project logic, (b) MEL plan and resources, (c) data and methods, (d) inclusion, (e) portfolio alignment, (f) long(er)-term evaluability, and (g) evaluability in practice. The implication for these multiple categories as a checklist is that, taken together, they make up the totality of what the AF-TERG identified as evaluable CCA projects. Within each category, we established a series of assessment criteria. The categories and criteria were a combination of (a) MEL standards already applied by the Adaptation Fund, and (b) standards not necessarily applied by the Fund but identified by the AF-TERG assessment team as standards and approaches of critical importance to credible MEL and/or evaluability, particularly for CCA-focused projects. Table 1 presents the categories, their criteria, and brief discussion of their relevance.

Process for Implementation

As part of the first phase, we reviewed the original proposal documentation for all 100 board-approved projects.[1] Proposal documentation was the main documentation for analysis in the sense of (a) making a consistent assessment of the evaluability of the portfolio at the project onset and of the structures already available and considered for evaluation in the project design, and (b) informing the MEL and proposal review processes in terms of evaluability in the Fund. We also analyzed available project inception reports, project performance reports, midterm evaluation reports, and terminal evaluations to develop an initial understanding of the evaluability in practice during project implementation.

Consistent with the intentions that the objectives of the assessment were clearly laid out and that the team would carry out a second phase of field verification for evaluability in practice, the first phase was exclusively desk based. The nature of the phased approach restricted the depth of analysis: Where gaps or uncertainties occurred within individual projects, we did not seek clarification through means such as follow-up interviews with implementing entities or project teams. However, limiting the work exclusively to a desk review allowed for broad coverage and

enabled analysis of the entire portfolio, while also shedding light on the Fund's existing review process. The second phase was intended to fill any information gaps identified during the first phase.

Based on a detailed set of guidance, reviewers assessed the evaluability of all 100 projects against each of the assessment criteria, providing narrative justification for assessments and allocating ratings where relevant (where logical, we applied rating scales to individual criteria). Table 2 provides an example of a criterion's rating system. All project-level assessments were recorded within a spreadsheet-based assessment tool that, in turn, supported portfolio-level analyses and in the longer term will serve as a transparent, accessible database of all the EA's underlying data.

In addition to the seven assessment categories and their underlying criteria, the tool also recorded descriptive detail for every project, such as project status, budget, country, and sector(s), for example. We subsequently used this detail to support cross-portfolio analyses, enabling the assessment to identify patterns or trends in evaluability according to criteria such as a project's age, the kind of implementing entity that leads the project, the context within which the project works, and so forth.

Analysis

The results of the portfolio evaluability assessment illustrate the common tensions already encountered in CCA monitoring and evaluation (M&E). More precisely, they concur with the three challenges identified by Fisher et al. (2015): assessing attribution, creating baselines, and monitoring climate change activities over longer time horizons. The assessment of various criteria allows us to pinpoint specific elements in the project design and implementation that contribute to these tensions. Finally, the results show the importance of undertaking evaluability assessments to help build stronger M&E in the field of CCA.

---

[1] The projects were at different points in the project cycle: approved but implementation had not started, under implementation, or completed at the time of assessment.

Table 1 The Adaptation Fund's evaluability assessment framework

Category	Project logic		
Criteria	Quality of project logic	Quality of evidence base	Clarity of project additionality

Project logic is integral to project evaluability because it describes the basis and justification for an intervention, the starting conditions and assumptions, the expected results, and the anticipated means or pathways through which the project will deliver those results. CCA-focused projects should also be based on a logic model that reflects—and identifies linkages between—both the human and natural systems being affected.

Category	MEL plan and resources		
Criteria	Quality of M&E plan	Quality of MEL resources	Quality of approach to learning

For a project to be evaluable, there must be a feasible approach to identifying the project's likely contribution to outcomes, including a clear direction as to what aspects of the project will be measured and how, a description of who will be involved (roles and responsibilities), the resources that will be available to deliver the proposed approach, and intended use for adaptive management.

Category	Data and methods			
Criteria	Quality of results statements	Quality of indicators or other measures	Quality of baselines	Quality of data

Project evaluability is often dependent on the relevance and quality of data generated during project implementation, with the gathered data typically (but not necessarily exclusively) defined by a project's indicators. Consequently, an assessment of project evaluability needs to consider the quality of project indicators, the methods through which progress against those indicators is assessed, and the quality of baseline against which project performance and results will be assessed.

Category	Inclusion		
Criteria	Quality of project logic	Quality of evidence base	Clarity of project additionality
	Quality of data disaggregation: gender		Quality of data disaggregation: other

The assessment considered the extent to which MEL activities involved and represented all of a project's interest groups, whether individuals, communities, sub-national institutions, or country governments. This includes assessment of the participation and representation of women, youth, and socially excluded and vulnerable groups. Recognition is growing that an intervention's approach to MEL—and the data that MEL generates—can only be credible if it is based on the consent and participation of the people and institutions that the intervention aims to support. This is particularly the case for CCA, where interventions are frequently focused on building individual, community, and institutional resilience.

Category	Portfolio alignment	
Criteria	Depth of alignment	Quality of monitoring and reporting against portfolio results

Data and learning generated through each project's MEL activity should also contribute to and strengthen the evaluability of the Adaptation Fund's overall portfolio. Project-level contributions to portfolio evaluability will be most readily achieved where clear alignment exists between a project's expected results and the Adaptation Fund's Strategic Results Framework.

(continued)

Table 1 (continued)

Category	Long(er)-term evaluability	
Criteria	Potential for postcompletion evaluation	

The kind of results and changes targeted by CCA interventions are mostly long(er)-term in nature, only identifiable and measurable well after a project has been implemented. But for an evaluation to be undertaken 5 years after a project's funding period, for example, resources must be available and—more important—sufficient foundations must be in place for an evaluation to be even plausible. The prospects for long(er)-term evaluability could be improved by, for example, a project having indicators that can be accurately measured over longer time horizons, a project logic model that extends beyond implementation, and the likely availability of institutions and individuals that participated in the original intervention.

Category	Evaluability in practice	
Criteria	Changes to MEL approach documented and justified	Quality of evaluability in practice

Documentation such as midterm and terminal evaluations can reveal how well a project's MEL strategy has performed in practice and, by extension, can provide insight into the actual strengths and challenges around project evaluability. Consequently, analyzing this documentation can improve understanding of the practical limitations (and opportunities) that projects face when it comes to ensuring interventions are evaluable. Comparing these documents to a project's original MEL design can also build understanding as to whether and how MEL strategies typically change. Once a project is under implementation, adjustments to MEL strategy are often required due to situations such as changes in plans, unanticipated monitoring challenges, or unforeseen resource constraints. Under such circumstances, project evaluability can be maintained and improved if changes to MEL strategy are well documented within progress reports.

Table 2 Example of a criterion's rating system

Criterion	Quality of project logic					
Guiding question	Does the project logic provide a sufficient basis against which performance and results can be evaluated?					
Scale	Very good quality	Good quality	Fair quality	Weak quality	Very weak quality	Logic not documented

Guidance	A project's logic is considered very good quality if it includes all of the following:
	    Explicit logic model, whether a theory of change, logical framework, or other explicit model
	    Clear description of the project's expected results
	    Clear description of the pathways and processes through which the project will deliver expected results
	    Clear description of the assumptions underpinning the project's logic
	    Clear description of the external influences that may affect project delivery
	    Intervention logic expresses a clear contribution to CCA
	    Project logic reflects both human and natural systems
	    Environmental and social risks are reflected in project logic
	The scale should then be applied according to the degree of alignment with the above elements and the clarity of discussion around project logic, with—at the lower end of the scale—very weak quality meeting only one of the above elements and/or logic is unclear.					

Logic and Additionality of Adaptation Projects

Relevance to Evaluability

To be evaluated effectively, a project must have clearly articulated logic. Clarity on additionality also strengthens evaluability by improving the potential for identifying (isolating) a project's influence and results. Within the project logic evaluability category, the assessment looked at the criterion clarity of project additionality (see Table 3).

Adaptation-Specific Evaluability Considerations

CCA projects invariably work within complex natural systems and target results that focus on human interactions within those systems. Those

Table 3 Guidance for assessing clarity of project additionality

Category	Project logic	
Criterion	Clarity of project additionality	Guiding question for assessors
		To what extent is it clear that the project's work is additional to business-as-usual and/or other initiatives that have been or are being delivered?

complex systems are influenced by multiple factors, many (if not most) of which will be completely outside the control of any given intervention. Similarly, to *attribute* higher-level results to a single project is often unrealistic or erroneous. The additionality of a CCA project is more often described in terms of the *contribution* to higher level results and the project's impact pathways toward those results. From an evaluability perspective, this necessitates a clear description not just of a project's additional contribution, but also the numerous other factors that will influence higher level results, and a project's interactions with those other factors.

Findings

The assessment found that clarity of project additionality was greatly supported by a project proposal template that required Fund applicants to provide detail on related initiatives and describe whether and how their proposed work duplicated those other initiatives. This requirement was primarily motivated by the Adaptation Fund's desire to ensure that their grants did not duplicate other funding sources. However, this simple, standard request also helped evaluability: It obliged applicants to think through and articulate how their project related to broader work on adaptation and, in doing so, helped to ensure that project proposals invariably contained clear descriptions of how interventions were (or were not) additional to external initiatives.

However, from an evaluability perspective, this created a common tension. Projects understandably and rightly sought to gain efficiencies and synergies with related interventions, often to the extent that the clarity of additionality was

reduced. Where the assessment identified a lack of clarity around an intervention's additionality, this was often because of programmatic strengths and efforts to contribute to a broader agenda. But when resources and efforts were combined across projects, perceiving how any eventual results could be directly attributable to a specific intervention or individual funding source became more difficult; at best, only a project's contribution to results would be evaluable.

As noted above, this scenario of multiple actors and influences working within complex systems is common (even standard) for CCA projects. The evaluability of such projects will benefit from clear, honest descriptions of how a project is distinct and—just as important—how a project intersects or even duplicates other work. Given that interdependencies between CCA projects are common, evaluability may be reliant on description and analysis of not just the main project's own results chains, but also the results chains of related initiatives. Furthermore, projects—and the funders that support those projects—may need to accept that the only measurement ever possible for a project may be its contribution to high-level results.

Evidence Base and Baselines: Natural vs. Human Systems

Relevance to Evaluability

When it comes to a project's evidence base, evaluability can be strengthened where there are clear lines and logical linkages between prior experiences and a proposed intervention. The evidence base and learning from previous work can also help to define how performance and results should be measured, including the design and setting of baselines. In turn, baselines are critical for evaluability: A project's progress can only be monitored, evaluated, and fully understood if comparisons can be made between a project's current position and a clearly described starting point that a baseline ideally establishes. The assessment looked at quality of evidence base and quality of baseline across two evaluability categories, project logic and data and methods (see Table 4).

Table 4 Guidance for assessing quality of evidence base and baseline

Category	Project logic	
Criterion	Quality of evidence base	**Guiding question for assessors:**
		Does the project logic reference and take into account data, evidence and learning from prior research, initiatives, and institutional experience?
Category	**Data and methods**	
Criterion	Quality of baseline	**Guiding question for assessors:**
		Does a baseline exist for all of the proposed project indicators/measures, or are clear plans in place for developing a baseline?

Adaptation-Specific Evaluability Considerations

Of particular relevance to CCA-focused work and to Adaptation Fund projects specifically, the assessment looked at how project evidence base and baselines took into account both natural and human (including institutional) systems. The conceptualization of—and separate emphasis on—natural and human systems is central to the Fund's definition (AF, 2017, p. 3) of an adaptation project:

> A concrete adaptation project/programme is defined as a set of activities aimed at addressing the adverse impacts of and risks posed by climate change. The activities shall aim at producing visible and tangible results on the ground by reducing vulnerability and increasing the adaptive capacity of human and natural systems to respond to the impacts of climate change, including climate variability.

A CCA project's evaluability will be partly determined by how well a project's evidence base and baseline describe the starting position of the targeted natural systems (e.g., forestry coverage, biodiversity, soil characteristics), the starting position of the targeted human systems (e.g., agricultural practices, economic incentives, government policy, institutions), and the current ways in which the two systems interact (e.g., agricultural practice reducing biodiversity, economic incentives accelerating deforestation).

Findings

With regard to additionality, the Adaptation Fund's proposal templates ensured that projects, in the main, presented clear, well-referenced descriptions of the preintervention evidence base. This requirement was principally used by Fund applicants to justify the case for a project, but it also served to strengthen evaluability: The evidence base and site-specific context inherently provided a basis against which project progress and results could be evaluated.

However, the assessment also found that the depth and quality of the preintervention evidence base varied according to whether evidence related to natural systems or to human systems. The use of detailed climatic and environmental baseline data (i.e., natural systems) was especially strong within most proposals. Conversely, proposals devoted less attention to describing and evidencing the nonenvironmental context within which an intervention was to operate, including the human and institutional aspects of a project. Beginning at the design/proposal stage, evaluability tended to be stronger for work relating to natural systems, less so for work relating to human systems (and, by extension, the linkages and interactions between natural and human systems).

While considering how the preintervention evidence base and baselines reflected human and natural systems, the evaluability assessment also reviewed the extent to which—once projects were under implementation—monitoring approaches (including results frameworks) measured change across both systems, and the interdependencies between them. Although the majority of projects did support some degree of measurement of both human and natural systems, direct monitoring approaches were very heavily geared toward only measuring change within human systems (e.g., agricultural infrastructure, institutional and individual capacities, legislation). Indeed, many projects that measured some aspect of natural systems did so through only one relatively high-level indicator (e.g., area of land restored). Moreover, only a handful of projects had results frameworks in place that would be capable of measuring both systems and the interdependencies between them. The comparatively strong baseline understanding of natural systems was often not being expanded (or even followed up) during implementation. Perhaps this is because accessing historical natural data (i.e., climate, environment, biodiversity) is comparatively easy, so a strong evidence base can be developed. Conversely, accessing historical human data could be more difficult (and such data may not be available), thus developing a detailed pre-implementation evidence base could be quite resource intensive.

Aside from the differing treatments of human and natural systems, the findings also illustrate that ensuring project evaluability is an ongoing process. Evaluability can't be achieved through project design alone: Even where evaluability appears strong at the pre-implementation stage, that level of evaluability has to be maintained throughout project delivery by, for example, designing and applying processes capable of gathering the breadth and quality of data necessary to measure results across both human and natural systems.

Resources Allocated to MEL: Direct vs. Indirect

Relevance to Evaluability

Project evaluability is partly dependent on sufficient institutional and/or financial resources being allocated toward MEL activity (with the definition of what comprises "sufficient" resources dependent on the nature of each project and its MEL strategy; see Table 5). To support the assessment of resource adequacy, we identified the level of financial resources allocated toward MEL for every project, recording two figures:

- Direct resources: Money allocated explicitly toward MEL activities.
- Indirect resources: Money allocated towards activities that, although not itemized or categorized as MEL, are likely to be of direct, substantive benefit to MEL.

Adaptation-Specific Evaluability Considerations

Many CCA interventions include—and are sometimes focused exclusively on—activity that

Table 5 Guidance for assessing quality of MEL resources

Category	MEL plan and resources	
Criterion	Quality of MEL resources	Guiding question for assessors:
		Are financial and institutional resources for MEL explicitly defined, and are these sufficient to support delivery of the MEL strategy?

can be considered indirect MEL. Examples of such activities are development of new climate monitoring approaches, consolidation of historical environmental data, research activities, and capacity development on the monitoring and interpretation of data. Project evaluability can be strengthened where these indirect activities are formally linked with the intervention's MEL strategy, in turn ensuring that project MEL (and evaluability) benefits from the widest possible range of data and resources.

Findings

During assessment of this criterion, our most notable finding was that the proportion of indirect financial resources associated with MEL-relevant activity was far larger than the proportion of resources directly allocated to MEL. Frequently, indirect MEL activity was a core component of projects, such as in projects focused almost exclusively on developing sub-national climatic monitoring infrastructure. Many projects also were strengthening institutions, capacities, and infrastructure required for longer term (postproject) monitoring and measurement of CCA-relevant indicators and results.

However, such obviously MEL-relevant activities were not recognized or presented as such by projects. This disconnect could have been due to project partners not conceptualizing the activity as MEL, or the common perception of MEL as a purely accountability-focused exercise, rather than as a tool that can also support learning and knowledge generation. Regardless of the reasons for the disconnect, where projects did not make the link between their direct MEL work and those indirect activities with clear relevance and value for MEL, the data and learning generated through indirect activities may have provided only a limited contribution or none at all to the project's formal MEL. This could have reduced the potential and strength of a project's evaluability.

Another notable finding was that the data collected through indirect MEL activities tended to be more focused on natural than on human systems. While some projects were actually gathering extensive data and/or developing infrastructure for the measurement of natural systems, this data was not being used to bridge the above-noted gap whereby formal, direct project MEL placed more (and sometimes exclusive) emphasis on the monitoring of human systems. Thinking beyond individual projects, we also see a risk that data could be overlooked by future research and meta-evaluations if not identified as being MEL relevant. In turn, this could reduce the potential contribution and value of data to longer-term learning and knowledge generation.

From an evaluability perspective, these findings highlight the importance of looking beyond a project's self-identified direct MEL activity, particularly where those projects are working to develop monitoring capacities and infrastructure. Moreover, the findings also demonstrate how the actual process of evaluability assessment can help to identify and uncover opportunities for strengthening not just evaluability, but a project's MEL more broadly.

Potential for Postcompletion Evaluation

Relevance to Evaluability

Although still a relatively underdeveloped approach, postcompletion evaluations are increasingly deployed as a tool to measure the longer term impact of interventions. However, for an evaluation to be undertaken several years after a project's closure, adequate resources must be available and, more important, sufficient foundations must be in place for an evaluation to be even plausible (see Table 6). The prospects for long(er)-term evaluability could be improved by, for example, a project having indicators that can be accurately measured over longer time horizons, a project logic model that extends beyond

Table 6 Guidance for assessing potential for postcompletion MEL

Category	Long(er)-term evaluability	
Criterion	Potential for postcompletion MEL	Guiding question for assessors:
		Are the project's logic, monitoring approach, and postcompletion institutional arrangements sufficiently long(er)-term in nature to support postcompletion MEL?

implementation, and the likely longer term availability of institutions and individuals that participated in the original intervention.

Adaptation-Specific Evaluability Considerations

The kind of results and changes targeted by CCA interventions are mostly longer term in nature, only identifiable and measurable well after a project has been implemented. Consequently, the justification for postcompletion evaluation is particularly strong in the adaptation arena. Therefore, considering the extent to which foundations are in place for longer-term evaluability—and postcompletion evaluation—is important for any assessment of an adaptation project's evaluability.

Findings

The assessment first sought to identify whether projects had formal plans in place for postcompletion evaluation. Only 3% of projects confirmed formal plans for postcompletion evaluations, with even the concept of postcompletion evaluation rarely being mentioned across the portfolio. Plans for postcompletion monitoring were far more prevalent, with several projects including components whose entire purpose was to establish organizational structures, systems, and capacities for longer-term (indefinite) monitoring of factors such as local meteorological data, water levels, or land use. However, the funding and institutional arrangements for these longer term systems were infrequently specified. All of these longer term monitoring systems focused on only one aspect of a project; we identified no examples of projects that planned broader longer term monitoring through approaches such as continued, wide-ranging monitoring against their original results framework.

The limited examples of longer term MEL are unsurprising; even within the MEL sector the concept of longer term MEL is still relatively new. Consequently, the assessment also considered the extent to which projects had in place certain foundations that at least strengthened the potential for postcompletion MEL, such as logic models that established pathways and results beyond the project's lifetime, project indicators that could plausibly be measured 5 years after project completion, indicators based on preexisting national data sets that are likely to be maintained over the longer term, and clear descriptions of postcompletion institutional ownership of the project's outcomes. Some instances of projects had promising building blocks in place, with several projects benefiting from clear exit strategies and descriptions of postcompletion ownership, whether institutional, community, or individual. A handful of projects also aligned their results frameworks with preexisting national results frameworks and data sets, with the explicit rationale being to support longer term monitoring of project results. However, these were exceptions; the broader portfolio was characterized by generally weak potential for postcompletion MEL.

Again, these findings are not surprising given the universally low application of longer term MEL and postcompletion evaluation. However, the process of evaluability assessment can help pinpoint gaps and opportunities to at least strengthen the foundations for any potential postcompletion evaluation.

Reflections on the EA Tool Development and Implementation

We undertook a reflective and consultative process to develop a tool specific to CCA programming and its particular challenges, aiming to fulfill our objective of an impartial and accurate assessment of the Adaptation Fund portfolio's evaluability. The tool was informed by a set of principles taking into account the Fund's current and historical approaches to evaluation and an understanding of our evaluand, CCA programs. We adopted the OECD definition of evaluability and set about a process akin to both Wholey's (1979) eight steps and Trevisan and Walser's (2015) four-stage approach: defining the evaluand, literature review, framework development, consultation and piloting, implementation, and presentation of results.

The process of our EA tool development was consultative, involving peer review by an evaluation consultant and by AF-TERG's advisory board of evaluation and climate change programming experts. We developed a 5-point scale with criteria of equal weighting. We piloted the tool, reflected on its results, further revised, and piloted again. Two colleagues undertook the review, aligning their interpretation of criteria and scoring. Although we did not determine an inter-rater reliability score, we systematically compared assessments to calibrate judgment.

The results of the EA identified challenges that are well documented about CCA programming (Fisher et al., 2015), as discussed above, and resonated with Adaptation Fund stakeholders, serving to promote discussion and decision making about the Fund's evaluation policy and processes for funding and partnership. A consideration for this first phase of the EA was the sole reliance on written documentation and the lack of opportunity to engage on site with partner agencies. In the second phase of our two-phased approach of assessment and verification, we will continue to collect data on site with partners and reflect on the tool's merit and utility as part of our effort to engage in formative meta-evaluation. With this positive EA experience and as we find additional use for the tool going forward, we may also meta-evaluate following field verification, using relevant criteria found in the Program Evaluation Standards (Yarbrough et al., 2011).

Conclusions

The framework that was developed and applied for the Adaptation Fund EA proved to be a useful tool for understanding evaluability strengths and gaps both within individual projects and across the Fund's whole portfolio. Moreover, the framework supported the identification of CCA-specific evaluability opportunities and challenges. Results of the assessment supported the three challenges related to climate change activities identified by Fisher et al. (2015): assessing attribution, creating baselines, and monitoring over longer time horizons.

The assessment confirmed the difficulty in identifying the additionality of CCA programs because of the complexity of such interventions. Clarifying the project logic through an EA helps conceptualize the project results in terms of contribution or attribution in a longer term perspective.

The study also highlighted the difficulty in setting baselines to measure the results of CCA in a comprehensive way, which would encompass impacts on both natural and human systems. It showed that, beyond the usual suspect of unavailability of nature-focused and human-focused data, the failure to integrate this data from both systems in the MEL system was the source of gaps in knowledge during project implementation.

Finally, the study highlighted the need to plan for and allocate resources for longer term M&E to be able to measure the delayed impacts of CCA interventions. The EA process can help pinpoint gaps and opportunities to at least strengthen the foundations for any potential postcompletion evaluation. The EA was incredibly useful in assessing the extent of the structures, processes, and resources in place capable of supporting credible M&E. Although we conducted this assessment retroactively, prospective use of the tool could inform and ensure that projects have the evaluability structures necessary at project design. The study of evaluability in practice, which verifies whether the evaluability is maintained throughout project delivery, could also be complemented with field verification to inform gaps in the processes related to project implementation.

The framework also offered a structured process for systematically thinking through what constitutes sound MEL more broadly. Indeed, a key conclusion of the assessment was that the strength and quality of a project's overall approach to MEL is a major determinant of a project's evaluability: If a project gets its MEL "right," the project is highly likely to also have strong evaluability. This suggests that, where improved evaluability is sought, focusing efforts specifically on strengthening evaluability may not be efficient or even necessary; instead, improvements to overall MEL strategy and processes should inherently deliver improvements to the quality of evaluability.

References

Adaptation Fund. (2011a). *Guidelines for project/programme final evaluations*. Author. https://www.adaptation-fund.org/document/guidelines-for-projectprogramme-final-evaluations/

Adaptation Fund. (2011b). *Results framework and baseline guidance – Project-level*. Author. https://www.adaptation-fund.org/wp-content/uploads/2015/01/Results%20Framework%20and%20Baseline%20Guidance%20final%20compressed.pdf

Adaptation Fund. (2012). *Evaluation framework*. Author. https://www.adaptation-fund.org/document/evaluation-framework-4/

Adaptation Fund. (2017). *Operational policies and guidelines for parties to access resources from the adaptation fund*. Author. https://www.adaptation-fund.org/document/operational-policies-guidelines-parties-access-resources-adaptation-fund/

Bours, D., McGinn, C., & Pringle, P. (2014). *Monitoring & evaluation for climate change adaptation and resilience: A synthesis of tools, frameworks and approaches* (2nd ed.). SEA Change Community of Practice and UKCIP. https://ukcip.ouce.ox.ac.uk/wp-content/PDFs/SEA-Change-UKCIP-MandE-review-2nd-edition.pdf

Bours, D., McGinn, C., & Pringle, P. (2015). *Monitoring and evaluation of climate change adaptation: A review of the landscape*. Wiley.

Davies, R. (2013). *Planning evaluability assessments: A synthesis of the literature with recommendations* (Working Paper 40). Department for International Development London, United Kingdom. https://www.gov.uk/dfid-research-outputs/planning-evaluability-assessments-a-synthesis-of-the-literature-with-recommendations-dfid-working-paper-40

Davies, R., & Payne, L. (2015). Evaluability assessments: Reflections on a review of the literature. *Evaluation, 21*(2), 216–231. https://doi.org/10.1177/1356389015577465

Fiala, N., Puri, J., & Mwandri, P. (2019). *Becoming bigger, better, smarter: A summary of the evaluability of Green Climate Fund proposals*. Green Climate Fund

(GCF) Independent Evaluation Unit (IEU). https://ieu.greenclimate.fund/documents/977793/985626/Working_Paper__Becoming_bigger__better__smarter_-_A_summary_of_the_evaluability_of_GCF_proposals.pdf

Fisher, S., Dinshaw, A., McGray, H., Rai, N., & Schaar, J. (2015). Evaluating climate change adaptation: Learning from methods in international development. In D. Bours, C. McGinn, & P. Pringle (Eds.), *Monitoring and evaluation of climate change adaptation: A review of the landscape* (pp. 13–35). Wiley.

Organisation for Economic Co-operation and Development, Development Assistance Committee. (2002). *Glossary of key terms in evaluation and results based management.* Author. https://www.oecd.org/dac/2754804.pdf

Rutman, L. (1980). *Planning useful evaluations: Evaluability assessment.* Sage.

Scriven, M. l. (1991). *Evaluation thesaurus* (4th ed.). Sage.

Scriven, M. l. (2007). *The logic and methodology of checklists* [unpublished manuscript]. Western Michigan University.

Smith, M. (1989). *Evaluability assessment: A practical approach.* Kluwer Academic.

Stufflebeam, D., & Coryn, C. (2014). *Evaluation theory, models, and applications* (2nd ed.). Jossey-Bass.

Trevisan, M. S. (2007). Evaluability assessment from 1986 to 2006. *American Journal of Evaluation, 28*(3), 290–303. https://doi.org/10.1177/1098214007304589

Trevisan, M. S., & Walser, T. M. (2015). Evaluability assessment: Improving evaluation quality and use. SAGE.

Wholey, J. S. (1979). *Evaluation: Promise and performance..* Urban Institute.

Wholey, J., Hatry, H. P., & Newcomer, K. E. (2010). *Handbook of practical program evaluation* (3rd ed.). Jossey-Bass.

Wilby, R. L., & Dessai, S. (2010). Robust adaptation to climate change. *Weather, 65*(7), 180–185. https://doi.org/10.1002/wea.543

Uitto, J. I., van den Berg, R. D., & Puri, J. (2017). *Evaluating climate change action for sustainable development.* Springer Open.

Yarbrough, D. B., Shulha, L. M., Hopson, R. K., & Caruthers, F. A. (2011). *The program evaluation standards: A guide for evaluators and evaluation users* (3rd ed). Sage.

Evaluating Transformational Adaptation in Smallholder Farming: Insights from an Evidence Review

Laura Silici, Jerry Knox, Andy Rowe, and Suppiramaniam Nanthikesan

Abstract

The literature on smallholder farming and climate change adaptation (CCA) has predominantly investigated the barriers to and determinants of farmer uptake of adaptation interventions. Although useful, this evidence fails to highlight the changes or persistence of adaptation responses over time. Studies usually adopt a narrow focus on incremental actions that provide limited insights into transformative adaptation pathways and how fundamental shifts in policy can address the root causes of vulnerability across different sectors and dimensions. Drawing on an evidence synthesis commissioned by the International Fund for Agricultural Development's Independent Office of Evaluation, this chapter outlines how lessons from CCA interventions can be transferred via three learning domains that are essential for transformational change:

scaling-up (in its multiple forms), knowledge management, and the human-environment nexus. We discuss the implications of our findings on monitoring, evaluation, and learning, highlighting the challenges that evaluators may face in capturing (a) the persistence or durability of transformational pathways, (b) the complexity of "super-wicked" problems, and (c) the relevance of context-dependent dynamics, within a landscape setting. We also address the contribution of evidence reviews to contemporary debates around development policy linked to climate change and agriculture, and the implications and value of such reviews to provide independent scientific rigor and robustness to conventional programmatic evaluations.

L. Silici (✉)
Independent Consultant, Rome, Italy

J. Knox
Water Science Institute at Cranfield University, Cranfield, UK
e-mail: j.knox@cranfield.ac.uk

A. Rowe
ARCeconomics, Maple Bay, BC, Canada

S. Nanthikesan
Independent Office of Evaluation of the International Fund for Agricultural Development, Rome, Italy

Introduction

Smallholder agriculture represents 75% of the world's farms (Lowder et al., 2016) and 80% of the source of food consumed in the developing world (International Fund for Agricultural Development [IFAD], 2020). Yet smallholder farmers constitute more than half of the world's undernourished people; they inhabit some of the most vulnerable and marginal landscapes, and many lack secure land tenure and water rights. These factors further exacerbate their exposure to

© The Author(s) 2022
J. I. Uitto, G. Batra (eds.), *Transformational Change for People and the Planet*, Sustainable Development Goals Series, https://doi.org/10.1007/978-3-030-78853-7_13

climate change (IFAD, 2020), alongside other sources of vulnerability such as population growth and land fragmentation. Indeed, smallholder agriculture is disproportionately threatened by unpredictable weather patterns, with the impacts of extreme events including floods, droughts, and heat waves having profound implications on both food security and poverty reduction, especially for rural communities dependent on rainfed agriculture (United Nations Environment Programme [UNEP], 2018).

Although extensive information is readily available on the projected agricultural impacts of climate change and on adaptation measures that could help minimize those impacts, assessments that specifically address the vulnerability of smallholder farmers to climate change are very limited (Donatti et al., 2019). Thus, there is a need to better understand how smallholder farmers perceive the risks of climate change, the factors that influence their decisions to adapt, and what adaptation strategies have been practiced and why (Belay et al., 2017).

This chapter draws on the findings of an evidence synthesis commissioned by the IFAD Independent Office of Evaluation (IOE) to inform a thematic evaluation of IFAD's support for smallholder farmers' adaptation to climate change. Evidence reviews aim to provide a transparent and robust assessment of what is known (and not known) in the literature regarding a specific topic, by adopting a systematic methodology to search, screen, and critically appraise a database of scholarly articles, including peer-reviewed and grey literature.

Executing the review involved several discrete stages following the well-established guidelines of the Collaboration for Environmental Evidence (CEE; Pullin et al., 2018), ensuring that the outcomes met quality standards with respect to robustness, transparency, and repeatability. The exercise provided a valuable narrative synthesis of the evidence produced over the last 10+ years on smallholder adaptation to climate change, focusing on how lessons from climate change adaptation (CCA) in smallholder farming could be transferred within three learning domains: (a) scaling-up, (b) knowledge management, and (c) the nexus interactions between the human and ecosystems. We briefly summarize the key findings from the review and discuss their policy and practice implications with respect to monitoring, evaluation, and learning (MEL) under the framework of transformational change, with a focus on the challenges that MEL faces. Specifically, these included addressing (a) the persistence, or durability, of transformational pathways; (b) the complexity of "super-wicked" problems (Levin et al., 2012) that lead to trade-offs across different intervention dimensions and respective goals; and (c) the relevance of context-dependent dynamics, within a landscape setting.

In smallholder farming, transformational adaptation entails adaptive strategies that address the different root causes of vulnerability, internalize aspects of gender, racial, and intergenerational equity (i.e., social and environmental aspects), and enable changes in the wider food system, including solutions at any point in the value chain and diversification into off-farm sectors. Transformative solutions exert influence beyond the boundaries of specific interventions, possibly reaching different sectors (outside agriculture) and dimensions such as gender roles and behavioral changes. More generally, they imply a conceptual shift toward transformative changes beyond incremental adaptive actions. Indeed, adaptation is not a discrete measure, but rather a dynamic and iterative process due to a changing context where the climate is one of many risks (Vermeulen et al., 2015; Wise et al., 2014). The pace of climate variability is also increasing alongside other sources of vulnerability such as population growth and land fragmentation and the emergence of alternative responses (off-farm employment, markets).

Successful implementation and expansion of transformation pathways in smallholder agriculture are underpinned by three mutually interacting factors. First, CCA solutions need to integrate the multiple nexuses between the socioeconomic sphere and the surrounding ecosystem; second, transformation entails scaling-up, in the intermediate or longer term; and third, an effective and usable climate adaptation knowledge base is needed to understand the nexus better in order to promote scaling-up.

Methodology

This review was originally conceived as a rapid evidence assessment (REA), an approach designed to deliver a relatively quick synthesis of evidence of existing research on a defined topic. The REA methodology involves defining an explicit research question(s) and following a strict procedure that is not dissimilar to the development of a protocol for conducting a systematic review. An REA is rigorous and has an approach that is tractable and transparent but generally lacks the depth of analysis often associated with a full systematic review. However, REAs are widely used for informing policy decision making. In this study, the overarching question that guided the review was: What interventions have been successful in building smallholders' adaptive capacity and responses to climate change and how have these been effectively transferred as learning outcomes in the three key dimensions of scaling-up, knowledge management, and ecosystem-human interactions? Our review followed the approach developed by CEE (Pullin et al., 2018) of first developing a protocol that guides the execution and describes how the literature searches will be conducted, including the bibliographic sources and search strings, the data extraction methods, quality assessment, and evidence synthesis. The review was constrained by a set of inclusion and exclusion criteria and the research question defined above.

We used a range of bibliographic databases including Scopus, Web of Science, and Science Direct, and web search engines (Google Scholar) and institutional websites for various international development organizations, think tanks, and research institutes. Following trial searches, we selected the search string "smallholder, AND agric* AND climat* AND change, AND adapt*". Collectively, the searches returned 806 sources of evidence. After a thorough screening of the titles and abstracts using a set of inclusion criteria, we selected 132 sources for full-text reading; of these, we then reviewed and synthesized 91. For a detailed explanation of the review methodology with information on the resources used, see Silici et al. (in press).

Key Messages Emerging from the Evidence Synthesis

Scaling Up Transformative Adaptation Pathways

Scaling up a process or an initiative may take multiple forms and does not imply only bringing an intervention to scale (to more people, larger areas) or adapting it to similar conditions in different locations (horizontal scaling-up). Scaling up a project can also relate to moving it forward into a more developed, complex phase, possibly including new components, configurations, and stakeholders (diagonal scaling-up). It can also consist of mainstreaming a certain approach into policy, leveraging and catalyzing policy and/or institutional change (vertical scaling-up; Neufeldt et al., 2015). In the latter case, scaling-up processes imply more than just physical or technical dimensions, but also social scaling-up (increasing social inclusiveness) and conceptual scaling-up in terms of moving beyond participation to embedding empowerment in the entire development process (Binswanger-Mkhize et al., 2009). With respect to adaptation, social and conceptual dimensions of scaling-up are particularly important to pursue genuinely transformative pathways.

The academic literature on CCA and smallholder farming has predominantly investigated the determinants of (or barriers to) farmer adaptation at a given point in time and for a defined location (Etana et al., 2020). In our review, about one fourth of the publications were quantitative studies addressing barriers and determinants of farmers' adoption of CCA. This evidence base provided valuable insights on the necessary preconditions and the enabling factors (or capacities) that determine smallholders' decisions to act in the face of climate change. Among the former are factors that underpin a farmer's decision to adapt: knowledge regarding climate change, farming and natural resource management, weather information services, and past experience of extreme climate events; collectively, these contribute toward building farmers' consciousness of the need to adapt. However, awareness alone does not trigger change. A series of enabling factors of social, economic, environmental, and institutional natures ultimately determine a smallholder's actual capacity and motivation to take action.

Planned adaptation strategies should thus ensure that the underlying determinants of adoption, such as access to knowledge and information, exist alongside other relevant enabling factors, including endowment with productive assets, human capital (education and skills), and institutional support (e.g., farmer groups and collective action; Atsiaya et al., 2019; Bedeke et al., 2019). Profiling different vulnerability categories is also extremely important because smallholders adopt different strategies according to their degree of sensitivity and adaptive capacity beyond their actual exposure to climate change. More well-off farmers, for instance, rely on a stronger economic buffer and usually have stronger connections with formal and informal institutions, which in turn affect their risk perception and adaptive decisions. Policy and institutions must build on this understanding to meet the particular needs and priorities for distinct smallholder groups, with the most notable disparities found in income and gender (Ruijs et al., 2011).

While this branch of research is very useful to understand farmers' decision making and how to support capacity building, it fails to show the changes or persistence of adaptation behavior over time (Etana et al., 2020). Moreover, because of a largely quantitative, linear approach and a predominant interest in mainstreaming technological solutions, these studies have maintained a rather narrow focus on incremental actions and adjustment approaches within the agricultural sector. The outcome is a limited perspective on agriculture-related solutions and a failure to detect innovative adaptation practices that incorporate local wisdom and creativity.

Other studies have adopted a more dynamic, longer term perspective in describing adaptation pathways. Here the concept of a pathway focuses more on an iterative process of decision making, rather than on the outcome, emphasizing the adaptive nature of the decision process itself in the face of high uncertainty and intertemporal complexity, rather than the achievement of a given objective (Wise et al., 2014). Some of the evidence also stresses the need to address the social roots of vulnerability and the necessity for political–economic change to achieve "transformative adaptation" (Bassett & Fogelman, 2013).

Several key points emerged regarding smallholder farming and transformational adaptation patterns. First, evidence of path dependence corroborates the need to provide financial and institutional support to overcome the behavioral and economic barriers of less endowed households, who could otherwise become locked in vicious cycles of inaction, thereby undermining overall community ability to adapt over time (Etana et al., 2020). In fact, we found that the sustainability of outcomes is better achieved when their distribution is more equitable and inclusive; this implies that all socioeconomic groups and different types of smallholders can respond to risk and vulnerability according to their adaptive capacity. In terms of planning, the review confirmed that no one blueprint or strategy works across all contexts; instead, adaptation should be perceived as a continuum of approaches, ranging from activities that aim to address the different drivers of vulnerability to measures explicitly targeting climate change impacts, including not only farming but also water and soil conservation and off-farm diversification. Planned responses should follow iterative processes, should set out all possible trade-offs (economic, temporal) between alternative strategies, and should be open to assess and envision all possible futures (Vermeulen et al., 2015, 2018). For example, the most viable pathway for some farmers may be to exit agriculture altogether, which itself requires careful management and planning of consequent rural transitions (Stringer et al., 2020).

Although identifying generic solutions is not possible, sustained adaptation and scaling-up do have some recurrent features. These include the need for integrated, multisector interventions; adopting participatory approaches in planning, implementation, and dissemination; and fostering knowledge exchange, peer learning and co-creation of knowledge. Participation and ownership of adaptive solutions are extremely important to ensure equitable scaling-up processes, because inter- and intra-household differences mean individuals react differently to multiple stressors. Institutional aspects are also critically important features in creating successful adaptation pathways. Sustainable adaptation only happens if farmers and their community members have the capacity to organize themselves. Institutional learning, or institutional capacity building, doesn't necessarily entail formalized structures (although linking with formal organizations does confer some advantages), but it is needed to foster adaptation through collective action and social learning. Institutional support, collective action, and participation are also critical in bridging local and scientific knowledge (Arouna & Akpa, 2019; Stringer et al., 2020; Vermeulen et al., 2018).

Transformative Knowledge Management

Access to knowledge constitutes one of the most important determinants of smallholders' decisions to respond to risk and is a critical element in building adaptive capacity. Creating the knowledge base that underpins its access and use, however, faces two challenges: Climate solutions have a shelf life, and they are very context specific. Therefore, the way knowledge regarding climate change and variability is produced, transferred, and exchanged is extremely relevant to securing equitable and inclusive scaling-up pathways (Popoola et al., 2020; Roncoli et al., 2010).

Building adaptive capacity does not rely solely on external, scientific knowledge. Knowledge embedded in farmers' experience and tradition is also critically important in raising awareness of climate risks and selecting appropriate responses. In fact, autonomous adaptation is based on farmer perception of climate change and variability. However, autonomous adaptations may be limited in scope and may be not entirely effective over the long run. They can even lead to maladaptation because farmers typically react to the actual threat experienced and do not consider likely future threats (Akinyemi, 2017; Makate, 2019). Knowledge based on local practices also may not be sufficient to prompt more transformative actions that take account of intergenerational

equity (Derbile et al., 2016), or to embark into more risky activities (Etana et al., 2020). Bridging local and external knowledge helps broaden farmers' knowledge base to include more forward-looking considerations (Makate, 2019; Shackleton et al., 2015). On the other hand, having a supply of scientific or external information does not imply that it is passed on, understood, and accepted; this depends on how it is communicated and, importantly, if it matches smallholders' needs.

School education, vocational training, and agricultural extension and advisory services (where available) are important enablers of adaptive capacity because farmers with more education and skills are able to understand, trust, and assimilate the information they receive and utilize it to support their needs (Guido et al., 2018; Henriksson et al., 2020). Joining formal groups such as producer organizations, cooperatives, or outgrower programs can allow farmers to gain access to technical skills, market information, weather forecasts, and other relevant information on credit, laws, and policies (Abass et al., 2018). These mechanisms for transferring knowledge help foster adaptive capacity and peer support, but they may not be effective for all smallholders or may even be exclusive of some groups if they have too narrow a technical focus and are conceived as a one-way, top-down transfer. The actual way in which knowledge is transferred to smallholders and how it then interacts with local knowledge is critically important.

Knowledge can be produced and circulated via more participatory ways. Partnerships and social learning (deep understanding and assimilation of concepts through social interaction) were identified in the literature as promising ways to link science, policy, and practice to tackle multiple and related challenges of agricultural development, food security, and climate change adaptation. The evidence showed that beyond empowering specific groups, inclusive learning processes (such as pro-poor research and gender transformative approaches) accelerate and improve development outcomes for everyone (Shaw & Kristjanson, 2014). Both formal groups and informal collective action, such as self-help neighborhood networks, work-sharing groups, and other community-based organizations, can foster synergies for social learning and capacity building. However, registered community-based organizations seem to be more significantly associated with adaptation (Khanal et al., 2019). Synergies between informal and formal structures also result in more effective organizations because they promote institutional learning, drive innovation and knowledge from external sources, and enlarge participation to more heterogeneous groups (Hulke & Revilla Diez, 2020).

Learning platforms based on participatory action research (PAR) that bring together different actors have been shown to be particularly effective in supporting adaptation strategies. They rely on social learning and collective action and help external actors (researchers, policymakers, and development institutions) assess community-based perspectives and existing adaptation options, using these as a starting point for further action (Asociacion Andes, 2016; Wekesa et al., 2017). This type of research can also help in situations where climate change is not the only externality but one of multiple sources of vulnerability, thereby providing a deeper understanding of the contextual barriers to adaptation. By fostering inclusion, social learning eases access to knowledge by women and people in other vulnerable categories who are not always targeted by institutional information sources in agriculture (Kerr et al., 2018; Mapfumo et al., 2013).

Several examples in the literature reported successful aspects of social learning through collaborative approaches that were broadly formalized into learning platforms or alliances. Farmer field schools (FFS) may also be considered as a relevant learning platform that integrates adaptation at different levels and scales, albeit the degree of participation in needs assessment and design of training modules is not as high as in PAR (Chandra et al., 2017; de Sousa et al., 2018). For learning platforms to be effective in promoting social learning and collective action in adaptation, strong facilitation is needed that minimizes power imbalances and builds trust among different stakeholders. One or more formal institutions should act as a broker to liaise between different interest groups and to mediate and coordinate partnerships. External actors may also have a role in creating or enhancing the demand for knowledge, thereby helping to overcome the limitations associated with relying only on local knowledge.

Researchers taking part in learning platform have to mind-shift into a learner role and need a deep understanding of local cultural norms and institutions to ensure inclusion of all views (e.g., to separate women and men in consultations, and to look beyond existing community groups that may include only stronger members). Time is also needed to build trust and a safe and comfortable interface environment through appropriate communication before the inception phase. Finally, economic incentives may also be needed to include marginalized categories who may not have time to engage in consultation activities (Shaw & Kristjanson, 2014). To work well, PAR and learning platforms also need a supportive institutional environment, reflected in economic support but also in an adequate policy and legislative framework (Chandra et al., 2017).

Ecosystem-Based Adaptation and Landscape Approaches in Smallholder Farming

Individual smallholder farms are typically small, but their collective share of contributions to the ecosystem burden cannot be ignored. Ecosystem goods and services are the backbone of farmers' agricultural economy. Common natural pool resources such as forests and wetlands provide complementary sources of livelihoods and, more generally, a healthy and productive landscape encourages households to remain in rural areas, slowing down outward migration from the countryside while increasing income opportunities and sustaining local adaptation (Arouna & Akpa, 2019; Food and Agriculture Organization of the United Nations, 2014). The interactions between farming and the environment also have negative connotations, with farmers often exposed to natural disasters and weather extremes. On the other hand, even if smallholder agriculture does not contribute to water and air pollution as much as large-scale intensive farming does, anthropogenic activities such as grazing, encroachment, and deforestation may severely undermine the natural resource base and ultimately smallholders' livelihoods. When taken collectively, these small units can still pose a heavy burden on the ecosystem; at the same time, just due to their large number and small scale, they can play an innovative role in forging beneficial ways of interaction between agriculture and the environment.

Despite these important connections, the review identified relatively few studies that focused explicitly on the nexus between humans and their surrounding ecosystems, or that used an environmental lens to critique adaptation within smallholder agriculture. These findings may well be influenced by the inclusion criteria defined in the review, but the fragmented evidence on this topic nevertheless reflects a reality where policy planning in agriculture, environment, and climate change still happen in silos, with limited exchanges between disciplines (El Chami et al., 2020). This is a key concern and should be recognized as a strategic development priority and considered explicitly in program evaluations.

Other areas of intervention can support stronger nexus integration between human and ecological systems in adaptation planning, implementation, and assessment. First, several studies stressed the need for a reframing of smallholder farming and agriculture more generally, and for an integrated system alongside natural resource management, energy, and climate change. Agricultural sustainability is a three-dimensional model that requires overcoming disciplinary boundaries. A transdisciplinary approach, which represents a step forward for interdisciplinarity with full integration of different disciplines, is much needed (El Chami et al., 2020). Such an approach should be pursued through policies that promote circular models of the economy and cyclical, rather than linear, growth-oriented systems. The economic analysis undertaken to inform such policies should reflect this complexity. That is, analysts should be able to value the costs and benefits beyond the conventional measures of productivity and efficiency to include, for instance, the economic value of

ecosystem goods and services and GHG mitigation, and thereby identify suboptimal equilibria where the different economic, social, and environmental objectives are consistently achieved (Reid et al., 2013).

Beyond stressing the importance of assessing adaptation responses through an integrated, cross-sectoral approach, the evidence review also highlighted the relevance of local and landscape dimensions. The ecological aspects of adaptation responses and land use changes are closely linked to local socioeconomic conditions: different locations, albeit close to each other, may take different adaptation paths and these should be contextualized within the agro-ecological landscapes in which they reside (López et al. 2020; Marquardt et al., 2020; Newsham & Thomas, 2009). Thus, a holistic approach to adaptation in smallholder agriculture should cut across not only sectors but also across multiple levels of intervention, from the household level through the community, right up to the landscape level.

In this context, ecosystem-based adaptation (EBA) has been proposed as an effective tool to achieve such an integrated vision (Abdelmagied & Mpheshea, 2020; El Chami et al., 2020; Vignola et al., 2015). Agro-forestry, for instance, has shown high potential to enhance smallholders' adaptive capacity (Lasco et al., 2014; Partey et al., 2018; Quandt, 2020). Other studies have highlighted that smallholders were aware of the links between the ecosystem and their economic activities, and many already use practices that could be classified as EBA (Chain-Guadarrama et al., 2019; Shah et al., 2019). However, the evidence also calls for greater involvement of researchers and policymakers to address some key challenges. These include, for example, integrating the farmers' local knowledge with results from scientific research and extension services to help manage and minimize the trade-offs at farm and landscape levels, such as competitive interactions between trees and crops, or pest infestations that can result from ecosystem stabilization (Lasco et al., 2014; Quandt, 2020). Scientific and economic research could also help quantify the economic returns from ecosystem goods and services, such as pest and disease control or the role of trees as buffer strips. Economic incentives based on these analyses are much needed because EBA approaches constitute long-term strategies with generally low returns (and hence incentives) in the short term.

Our review also showed that institutional support to social networks and collective action was very important in smallholders' pursuing adaptation pathways that assimilate the nexus between human and ecological systems. Being part of a network helps farmers to coordinate collective action around common pool resources. Strong social networks, coupled with the presence of environmental champions, can help facilitate the spread of knowledge-intensive practices for adaptation, such as agro-ecology, and lead to higher environmental sustainability in smallholder farming (Nyantakyi-Frimpong et al., 2019; SaintVille et al., 2016).

Some authors advocated for combining EBA and community-based adaptation and mainstreaming them into large-scale planning. According to Reid (2016, p. 5) "good EBA should (but does not always) have a strong community focus." The participatory dimension of EBA is easily explained by the fact that many (although not all) natural resource and environmental practices transcend administrative boundaries and go beyond field limits, requiring collective agreement and action. Indeed "good" community-based adaptation often considers ecosystems and ecosystem services to a strong degree. Local adaptation activities in the field tend to combine the two approaches and compensate for the shortcomings of top-down initiatives based on hard infrastructure (Reid, 2016). However, to achieve fruitful integration of these two approaches at scale, an important component is addressing the policy context to assess whether governance is actually inclusive for all community members, whether there is devolution of rights and responsibilities to local institutions, and how this works in practice. Incentives are needed both for individuals and community groups. These could be distributed via external funding for public projects and initiatives, as financing tools such as offering microcredit or revolving funds, or as payments for ecosystem services (PES) and direct compensation (Reid, 2016).

Policy Shortcomings

Policymakers play a key role in supporting smallholders' adaptation to climate change. Policies at national and international levels define and influence adaptation pathways, first and foremost by setting the vision that underlies the legal and the institutional frameworks. High-level, public intervention is needed to ensure that autonomous adaptations (that may be guided by short-term perspectives) and planned adaptations (more forward looking to intergenerational sustainability) work consistently toward virtuous patterns of transformative adaptation. Evidence showed that path dependence in farmers' choices to uptake adaptation calls for higher level coordination and investments across three intervention scales—household, community, and landscape. Institutional and financial support is necessary to preserve local adaptation efforts (e.g., through infrastructure) and to secure an enabling policy and legal environment (e.g., land tenure, access rights to natural resources). External institutions such as government and development actors should also provide economic incentives to ensure that environmental considerations are fully integrated into human responses (such as compensation for investments in agroforestry that don't deliver immediate returns) and, more generally, to reconcile the unavoidable trade-offs between competing objectives in the economic, social, and environmental domains.

The narrative that emerged from the review highlights that policies would need to undergo some fundamental shifts to support adaptation pathways that are genuinely inclusive and sustainable. First, adaptation strategies should address the root causes of vulnerability, which in the farming sector may lie in complex situations where climate change is just one of multiple stressors. This implies pursuing transformative strategies that have a holistic, intersectoral approach, keeping open all the options for response (including, for example, exiting agriculture). Within the agricultural sector, this means welcoming system-oriented solutions that go beyond technological interventions at plot and farm level; in fact, such interventions may prove to be only coping strategies and, if not well targeted, might also generate or exacerbate inequalities at the expense of poorer, less knowledgeable, or less connected farmers. Indeed, addressing the root causes of vulnerability also requires assessing and, if necessary, rebalancing power relationships and politics of knowledge at the community level to avoid asymmetries in access to knowledge and information, the exclusion of most vulnerable groups, and elite capture of benefits. Finally, policy plays a fundamental role in reorienting science and research toward more participative, multistakeholder, and transdisciplinary approaches, according to the principles discussed above.

However, the evidence presents a largely empirical approach and we often perceived a seeming disconnection between policy, research, and practice. Although extensive empirical literature exists on the impacts of climate change, policymakers do not systematically use the evidence available (Khan & Akhtar, 2015). A further problem is that while many studies identify barriers to adaptation in farming, the appreciation of their interactions and impacts remains scarce. At the policy level, perceptions of climate change mainly as an environmental issue rather than a broad development issue results in it often being sidelined, constituting a barrier to planning and implementing effective actions. Other political barriers include a shortfall of funding and insufficient coordination (Shackleton et al., 2015). Poor interaction between different disciplines, both in research and in policy, was identified as a further gap. Although recognition is growing that smallholder farming and the surrounding ecosystem are tightly coupled, most climate change interventions in agriculture have still focused on farm-related technological solutions. A technocentric approach risks neglecting environmental and natural resource dimensions because it does not solve problems from a holistic perspective. It also risks exacerbating income and gender inequalities because access to assets and skills is often gender biased and generally interdicted to the most vulnerable.

Pursuing the idea of adaptive development may help overcome some of these limitations and promote a truly transdisciplinary approach where reducing societal inequality and injustice is a pre-requisite for adaptation that works for everyone (Lemos et al., 2013). However, several fundamental barriers need to be overcome. At national levels and in international forums, an original disconnection exists between farmers, technical experts, and policymakers and coordination remains limited. True participation by smallholders in the decision-making process is scarce, with "participation" often reduced to consultation, and the devolution of power is weak. As a result, policies often do not reflect smallholders' real needs and preferences, which risks reinforcing vicious circles of power imbalance and inequitable adaptation outcomes (Hameso, 2017; Nigussie et al., 2018; Sova et al., 2015; Taylor, 2018).

Further challenges lie in the difficulties of mainstreaming successful local adaptation solutions into large-scale planning; because climate impacts are often locally specific, large-scale initiatives to support smallholder farmers must consider local priorities and integrate lessons from successful autonomous adaptation efforts. Scaling up climate adaptation into large-scale agricultural initiatives requires not only integration of lessons from community-based adaptation (i.e., learning local priorities, capacities, and lessons), but also the building of inclusive governance. Mainstreaming community-based adaptation, in particular, is more than scaling-up of specific adaptation practices or knowledge; it is about mainstreaming institutional and organizational approaches that allow this knowledge to be generated (Wright et al., 2014). According to this vision, securing impact at scale in practice involves providing the lawful circumstances to embed local institutions into broader governance frameworks, to legally devolve rights and responsibilities to local groups and communities (especially with regard to natural resource management) and to reform land tenure and rights to access natural resources as required (Reid, 2016).

One major opportunity to bring social learning on adaptation to the national level is via existing extension and advisory services, but in many lower-middle income countries, extension services are limited and often biased to target relatively wealthier households (Wright et al., 2014).

Another way to support effective knowledge sharing is through stakeholder platforms. Stakeholder consultations in policy and in research help consider the multiple and sometimes competing objectives among different sectors and interest groups, and the resulting trade-offs in the distribution of benefits and costs (Abegunde et al., 2019). Stakeholder-centered methods are increasingly used alongside other trade-off analysis methods oriented to better management, that is, with the aim to improve the relative situation for different groups, acknowledging that eliminating trade-offs and generating win-win solutions for all is not possible (Gusenbauer & Franks, 2019).

Implications for Monitoring, Evaluation, and Learning

Evidence assessment and continuous learning are critical to inform policy makers about what works for whom. In the context of the three learning outcomes of relevance to this study, monitoring, evaluation, and learning (MEL) should be able to capture the success of both autonomous and planned adaptation strategies in pursuing transformative pathways in smallholder farming. The outputs from the review highlighted a number of challenges for MEL in attaining this goal, and showed opportunities for innovation and cross-learning from different disciplines and methodological approaches.

One challenge lies in the need to evaluate the durability and the persistence over time of non-linear development processes. The evidence in the literature on scaling-up and knowledge management showed that social equity remains a key factor of transformational patterns; on the other hand, the reiteration of inequitable starting conditions coupled with scarce opportunities for building adaptive capacity causes path dependence in farmer decision making, leading to technology lock-in situations and possibly maladaptation. Enabling countries and organizations to better assess and evaluate transformative adaptation thus requires an inclusive approach built on effective knowledge sharing and atten-

tion to the behavioral and institutional changes that address the needs of the most vulnerable (Anderson, 2011). Without understanding how vulnerability to climate change is produced and distributed, planned adaptations may inadvertently promote injustice and deepen inequality. Inequality of outcomes may be channeled through wealth, gender, and local relationships that influence dynamics of power, knowledge flow, and uptake of certain technologies. Therefore, at a higher level, evaluation should seek to consider not only the outcomes of adaptation but also the processes behind them, to understand whose preferences and priorities have guided the decisions and the actions that have been implemented. At the operational level, a fundamental challenge to monitoring is that adaptive strategies also take time to implement and deliver impacts that can be monitored; key milestones are therefore needed to track actual changes over time.

Another related challenge for those engaged in evaluation is that complex, iterative, adaptive processes may often cut across different sectors beyond agriculture. MEL should be able to assess outcomes cutting across sectors (transdisciplinary) and take into account these relationships—and the trade-offs—across different intervention domains with different and sometimes competing goals (e.g., food security and conservation). The evidence showed that failing to fully acknowledge the interactions between the human and environmental systems, and continuing to treat agriculture separately from the environment, may undermine a full understanding of adaptation solutions both in agriculture and beyond. The ecological aspects of adaptation responses in agriculture, and more generally in land use changes, are closely linked to the social and economic dimensions of migration, labor, and shifting production patterns. These interlinkages are highly dependent on context and framed within the scope of landscapes that do not necessarily overlap with administrative boundaries and/or productive areas. These considerations call for MEL in agriculture to redefine both its focus (beyond the farming sector) and its scope (across traditional boundaries): Evaluators should go beyond the project or program scale and use the community and landscape dimensions as the scope for evaluating context-dependent dynamics. At a higher level, a conceptual shift is needed in what is conventionally valued as productive and efficient in agriculture in order to embed considerations of inter- and intragenerational equity (including ecological sustainability) into the assessment of adaptation outcomes.

Not only is a conceptual shift required, but changes and improvements in methodology are also necessary to assess and evaluate adaptation to climate change in agriculture. One operational challenge lies in the fact that adaptive strategies take time to bear outcomes that can be monitored, so evaluators need to identify and track key milestones toward the actual change. Another issue is that current M&E approaches mostly rely on purely quantitative methods, which are not suited to understanding complex processes and interactions or to including more nuanced perspectives on equity and vulnerability. More expertise is needed on evaluation applied to adaptation and on the combined application of quantitative and qualitative approaches (Anderson, 2011). Evaluators should also use innovative approaches in economics to value ecosystem goods and services; to assess trade-offs more systematically across different sectoral, spatial, and time scales; and to better evaluate social costs and benefits in the calculation of PES and other economic incentives for farmers.

Finally, improvements in governance are inherently difficult to assess, and so too is understanding how benefits are distributed among different groups. Indeed, even agreeing on a definition of successful adaptation may be difficult because different interest groups will evaluate the same outcome differently. Although using participatory approaches and mutual learning is critical to overcome these issues, they create difficulty in producing impartial and definitive assessments (Anderson, 2011). In this context, stakeholder platforms provide a promising tool (alongside other analytical methods) to encourage mutual learning, communication, and governance in adaptation. These platforms also may be useful to inform the development of innovative MEL approaches.

Future Role of Evidence Reviews in Programmatic Evaluation

Beyond providing valuable insights and new evidence on climate change adaptation in smallholder agriculture, this type of review can also feed into broader discussions on how evidence from a range of sources (including systematic and rapid evidence reviews) can be usefully integrated into thematic evaluations and what added value such syntheses can provide for MEL. The value to MEL can be explained in relation to a pyramid of evidence (Fig. 1) where the base represents the degree of uncertainty and the different levels represent increasing consensus for different types of enquiry. In evaluation exercises, one task is usually to gather all available data and information from a range of sources (including project design and progress reports, and semistructured interviews with key informants and stakeholders) to corroborate and/or triangulate expert opinion. The rationale is to realize the potential of the data gathered to inform the evaluation while minimizing any bias and uncertainty in the final judgement. Although evaluation panels can access different target audiences, the information they acquire is subject to uncertainty. This could include, for example, the timing and scale of enquiry; the resources available to conduct in-depth, in-country technical missions; access to independent key informants; the relevance of monitoring data; and personal subjectivity. Conclusions also tend to rely on expert judgement that depends on many factors including the evaluation experience and discipline expertise of individual team members.

Traditional literature reviews also can provide valuable insights for evaluation exercises but are contingent on the sources available (and used), the skills of the researcher in framing and executing the searches, the scope of the review, and the method of analysis with inferences based on only a sample of the available evidence. In contrast,

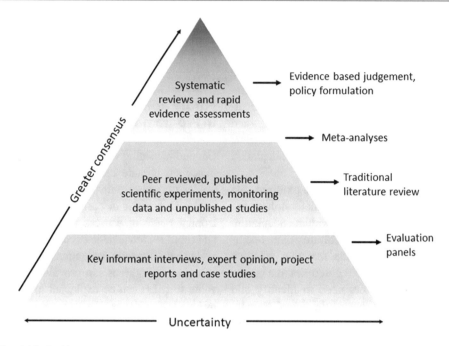

Fig. 1 Pyramid of evidence to support project and program evaluation

systematic reviews are often considered to be at the top of the pyramid of evidence hierarchy, providing the highest degree of consensus and lowest amount of uncertainty, given their highly structured and systemic methodology. Rapid evidence assessments (REAs) are consistent with many of the systematic review principles and popular for informing decision making and policy formulation.

This synthesis of evidence was based on the structured methodology for an REA, although it ultimately presented a much deeper and wider ranging level of analysis more usually associated with a systematic review. This narrative synthesis required a hybrid approach articulated across three related yet distinct learning domains, with each characterized by a high degree of complexity. We chose the REA framework because it can be designed and executed in a relatively short time span (1–3 months) depending on the scope of the research question and available literature, compared to 3–6 months for a systematic review. As such, it is more appropriate for informing programmatic evaluations where a relatively quick synthesis of evidence is required. An REA can also

provide a deep understanding of the volume and characteristics of evidence that is available on a certain topic and make it more accessible for further scrutiny. Hence, REAs can facilitate answering targeted, research-based evaluation questions by maximizing use of the existing evidence base, while also providing a clear picture of the adequacy of that evidence (Collins et al., 2015).

Evidence reviews can also help evaluators contextualize projects and programs within a broader picture for use in comparing their results with findings drawn from international evidence. Thus, they can help evaluators understand how conceptual issues are addressed across different disciplines and timeframes, with support from a wider pool of case study material. In some instances (for example, when exploring nexus issues between the human system and ecosystem in the agricultural literature), REAs can help identify existing gaps in knowledge and emerging trends in research, policy, and practice to inform how these gaps and trends might then affect intervention outcomes. Finally, evidence reviews can also help connect research communities with practitioners by providing valuable and informed access to the peer-reviewed scientific literature that is either overlooked or not known to exist within the project and program evaluation community.

Acknowledgements This research was funded by the International Fund for Agricultural Development (IFAD) as part of a broader thematic evaluation of their global development program on smallholder adaptation to climate change. The authors acknowledge colleagues from the thematic evaluation team who provided valuable critique and feedback on this review, including Nurul Alam, James Gasana, Margarita Borzelli Gonzalez, Christian Hergarten, Prashanth Kotturi, Susanne Leloup, Carsten Schwensen, and Horst Weyerhaeuser. The opinions expressed in this chapter are those of the authors and do not constitute a formal view from IFAD.

References

Abass, R., Mensah, A., & Fosu-Mensah, B. (2018). The role of formal and informal institutions in smallholder agricultural adaptation: The case of Lawra and Nandom Districts, Ghana. *West African Journal of Applied Ecology, 26*, 56–72. https://www.ajol.info/index.php/wajae/article/view/182466

Abdelmagied, M., & Mpheshea, M. (2020). *Ecosystem-based adaptation in the agriculture sector – A nature-based solution (NbS) for building the resilience of the food and agriculture sector to climate change.* Food and Agriculture Organization of the United Nations. http://www.fao.org/3/cb0651en/CB0651EN.pdf

Abegunde, V. O., Sibanda, M., & Obi, A. (2019). The dynamics of climate change adaptation in sub-Saharan Africa: A review of climate-smart agriculture among small-scale farmers. *Climate, 7*(11), 132. https://doi.org/10.3390/cli7110132

Akinyemi, F. O. (2017). Climate change and variability in semiarid Palapye, Eastern Botswana: An assessment from smallholder farmers' perspective. *Weather, Climate, and Society, 9*(3), 349–365. https://doi.org/10.1175/WCAS-D-16-0040.1

Anderson, S. (2011). *Assessing the effectiveness of climate adaptation.* International Institute for Environment and Development.

Arouna, A., & Akpa, A. K. (2019). Water management technology for adaptation to climate change in rice production: Evidence of smart-valley approach in West Africa. In A. Sarkar, S. R. Sensarma, & G. W. vanLoon (Eds.), *Sustainable solutions for food security* (pp. 211–227). Springer International.

Asociacion Andes. (2016). *Resilient farming systems in times of uncertainty: Biocultural innovations in the Potato Park, Peru.* International Institute for Environment and Development. https://pubs.iied.org/14663IIED

Atsiaya, G. O., Ayuya, O. I., Nakhone, L. W., & Lagat, J. K. (2019, 7). Drivers and responses to climate variability by agro-pastoralists in Kenya: The case of Laikipia County. *SN Applied Sciences, 1*, 827. https://doi.org/10.1007/s42452-019-0849-x

Bassett, T. J., & Fogelman, C. (2013). Déjà vu or something new? The adaptation concept in the climate change literature. *Geoforum, 48*, 42–53. https://doi.org/10.1016/j.geoforum.2013.04.010

Bedeke, S., Vanhove, W., Gezahegn, M., Natarajan, K., & Damme, P. V. (2019). Adoption of climate change adaptation strategies by maize-dependent smallholders in Ethiopia. *NJAS – Wageningen Journal of Life Sciences, 88*, 96–104. https://doi.org/10.1016/j.njas.2018.09.001

Belay, A., Recha, J. W., Woldeamanuel, T., & Morton, J. F. (2017). Smallholder farmers' adaptation to climate change and determinants of their adaptation decisions in the Central Rift Valley of Ethiopia. *Agriculture and Food Security, 6*, 24. https://doi.org/10.1186/s40066-017-0100-1

Binswanger-Mkhize, H. P., de Regt, J. P., & Spector, S. (2009). *Scaling up local and community driven development: A real world guide to its theory and practice.* World Bank. https://openknowledge.worldbank.org/handle/10986/28252

Chain-Guadarrama, A., Martínez-Rodríguez, R. M., Cárdenas, J. M., Mendoza, S. V., & Harvey, C. A. (2019). Uso de prácticas de Adaptación basada en Ecosistemas por pequeños cafetaleros en Centroamérica. *Agronomía Mesoamericana, 30*(1), 1–18. https://doi.org/10.15517/am.v30i1.32615

Chandra, A., Dargusch, P., McNamara, K. E., Caspe, A. M., & Dalabajan, D. (2017). A study of climate-smart farming practices and climate-resiliency field schools in Mindanao, the Philippines. *World Development, 98*, 214–230. https://doi.org/10.1016/j.worlddev.2017.04.028

Collins, A. C., Coughlin, D., Miller, J., & Kirk, S. (2015). *The production of quick scoping reviews and rapid evidence assessments: A how to guide.* United Kingdom Department for Environment Food and Rural Affairs and Natural Environment Research Council. https://assets.publishing.service.gov.uk/government/uploads/system/uploads/attachment_data/file/560521/Production_of_quick_scoping_reviews_and_rapid_evidence_assessments.pdf

de Sousa, K., Casanoves, F., Sellare, J., Ospina, A., Suchini, J. G., Aguilar, A., & Mercado, L. (2018). How climate awareness influences farmers' adaptation decisions in Central America? *Journal of Rural Studies, 64*, 11–19. https://doi.org/10.1016/j.jrurstud.2018.09.018

Derbile, E., Jarawura, F., & Dombo, M. (2016). Climate change, local knowledge and climate change adaptation in Ghana. In J. H. Joseph & A. Yaro (Eds.), *Adaptation to climate change and variability in rural West Africa*. Springer.

Donatti, C., Harvey, C. A., Martinez-Rodriguez, M. R., Vignola, R., & Rodriguez, C. M. (2019). Vulnerability of smallholder farmers to climate change in Central America and Mexico: Current knowledge and research gaps. *Climate and Development, 11*(3), 264–286. https://doi.org/10.1080/17565529.2018.1442796

El Chami, D., Daccache, A., & Moujabber, M. E. (2020). How can sustainable agriculture increase climate resilience? A systematic review. *Sustainability, 12*(8), 3119. https://doi.org/10.3390/su12083119

Etana, D., Snelder, D. J., van Wesenbeeck, C. F., & de Cock Buning, T. (2020). Dynamics of smallholder farmers' livelihood adaptation decision-making in Central Ethiopia. *Sustainability, 12*(11), 4526. https://doi.org/10.3390/su12114526

Food and Agriculture Organization of the United Nations. (2014). *Adapting to climate change through land and water management in Eastern Africa: Results of pilot projects in Ethiopia, Kenya and Tanzania.* http://www.fao.org/family-farming/detail/en/c/853718/

Guido, Z., Finan, T., Rhiney, K., Madajewicz, M., Rountree, V., Johnson, E., & McCook, G. (2018). The stresses and dynamics of smallholder coffee systems in Jamaica's Blue Mountains: A case for the potential role of climate services. *Climatic Change, 147*(1–2), 253–266. https://doi.org/10.1007/s10584-017-2125-7

Gusenbauer, D., & Franks, P. (2019). *Agriculture, nature conservation or both? Managing trade-offs and synergies in sub-Saharan Africa.* International Institute for Environment and Development. https://pubs.iied.org/14675IIED

Hameso, S. (2017). Farmers and policy-makers' perceptions of climate change in Ethiopia. *Climate and Development, 10*(4), 347–359. https://doi.org/10.1080/17565529.2017.1291408

Henriksson, R., Vincent, K., Archer, E., & Jewitt, G. (2020). Understanding gender differences in availability, accessibility and use of climate information among smallholder farmers in Malawi. *Climate and Development.* https://doi.org/10.1080/17565529.2020.1806777

Hulke, C., & Revilla Diez, J. (2020). Building adaptive capacity to external risks through collective action – Social learning mechanisms of smallholders in rural Vietnam. *International Journal of Disaster Risk Reduction, 51*, 101829. https://doi.org/10.1016/j.ijdrr.2020.101829

International Fund for Agricultural Development. (2020) *Adaptation for smallholder agriculture programme: ASAP.* https://www.ifad.org/documents/38714170/40213192/asap.pdf/b5a8c1f9-f908-4a68-ad30-e3d5eeb17c31

Kerr, R. B., Nyantakyi-Frimpong, H., Dakishoni, L., Lupafya, E., Shumba, L., Luginaah, I., & Snapp, S. S. (2018). Knowledge politics in participatory climate change adaptation research on agroecology in Malawi. *Renewable Agriculture and Food Systems, 33*(3), 238–251. https://doi.org/10.1017/S1742170518000017

Khan, M. A., & Akhtar, M. S. (2015). Agricultural adaptation and climate change policy for crop production in Africa. In K. R. Hakeem (Ed.), *Crop production and global environmental issues* (pp. 437–541). Springer International.

Khanal, U., Wilson, C., Hoang, V., & Lee, B. (2019). Impact of community-based organizations on climate change adaptation in agriculture: Empirical evidence from Nepal. *Environment Development and Sustainability, 21*(2), 621–635. https://doi.org/10.1007/s10668-017-0050-6

Lasco, R. D., Delfino, R. J., & Espaldon, M. L. (2014). Agroforestry systems: Helping smallholders adapt to climate risks while mitigating climate change. *Wiley Interdisciplinary Reviews: Climate Change, 5*, 825–833. https://doi.org/10.1002/wcc.301

Lemos, M., Agrawal, A., Eakin, H., Nelson, D., & Engle. (2013). Building adaptive capacity to climate. In A. Ghassem & J. W. Hurrell (Eds.), *Climate science for serving society: Research, modelling and prediction priorities.* Springer.

Levin, K., Cashore, B., & Bernstein, S. (2012). Overcoming the tragedy of super wicked problems: Constraining our future selves to ameliorate global climate change. *Policy Sciences, 45*, 123–152. https://doi.org/10.1007/s11077-012-9151-0

López, S., López-Sandoval, M., & Jung, J.-K. (2020). New insights on land use, land cover, and climate change in human–environment dynamics of the Equatorial Andes. *Annals of the American Association of Geographers.* https://doi.org/10.1080/24694452.2020.1804822

Lowder, S. K., Skoet, J., & Raney, T. (2016). The number, size and distribution of farms, smallholder farms and family farms worldwide. *World Development, 87*, 16–29. https://doi.org/10.1016/j.worlddev.2015.10.041

Makate, C. (2019). Local institutions and indigenous knowledge in adoption and scaling of climate-smart agricultural innovations among sub-Saharan smallholder farmers. *International Journal of Climate Change Strategies and Management, 12*(2), 270–287. https://doi.org/10.1108/IJCCSM-07-2018-0055

Mapfumo, P., Adjei-Nsiah, S., Mtambanengwe, F., Chikowo, R., & Giller, K. E. (2013). Participatory action research (PAR) as an entry point for supporting climate change adaptation by smallholder farmers in Africa. *Environmental Development, 5*, 6–22. https://doi.org/10.1016/j.envdev.2012.11.001

Marquardt, K., Pain, A., & Khatri, D. B. (2020). Re-reading Nepalese landscapes: Labour, water, farming patches and trees. *Forests, Trees and Livelihoods, 29*(4), 238–259. https://doi.org/10.1080/14728028.2020.1814875

Neufeldt, H., Negra, C., Hancock, J., & Foster, K. (2015). *Scaling up climate-smart agriculture: Lessons learned from South Asia and pathways for success.* World Agroforestry Centre. https://doi.org/10.5716/WP15720.PDF%20

Newsham, A., & Thomas, D. (2009). *Agricultural adaptation, local knowledge and livelihoods diversification in North-Central Namibia*. Tyndall Centre for Climate Change.

Nigussie, Y., van der Werf, E., Zhu, X., Simane, B., & van Ierland, E. C. (2018). Evaluation of climate change adaptation alternatives for smallholder farmers in the Upper Blue-Nile Basin. *Ecological Economics, 151*, 142–150. https://doi.org/10.1016/j.ecolecon.2018.05.006

Nyantakyi-Frimpong, H., Matouš, P., & Isaac, M. E. (2019). Smallholder farmers' social networks and resource-conserving agriculture in Ghana: A multicase comparison using exponential random graph models. *Ecology and Society, 24*(1), 5. https://doi.org/10.5751/ES-10623-240105

Partey, S. T., Zougmoré, R. B., Ouédraogo, M., & Campbell, B. M. (2018). Developing climate-smart agriculture to face climate variability in West Africa: Challenges and lessons learnt. *Journal of Cleaner Production, 187*, 285–295. https://doi.org/10.1016/j.jclepro.2018.03.199

Popoola, O. O., Yusuf, S. F., & Monde, N. (2020). Information sources and constraints to climate change adaptation amongst smallholder farmers in Amathole District Municipality, Eastern Cape Province, South Africa. *Sustainability, 12*(14), 5846. https://doi.org/10.3390/su12145846

Pullin, A. S., Frampton, G. K., Livoreil, B., & Petrokofsky, G. (Eds.). (2018). *Guidelines and standards for evidence synthesis in environmental management, version 5.0*. Collaboration for Environmental Evidence https://environmentalevidence.org/information-for-authors/

Quandt, A. (2020). Contribution of agroforestry trees for climate change adaptation: Narratives from smallholder farmers in Isiolo, Kenya. *Agroforestry Systems, 94*(6), 2125–2136. https://doi.org/10.1016/j.gloenvcha.2006.02.006

Reid, H. (2016). Ecosystem- and community-based adaptation: Learning from community-based natural resource management. *Climate and Development, 8*(1), 4–9. https://doi.org/10.1080/17565529.2015.1034233

Reid, H., Chambwera, M., & Murray, L. (2013). *Tried and tested: Learning from farmers on adaptation to climate change*. International Institute for Environment and Development. https://pubs.iied.org/14622IIED

Roncoli, C., Okoba, B., & Gathaara, V. (2010). *Adaptation to climate change for smallholder agriculture in Kenya: Community-based perspectives from five districts* (Project note from adaptation of smallholder agriculture to climate change in Kenya project). International Food Policy Research Institute.

Ruijs, A., de Bel, M., Kononen, M., Linderhof, V., & Polman, N. (2011). *Adapting to climate variability: Learning from past experience and the role of institutions*. World Bank. https://openknowledge.worldbank.org/handle/10986/26896

SaintVille, A. S., Hickey, G. M., Locher, U., & Phillip, L. E. (2016). Exploring the role of social capital in influencing knowledge flows and innovation in smallholder farming communities in the Caribbean. *Food Security, 8*(3), 535–549. https://doi.org/10.1007/s12571-016-0581-y

Shackleton, S., Ziervogel, G., Sallu, S., Gill, T., & Tschakert, P. (2015). Why is socially-just climate change adaptation in sub-Saharan Africa so challenging? A review of barriers identified from empirical cases. *Wiley Interdisciplinary Reviews: Climate Change, 6*(3), 321–344. https://doi.org/10.1002/wcc.335

Shah, S. I., Zhou, J., & Shah, A. A. (2019). Ecosystem-based adaptation practices in smallholder agriculture: Emerging evidence from rural Pakistan. *Journal of Cleaner Production, 218*, 673–684. https://doi.org/10.1016/j.jclepro.2019.02.028

Shaw, A., & Kristjanson, P. (2014). A catalyst toward sustainability? Exploring social learning and social differentiation approaches with the agricultural poor. *Sustainability, 6*(5), 2685–2717. https://doi.org/10.3390/su6052685

Silici, L., Rowe, A., Suppiramaniam, N., & Knox, J. W. (in press). Building adaptive capacity of smallholders to climate change: Review and synthesis of evidence on learning outcomes. *Environmental Research Letters*.

Sova, C., Vervoort, J., Thornton, T., Helfgott, A., Matthews, D., & Chaudhury, A. (2015). Exploring farmer preference shaping in international agricultural climate change adaptation regimes. *Environmental Science & Policy, 54*, 463–474. https://doi.org/10.1016/j.envsci.2015.08.008

Stringer, L., Fraser, E. D. G., Harris, D., Lyon, C., Pereira, L., Ward, C. F. M., & Simelton, E. (2020). Adaptation and development pathways for different types of farmers. *Environmental Science and Policy, 104*, 174–189. https://doi.org/10.1016/j.envsci.2019.10.007

Taylor, M. (2018). Climate-smart agriculture: What is it good for? *The Journal of Peasant Studies, 45*(1), 89–107. https://doi.org/10.1080/03066150.2017.1312355

United Nations Environment Programme. (2018). *The adaptation gap report 2018*. https://www.unep.org/resources/adaptation-gap-report-2018

Vermeulen, S., Mason, M., Dinesh, D., & Adolph, B. (2015). *Radical adaptation in agriculture: Tackling the roots of climate vulnerability*. International Institute for Environment and Development. https://pubs.iied.org/17309IIED

Vermeulen, S. J., Dhanush, D., Howden S., M., Cramer, L., & Thornton, P. K. (2018). Transformation in practice: A review of empirical cases of transformational adaptation in agriculture under climate change. *Frontiers in Sustainable Food Systems, 2*, 65. https://doi.org/10.3389/fsufs.2018.00065

Vignola, R., Harvey, C. A., Bautista-Solis, P., Avelino, J., Rapidel, B., & Camila Donatti, R. M. (2015). Ecosystem-based adaptation for smallholder farmers: Definitions, opportunities and constraints. *Agriculture, Ecosystems & Environment, 211*, 126–132. https://doi.org/10.1016/j.agee.2015.05.013

Wekesa, C., Ongugo, P., Ndalilo, L., Amur, A., Mwalewa, S., & Swiderska, K. (2017). *Smallholder farming systems in coastal Kenya: Key trends and innovations for resilience.* International Institute for Environment and Development. https://pubs.iied.org/17611IIED

Wise, R. M., Fazey, I., Stafford Smith, M., Park, S. E., Eakin, H. C., Archer Van Garderen, E. R. M., & Campbell, B. (2014). Reconceptualising adaptation to climate change as part of pathways of change and response. *Global Environmental Change, 28*, 325–336. https://doi.org/10.1016/j.gloenvcha.2013.12.002

Wright, H., Vermeulen, S., Laganda, G., Olupot, M., Ampaire, E., & Jat, M. L. (2014). Farmers, food and climate change: Ensuring community-based adaptation is mainstreamed into agricultural programmes. *Climate and Development, 6*(4), 318–328. https://doi.org/10.1080/17565529.2014.965654

Introduction

Carlo Carugi (✉) e-mail: ccarugi@thegef.org
Global Environment Facility Independent Evaluation Office, Washington, DC, USA

This part discusses evaluation approaches for assessing integrated environmental and socioeconomic co-benefits. It focuses on how best to approach the socioeconomic and environment nexus from an evaluation point of view in today's Sustainable Development Goals (SDGs) context. To set the stage of this discussion, the part starts with "Evaluation at the Endgame: Evaluating Sustainability and the SDGs by Moving Past Dominion and Institutional Capture" and Rowe's provocative call for attention to the fact that while the SDGs have brought the social, the economic, and the environmental closer to each other, they still treat them as separate, which does not bode well for sustainability. Evaluation should be aware of that. According to Rowe, there is no disputing the serious threats to sustainability faced by all on this planet, or that humans have contributed significantly to this. The good news is that a potential exists to diminish the level of threat. Although thus far evaluation has not shown much interest in environmental sustainability, it can still assume a more activist role and contribute value to checkmating extinction. The chapter first outlines the current stance of evaluation with respect to sustainability and then turns to how to design evaluations that contribute to efforts to reverse the "end game" environmental trends.

The part continues with Miyaguchi's chapter, "Importance and Utilization of Theory-Based Evaluations in the Context of Sustainable Development and Social-Ecological Systems," in which he emphasizes the importance of evaluation theories and practices considering the nexus between the natural and human systems. It is hard to disagree with him. Miyaguchi rightly notes that the breadth of the partnerships and collaborations brought about by the SDGs among non-state actors, including the private sector, nongovernmental organizations, and nonprofit organizations, has been unprecedented. But the inconsistencies and incompatibility among 169 SDG targets and their indicators represent a challenge in light of the abilities and capacities of many states, especially those of developing countries, to adequately monitor and evaluate the current and future SDG achievement status. A related major challenge lies in evaluating the nexus of human and natural systems. The author discusses the utility of theory-based evaluation to address that challenge, and introduces a holistic framework called CHANS (coupled human and natural systems) as a useful analytical framework in evaluating complex, social-ecological systems.

Finland is committed to implementation of the United Nations 2030 Agenda for Sustainable Development and its goals. As a demonstration of this commitment, the Finnish Government established that the implementation of the 2030 Agenda should be evaluated once during every 4-year electoral term. In "Pathway to the Transformative Policy of Agenda 2030: Evaluation of Finland's Sustainable Development Policy," Räkköläinen and Saxén describe the independent evaluation of Finland's national implementation of the 2030 Agenda. The purpose of the evaluation was to support evidence-based decision making by informing the government and the public on the efficiency of the 2030 Agenda implementation. This requires an entirely new kind of comprehensive approach and coherence in the policies of the various Finland administrative branches. According to the authors, the evaluation was very timely as it strengthened the knowledge base for updating the implementation plan for the 2030 Agenda after the parliamentary elections in 2019, contributing to the preparation of the new Government program. Another vital role of the evaluation was providing content for the social policy debate that preceded the parliamentary elections.

Sustainable development is characterized by new demands for development actors to mainstream and incorporate in their work as they emerge. Recent noteworthy examples include the environment and climate change. Evaluators must consider how to reflect such emerging priorities in their work and do so in a way that allows measuring progress and achievements. In "Evaluating for Resilient and Sustainable Livelihoods: Applying a Normative Framework to Emerging Realities," Kotturi discusses the experience of the Independent Office of Evaluation (IOE) of the International Fund for Agricultural Development (IFAD) in incorporating concerns on environment and climate within its evaluations. The IOE did this in two ways, namely by adapting its existing methodologies to address environmental concerns while continuing to meet the demands placed on IFAD's evaluation function, and by introducing in its evaluations new methodologies that allow assessment of environmental change, such as geospatial analysis. IFAD has started geotagging its project/program sites more systematically and plans to introduce this practice as a standard part of its operational procedures. Geospatial data are being harmonized across IFAD regional divisions and made available on a web-based platform called IFAD Geonode. To be able to consistently use these data in its evaluations, the IOE is exploring ways to further build its geospatial analysis skills and capacity. This reciprocal adaptation between IFAD operations and evaluation function is an example to follow.

Change in forest cover over time is one of the environmental changes that are most easily observable with geographic information systems (GIS). In "Measuring the Impact of Monitoring: How We Know Transparent Near-Real-Time Data Can Help Save the Forests," Shea of the World Resources

Institute describes Global Forest Watch (GFW), an online platform that can support monitoring and evaluation of conservation projects, international commitments to reduce deforestation, and private sector zero-deforestation plans. GFW is based on the idea that transparent, publicly available data can support the greater good, which in this case is reducing deforestation. However, by its very nature, the use of freely available data can be difficult to track and its impact difficult to measure. This chapter provides a framework for other open-data platforms to monitor outcomes and measure impact by exploring four different options for measuring GFW's reach and impact. GFW's experience indicates that while quantitative methods are capable of measuring outcomes and impact, they are not sufficient. Qualitative methods are also necessary to understand the mechanisms of adoption and produce lessons for furthering the reach and impact of open-data solutions.

Anand and Batra close the part with "Application of Geospatial Methods in Evaluating Environmental Interventions and Related Socioeconomic Benefits," discussing the extensive and diversified experience gained by the Independent Evaluation Office (IEO) of the Global Environment Facility (GEF) in blending geospatial analysis and remote sensing with qualitative analysis in GEF IEO evaluations. Starting from the consideration that environmental interventions such as those the GEF funds are essential for achieving the objectives laid out in the SDGs and the international environmental conventions the GEF is mandated to serve, Anand and Batra convincingly point at the complexities inherent in the assessment of expected environmental results and related socioeconomic benefits. This complexity originates from the interlinkages, be these synergies or trade-offs, between environmental and developmental goals, and the frequent lack of data, especially in developing countries, that characterizes these interventions. Geospatial approaches and tools allow assessment of environmental and socioeconomic outcomes—in other words, measuring the progress of initiatives over time while addressing their complexity. Drawing on GEF interventions in biodiversity (SDG 15), land degradation (SDG 15), and climate change (SDG 13), the authors discuss the application of geospatial approaches to assess GEF interventions' relevance, results, and sustainability in terms of their environmental outcomes as well as observable socioeconomic (SDG 1, 2) and health (SDG 3) co-benefits.

Evaluation at the Endgame: Evaluating Sustainability and the SDGs by Moving Past Dominion and Institutional Capture

Andy Rowe

Abstract

Three facts underlay this chapter. First, the human system and all our ambitions for improving the human system depend on sustainable natural systems. Second, we do not have much time. On track to fall well short of all sustainability goals, the climate and sustainability crises grow and extinction looms. Third, up to this point evaluation has shown little interest in sustainability, yet evaluation potentially addresses the very questions that are central to informing and guiding rapid adaptation of human behavior to successfully surmounting extinction.

Business-as-usual evaluation will not suffice. At the endgame with extinction looming, we need an evaluation that is more nimble, keeps up with rapidly accelerating knowledge, is relentlessly use-seeking and that guides the way to joined-up approaches. The evaluation we need will systematically mainstream sustainability across all evaluations and interventions, in all evaluation criteria and standards. For this, all evaluations will always address nexus where human and natural systems join and incorporate knowledge and methods from both systems. Existing evaluation knowledge is well suited to this task, as are knowledges in biophysical sciences. We know and promote knowledge processes for integrative evaluation and are starting to shift toward the requirements for evaluation at the nexus. As this chapter shows, the anchors holding us back are political, not technical.

Introduction

Every line of evidence leads us to conclude that the threats to sustainability of the planet and the life it supports are very real, large, multi-faceted and imminent. And yet globally we are falling well short on milestones such as the 2030 Agenda for Sustainable Development and 2050 carbon reduction goals. We have pushed natural systems beyond their capacity to adapt and continue to provide the services on which we depend. We are at the endgame on this planet.[1]

With some important exceptions, evaluation globally has not recognized the overwhelming evidence that sustainability is a matter worthy of our attention. Sustainability is a materially differ-

---

[1] In revising this chapter, I came across the work of Robert Nadeau that seems to presage some of my own work. He wrote a decade earlier about the shortcomings of neoclassical economics in not addressing the connection between environment and economics that he attributes to a two culture (economics and environment) thinking (Nadeau, 2008) identified initially by C. P. Snow (Nadeau, 2006).

A. Rowe (✉)
ARCeconomics, Maple Bay, BC, Canada
e-mail: andy@arceconomics.com

J. I. Uitto, G. Batra (eds.), *Transformational Change for People and the Planet*, Sustainable Development Goals Series, https://doi.org/10.1007/978-3-030-78853-7_14

ent matter than those that evaluators are accustomed to addressing because there is a hard stop if we fall short; absent significant improvements in our performance, that hard stop is a clear pathway to extinction. Meaning that evaluation at the endgame is different from business-as-usual evaluation. As with chess, the sustainability endgame needs to be fully goal focused and must fully commit all resources to strategies to achieve checkmate.

This chapter is concerned about the character of evaluation that will enable the field to make useful contributions at the endgame. The most fundamental change is from evaluations' almost monastic focus on the human system to systematic consideration of all interventions (projects, programs, strategies, policies) in their nexus location where both human and natural systems are present, have influence, provide value, and are affected.

The underlying mechanism for this monastic, human-centered worldview lies in the rootstock of evaluation that is said to be provided by Western social and management sciences with accountability, social inquiry, and social research methods as the trunk of the tree (Christie & Alkin, 2008; Alkin, 2004). That evaluation rootstock is embedded in and draws nutrition from the accumulated soils of Judeo-Christian society strongly infused with dominion, a worldview in which humans have ascendancy over other living and nonliving things, and over other peoples (Rowe, 2018). Humans, and of course especially those of European origin (i.e., white), are at the top of the heap; all else serves. Nonhuman living and nonliving things that constitute the natural systems on which all life depends are regarded as resources to be freely extracted to support humans. And while social sciences and evaluation are adapting to recognize and address how dominion has shaped thought and practice (e.g., gender bias, racism) the presumption that only humans have value and therefore merit consideration continues virtually unchecked in evaluation.

Accountability is one of the stems of the evaluation tree effectively partitioning governance structures from interventions at all levels so that connectivity to public policy goals is truncated (Chelimsky, 2012). It is an important mechanism for the observable, inverse relationship between public expenditures and the status and trends on conditions targeted by public policy such as public health and education (Williams, 2019). Sustainability is about connected systems while accountability is about partitioned systems, making pursuit of sustainability at odds with contemporary approaches to accountability. Accountability is an important authorizing mechanism bringing dominion into evaluation with the unintended effect of imparting a systematic positive bias to evaluation (Rowe, 2019b).

The COVID-19 pandemic provided dramatic evidence that human and natural systems are connected (Patton, 2020b). The virus reached us along pathways created by our relentless incursions into natural systems. The inverse and causal relationship between contemporary forms of economic growth and environmental health have been starkly shown with the slowing of economic and social activities causally linked to reduced incidence of some important health conditions such as asthma and reductions in GHG emissions from economic downturn. The economic downturn has resulted in falling petroleum prices, making it less expensive to produce virgin plastic from fossil fuels as compared to recycling. At the same time, demand for disposable (plastic) protective equipment has increased manyfold. For example, daily single-use plastic medical waste (gloves, masks, and gowns) in Wuhan at the peak of the pandemic there increased sixfold compared to prepandemic averages (Adyel, 2020), all of which is disposed in landfills. Demand for plastic packaging is estimated to have increased by 5.5%, strongly related to the increased consumption of take-out foods; plastic deposits in landfills increased by 1400 tons during the 8-week shutdown in Singapore (Adyel, 2020). That is, the pandemic reduced economic activity, decreasing demand for fossil fuels and lowering their price, leading to increased fossil fuel use to produce single-use plastic commodities. This resulted in increased deposits in landfills and in unmanaged streams of disposal, a good example of connectivity from public health to economy to environment.

The evaluation worldview must shift to acknowledge that human life is intrinsically contingent on healthy natural systems with which we are coupled[2] and that we must end the unnecessary harm we cause and move to restoring critical environmental values. Indigenous worldviews are instructive; for example, we might take direction from Daniel Wildcat from Haskell Indian Nations University:

> Think of how our worldview changes if we shift from thinking that we live in a world full of resources to a world where we live among relatives. (Zak, 2019)

Evaluation does not address the natural system for social and political reasons, but we have the knowledges, tools, and methods needed to renovate evaluation by drawing on a broad palate including evaluation, social and biophysical sciences, conflict resolution, law, and other fields (Patton, 2020b; Rowe, 2018). And emerging efforts by evaluators are starting to build foundations for incorporating sustainability into evaluation, such as in Blue Marble Evaluation (Patton, 2020a) and Better Evaluation (2020). The need for these efforts is amply demonstrated by two recent stocktakings showing evaluation to be only in the early stages of addressing nexus, that development evaluation appears to lead national and sectoral efforts, and that the intellectual infrastructure for nexus evaluation can only be described as weak (Sustainability Working Group, Canadian Evaluation Society [CES], 2020; United Nations Evaluation Group Working Group on Integrating Environmental and Social Impact into Evaluations [UNEG Working Group], 2020).

This chapter's focus is on evaluation at the endgame. I begin with the findings of the two sustainability stocktakings to describe where evaluation is now with respect to systematically

incorporating sustainability into evaluation—effectively our starting point for the endgame. The findings clearly point to evaluation's almost singular focus on human systems and to an intellectual infrastructure that is not fit for the purpose of incorporating sustainability. I then briefly reprise my arguments that the cause for this state of affairs lies in a worldview of dominion whereby humans, and especially white humans, hold dominion over all other living and nonliving things. This worldview is pervasive in social science and evaluation, with accountability serving as a key mechanism authorizing disregard of the natural system in evaluation. To these earlier arguments I add institutional capture as a further mechanism separating human and natural systems in evaluation and use the example of the SDGs to illustrate this. I then return to the endgame, illustrating some fundamental differences between evaluation needed for the endgame and the evaluation we have now.

Taking Stock on Evaluation Practice and Resources on Sustainability

Two recent and complementary stocktaking efforts have assessed current evaluation practice and resources to incorporate sustainability. The UNEG Working Group on Integrating Environmental and Social Impact into Evaluations completed a stocktaking of evaluation policy and guidance on social and environmental considerations and of practices of UNEG member evaluation offices in addressing social and environmental considerations (UNEG Working Group, 2020).[3] The stocktaking is to contribute to deliberations about a common UN-wide approach for incorporating environmental and social considerations into all evaluations (whether or not the evaluand is an environmental program). The second stocktaking was conducted by the Sustainability

---

[2] I use the concept of *coupled systems* to refer to the dynamic complex relationships across intimately connected human and natural systems (Liu, 2007; Ostrom, 1990; Rowe, 2019b) and *nexus* to refer to evaluation of sustainable development with interlinked social, economic and environmental dimensions (Uitto, 2019).

[3] UNEG is a professional network that brings together the evaluation units in the UN system, including the various UN departments, specialized agencies, funds and programs (http://unevaluation.org/) as well as non-UN organisations such as the GEF IEO.

Working Group of the Canadian Evaluation Society (CES) for two purposes: to assess the extent to which sustainability has been addressed in federal evaluations and by other governments and organizations in Canada and by Canadian evaluators working internationally, and to assess the intellectual infrastructure for evaluating sustainability in Canada and the United States (CES, 2020). The CES stocktaking is informing consideration of how the CES can mainstream sustainability in its own work and in evaluation in Canada. The CES stocktaking report was completed in 2020 with much of the work undertaken on a pro-bono basis by four leading Canadian consulting firms.[4]

These two undertakings cover a wide swath of global evaluation with UNEG addressing development evaluation and the CES addressing evaluation at national and sub-national levels while also assessing the Canadian and U.S. intellectual infrastructure for mainstreaming sustainability. Together, these two stocktaking efforts provide powerful evidence that the evaluation field is, at best, mildly and only recently addressing sustainability and that the social dimension is the priority for evaluation.

The two stocktaking efforts clearly showed that sustainability is largely missing in action from evaluation in the UN system and in Canada, and from the intellectual infrastructure for evaluation in the United States and Canada.

- The UNEG stocktaking also revealed that, first, coverage of the social system is also only partial and, despite heightened awareness of social–natural systems interaction, evaluation guidance on environment is extremely limited; and second, that the over-arching need emerging from documentary analysis and survey responses of UNEG member agencies is for a comprehensive document providing advice on how to evaluate the interactions among social and environmental considerations within the framework of UN activities

in support of the SDGs (UNEG Working Group, 2020, p. 6).
- The CES stocktaking showed sustainability and consideration of the natural system to be largely missing from federal evaluations conducted in 2016–2018, with Global Affairs Canada being a notable exception, and that the intellectual infrastructure in Canada and the United States for evaluation in the natural system is very limited.

The Canadian stocktaking is worth highlighting given the strong and long-standing evaluation infrastructure:

- The CES is the elder national evaluation organization among its global peers, membership per capita is highest relative to peer organizations, national training programs have been in place since the mid-1990s, and the CES developed the first evaluator credentialing in 2009.
- The Canadian government enacted a government-wide measurement and evaluation system in 1977 and the National Evaluation Policy in 1994 and 2001, requiring all federal programs and initiatives of material importance (roughly greater than $5 million CDN) to be evaluated at least once every 5 years. This ensured that all federal departments have a strong evaluation function and that supporting evaluation in their departments and responding to evaluations is an important part of the performance criteria of federal senior managers.
- Provinces and territories also have evaluation functions and requirements, as do other levels of government such as school boards and health agencies.

For evaluation function and infrastructure, Canada is a global leader. Canada also has signed most international climate and sustainability protocols and agreements and the elected government platform and positions have, since 2015, accorded sustainability and climate a strong priority.

Given the relative strength of evaluation in Canada and wide acceptance of the importance

---

[4]Baastel, Goss Gilroy, Prairie Research, and Universalia.

of climate and sustainability, it is reasonable to expect more positive observations than the sustainability stocktaking showed. The stocktaking had four elements:

1. A review of all federal evaluations from 2016–2018 revealed only a very tiny portion addressing nexus or sustainability. Global Affairs Canada was the leader, associated with its responsibilities for international climate and sustainability agreements. Natural resource-focused departments only evaluated human system effects; that is, departments in the Canadian government whose mandates included natural resources conducted evaluations from an extraction stance.

2. A review of Canadian philanthropic, nongovernmental, and First Nation evaluations did not identify much in the way of evaluations addressing nexus, although they did address natural systems when this was the focus of funding. Evaluations from these sectors rarely considered both human and natural systems.

3. Examination of whether Canadian-based evaluators working internationally considered the natural system and nexus did identify international examples where this occurred.

4. And perhaps most concerning, the intellectual infrastructure for nexus evaluation or even just evaluation of natural system effects is almost asymptotic to zero; that is, the natural system does not appear in peer-reviewed evaluation literature in Canada and the United States,[5] conference presentations, gray literature, and professional and university-based training. For example, just 4% of published papers in the four leading North American evaluation journals addressed natural system matters and only a few of these addressed nexus.

The findings of the two stocktakings are sobering but also encouraging. They are sobering in their confirmation that the evaluation field has little or no presence and little existing capacity in

contributing to sustainability, the leading issue of the day. But we can find encouragement because they clearly point to a growing recognition that sustainability is a top matter and to an interest in addressing sustainability as a priority.

Given the similarity of findings of the UNEG and Canadian efforts, a search for the systematic origins for the clear prioritization in evaluation of the human over the natural system, and the separation of the two systems, is reasonable. The next section proposes that the origins lie in a dominion-infused worldview asserting that humans are imbued with rights over all else—basically colonization of the planet to serve humans. Accountability structures have served as an important mechanism framing evaluation from a dominion perspective, and global and national governance units have sought to capture the resulting siloed landscape.

Dominion, Accountability, and Institutional Capture

The two stocktaking efforts clearly show that evaluation strongly prioritizes social matters, has very limited capacity to address natural systems,[6] and only rarely, across the vast landscape of evaluations covered by the two stocktaking efforts, are the two systems, human and natural, considered together.

I offer an explanation that evaluation rests on knowledge that itself rests on a worldview of dominion in which humans, and especially humans of European origin, have dominion over all other living and nonliving things and regard these as resources for use as humans see fit. Social inquiry and social research methods are said to be the rootstock of evaluation (Alkin, 2004), but I argue that they draw their nutrition from the terroir of dominion (Rowe, 2019b). The other rootstock of evaluation is said to be accountability. This management construct is layered on top of dominion and is the second causal force that has contributed to an almost monastic focus

---

[5]The stocktaking was limited to Canada and the United States for purposes of feasibility only.

[6]Natural systems are inclusive of environmental impacts as addressed by the UNEG stocktaking.

on the human system by bounding accountability, and consequently evaluation, to the intent and boundaries of interventions severing or at least loosening connection to the public policy goals for which they exist and to other efforts addressing those goals. Third, without structure that recognizes the connectivity between human and natural system goals and the dependence of the social system on the national systems, the SDGs offer a goal structure, initially and still today, in which the natural system does not need to be considered. This section provides a brief overview of how dominion, accountability, and institutional capture contribute to evaluation's overwhelming focus on the social system and neglect of the natural system, as reflected by the UNEG stocktaking.

Dominion

Evaluating sustainability first requires systematically recognizing and addressing those elements of both the human and natural systems that influence and are influenced by the evaluand. The stocktaking efforts showed evaluation to have an overwhelming focus on the human system, reflecting a dominion worldview where humans are ascendant and all other things, living and natural, can be extracted and deployed for human use. This is an implausible position: If human life depends on what we draw from the natural system, then the natural system must have value to the human system. The position that the natural system has no value and need not be considered has deep roots in social science and economics, which in turn are rooted in Judeo Christian worldviews and associated beliefs about dominion. Dominion is quite a simple concept whose existence is undeniable but, like any deeply embedded concept, it can be challenging to recognize and address. Dominion also provides a causal connectivity between the treatment of colonized and subjugated peoples and the treatment of other species and elements in the natural system. Indeed, one of the rationales for the actions of colonizers was the superiority of their worldview over the very different worldviews of many

of the colonized peoples who regarded themselves and other living and nonliving things as equal and part of a whole.

Dominion means that other living and natural things do not have value, that they exist to serve humans, and any monetary value ascribed to them results from ownership or regulated rights that provide the ability to control access and use. A classic example of dominion in action was the construction of massive dams for electrical generation in pursuit of industrial and economic development. Early critiques and resulting modification of cost benefit and other analysis of dams recognized and evaluated the direct losses to humans above and below the dams. But only recently have the ecosystem losses from flooding above the dam and water loss below the dam begun to be imputed, although on a limited basis. Because living and natural things other than humans were not valued, no mechanism was in place to recognize their importance and scarcity, directly causing relatively unfettered extraction and destruction—the fundamental cause of the sustainability crisis and climate change.

The issue of temporal and spatial scales is another way that dominion and accountability have led to evaluation's monastic focus on the human system. Systems are by their nature coupled, extensive, and dynamic, each with a wide range of temporal and spatial scales and often very diverse units of account (Rowe, 2012). Human temporal and spatial scales differ significantly from scales relevant to the natural system, and, of course, with a dominion-infused worldview, the units of account that matters are human. When the natural system is considered, it is usually from an extraction perspective in terms of utility to humans, not as a coupled system meriting its own place in evaluations.

Evaluation is a human system activity usually conducted from temporal scales meaningful to the aspects of the human system that is commissioning and undertaking the evaluation. By their nature, effects of a human or natural intervention have broad reach, well beyond the temporal and spatial reach of the intervention. Evaluations are aligned with the programmatic schedules of interventions and usually extend backward to the start of the intervention and forward to some programmatic or arbitrary time, usually less than 10 years from their start. These temporal scales bear no relevance for the temporal scales of natural system elements that can range from centuries to moments.

The value and function of natural systems is not the only consequence of dominion. Clearly, racism and misogyny are causally linked to the dominion of white, European-origin males. To illustrate, a 1987 synthesis of two national 1986 studies in the United States found that race was the was the most significant factor in locating toxic landfills and that 3 of 5 Black and Hispanic Americans, and approximately half of all Asians, Pacific Islanders, and American Indians, lived in communities with uncontrolled toxic waste sites (Gilio-Whitaker, 2019). And while the roots of racism, misogyny, and extraction are commonly and firmly planted in dominion, actions on these matters are often pitted against one another using class, religion, nationality, and other constructs, and all and each constrained by what is deemed possible within capitalism and not overly deleterious to economic growth.

Accountability

Accountability is cited as one of the main stems of evaluation (Alkin, 2004); from the perspective of sustainability, accountability can be described as a highly evolved contagion. It is a management construct designed to enable monitoring and improvement of agreed outcomes and is usually linked to program, management, and personnel performance. Managers and programs seek to constrain risk of falling short on accountability metrics by focusing on what they have the authority, resources, and capacities to be able to likely achieve. This provides incentives to narrow the programmatic box for which they are accountable and to resist being accountable for contributing usefully to other boxes.

The two stocktakings observe that the natural system, sustainability, and the nexus are systematically absent from evaluations. However, the remit of some agencies does address the natural

system, such as the Global Environment Facility (GEF), UN Environment, national government departments such as environment and natural resources, and environmental NGOs. The evaluation record of these is mixed; while the GEF Independent Evaluation Office addresses both systems and incorporates nexus, other natural system agencies focus almost exclusively on the natural system. We can generalize this by framing evaluations as single system (either human or natural) or two system (Rowe, 2012).

Evaluations are overwhelmingly single system, a situation to which accountability frames contribute. Since natural system values are infrequently considered, accountability reinforces ignoring the natural system. We know that even within the human system we must recognize and incorporate connectivity to reach to public policy goals. Reinforcing and incentivizing partitions between human and natural systems and within human systems accountability reinforces silos, the opposite of the silo busting required for evaluation at the nexus and for evaluation more generally.

Evaluating sustainability requires evaluation practices and methods that (a) recognize and operate at the nexus where both human and natural systems are present and (b) address the intrinsic coupling between and within human and natural systems. It is bad enough that the natural system is not valued and that systems approaches and understanding are unlikely with political and administrative partitioning. Accountability reinforces and further constrains possibilities of addressing sustainability in programming and evaluation with its focus on "accountability scales" that rarely reach beyond the accountability frame of the intervention.

One result is that the responsibility and remit of the intervention and reach of its direct effects frame the spatial scale for the evaluation within the larger framing of governance structures such as local area, province, or country, or within the remits of the responsible government organization. Ecosystems and landscapes provide more relevant spatial framing for natural systems; there is no reason to expect the boundaries and shapes of ecosystems to align with human system political and administrative boundaries or program areas. And ecosystems are not always appropriate for the territory of an organism—for example, a wolf, whale, or snail function across or entirely within an ecosystem. At a minimum, the relevant spatial scales for the natural system can be thought of as an ecosystem, often highly coupled with other ecosystems. This presumption that boundaries and territories in the natural system will align with the political and administrative boundaries of the human system is not limited to the natural system. For example, the same assumption is made about boundaries of Indigenous traditional lands, and that the Canada/United States border is relevant or appropriate where it crosses traditional lands. For many Indigenous peoples, the relevant spatial boundaries are their traditional territories from which food; medicine; and spiritual, ceremonial, and community values are drawn (Gilio-Whitaker, 2019). Instead, evaluation is likely to address the spatial scales defined by colonial occupation such as a reserve or First Nation territory; these are always and importantly smaller than traditional territories and often exclude areas of high importance to Indigenous peoples.

Program managers, evaluators, and especially evaluation commissioners often insist that an evaluation be conducted within the frame of the stated goals and operations of (accountability of) the intervention. This severs interventions from each other and limits the reach of evaluation, falling well short of the critically important public policy goals such as ending poverty or achieving sustainability. As Williams (2019) observed, such a frame establishes a program and evaluation ecosystem where programs systematically are assessed as providing positive contributions to the broad goal and where no progress is visible toward achieving the goal itself. It also creates a systematic positive bias in evaluation (Rowe, 2019b).

Institutional Capture

Institutional capture is the process by which identified needs and demands for major structural change are captured by existing structures, policies, and approaches. The SDGs were such a moment when sustainability was recognized as an overriding priority requiring major structural change to address. By and large, responsibilities for individual SDGs were assigned without changing the partitioned structure of organizations. But successfully addressing sustainability programmatically or in evaluation requires platforms suited to the task; the partitioned structures, policies, and approaches are not well suited to pursuit and evaluation of sustainability. Understandably, the UN and other multilateral organizations staked claims on specific SDGs, pursuing the assurance this provided to their futures; some such as the United Nations Development Programme (UNDP), the United Nations Industrial Development Organization (UNIDO), and the International Fund for Agricultural Development (IFAD) now explicitly recognize this connectivity, while others are on the pathway to do so.

The evaluation criteria of the Development Assistance Committee of the Organisation for Economic Co-operation and Development (OECD DAC) Network on Development Evaluation (2019) address sustainability as sustaining interventions and achievement of impacts, and not, as most think of sustainability, as a nexus concept of human and natural systems together with emphasis on sustaining the capacity of the natural system to enable life. In this, evaluation is somewhat distinct—elsewhere sustainability is recognized as a science "with a room of its own" (Clark, 2007); the 2009 Nobel Prize in Economics was awarded to Elinor Ostrom for her work on the commons and as one of the founders of coupled human and natural systems (CHANS) analysis. And, as the UNEG and CES stocktaking efforts have shown, evaluation has also been largely captured by the institutions it serves.

Sustainability-Ready Evaluation

Evaluators are good observers and place confidence in good evidence. They will increasingly be persuaded by the emerging knowledge on sustainability and climate, and increasingly recognize how these have affected the human issues and populations that have been evaluators' primary concern. They will also recognize how their long-preferred interventions and methods in the human system can contribute to worsening climate and sustainability. The underlying premise of sustainability-ready evaluation is that evaluators will recognize the need to address effects in the natural as well as the human system and take evaluation to a place where existing capacities are insufficient. Evaluators will need to, for example: recognize, speak, and hear representatives of natural system knowledge; learn how to feasibly address dynamically coupled systems (Liu, 2007); incorporate effects that have widely differing temporal and spatial scales and very differently framed units of account; and be open to and advocate for shared evaluation functions (see Carugi & Bryant, 2019; Rowe, 2012; Uitto, 2019).

Other fields of inquiry and assessment will be important contributors to developing and implementing evaluation at nexus settings. Evaluation is a cross-disciplinary field accustomed to drawing from other fields of inquiry, and this is fortunate because evaluating sustainability will require knowledge from and engagement from more system sciences. Climate and materials sciences, ecology, and geography will be important as will knowledge from more focused fields such as energy engineering, biology, agriculture, forestry and fisheries, and areas of public administration such as procurement. Two connected fields concerned with understanding and assessing nexus will likely be critically valuable fellow travelers: sustainability science (Kates, 2011; Clark et al., 2016) and CHANS work and networks (Liu, 2007; Ostrom, 1990).

Strongly siloed culture, structures, and practices of evaluation and programs create challenges to mainstreaming sustainability in the nexus sense of human and natural systems. To truly incorporate the natural system into long-standing and newer interventions whose primary focus is in the human system is proving difficult; likewise, to get evaluations to address the natural system is challenging, as shown by the UNEG stocktaking. However, the effort does appear to be gaining some momentum, such as in research on environmental effects of refugee camps (Braun et al., 2016), although discarded Covid-19 face masks are already finding their way to landfills and water bodies. As Fabien Cousteau (2020) wrote recently,

> We live in a closed-loop system. We can't actually throw things "away." The plastic we toss in the garbage often just ends up inside the bodies of marine animals, before finding its way back inside of us. (para. 12)

This means that what are usually classed as unexpected or unintended effects, or effects that were known but ignored because they lay outside the accountability frame of the intervention, now have to be recognized as a direct effect of the intervention. I have shown (Rowe, 2018, 2019a, b) that ignoring direct effects in the natural system imparts a systematic positive bias to evaluations. To make the point clear, evaluation conducted in silos has a systematic positive bias favorable to the intervention and, importantly, arising because of the accountability frames that are applied as discussed above.

Sustainability-ready evaluation is an evaluation function that is ready to recognize these connections and able to cross them. It is an evaluation function with individual evaluators and evaluation organizations that are enthused by contributing to a future we choose (Figueres & Rivett-Carnac, 2020). There are many strongly held visions of what that future should be, with associated and strongly held views of what we need to change to get there. It is not the job of evaluation to pick a pathway or end point; our job is to be enthused and capable of contributing to improvement, including sorting and valuing the competing pathways and desired new ways. Evaluation today is appropriately described as close to sustainability-ignorant and far from sustainability-ready.

How Can Evaluation Contribute to Checkmating Extinction?

An evaluation able to contribute to the defeat of extinction requires some relatively simple changes in how we frame and undertake our work, but these simple changes will significantly alter the stance and thus the politics of evaluation. Here I briefly sketch some important changes in stance for an evaluation fit for purpose for the endgame.

Checkmating extinction will only be possible if evaluation shifts from a singular focus on the human system to mainstreaming nexus in all evaluation. We are right now at a juncture where urgently needed changes seem possible. A second major change in the stance of evaluation relates to expectations of goal achievement: The current standard of progressing toward goals merely draws out a checkmate in favor of extinction. Instead, overcoming extinction requires a stance at the endpoint and assesses achievement of these goals with evaluation providing guidance to improve performance. Of course, achieving these goals requires joined-up, system-wide efforts for which we need to join evaluation stances with systems approaches. Conditions are worsening faster than expected and efforts to understand status and trends in natural systems and options for mitigation and adaptation are generating new knowledge at a rapid pace. This means that the stance of evaluation must be nimble and adaptive to integrate these changes, and be undertaken with sufficient rapidity to align with significantly accelerated decision cycles. Together, all of this means that evaluation for the endgame must be relentlessly use seeking and forward looking.

These are but some of the features needed for an evaluation function and practice that is an ally in efforts to checkmate extinction. Consideration of this stance will identify additional necessary features and perhaps diminish the importance of some that are discussed below. This chapter is only an early step in identifying the stance needed for an evaluation that contributes to the endgame.

Recognizing Natural Systems as the Foundation for the Human System Means Adding the Natural System Perspective to All Evaluation Criteria

The opponent at the endgame is continued destruction of the natural system by humans, meaning that both systems must be considered and addressed by evaluation at the endgame. That is, nexus is the required position for evaluation at the endgame.

Think of the relationship between human and natural systems with the natural system as a bank account. The human system has well exceeded its overdraft limit so that now every draw we make must have a repayment schedule that not only matches current withdrawals but also systematically and strategically starts to reduce the overdraft.

Environmental and social safeguards and policies have been enacted by most development donors with the requirement that they are applied in project development, funding, implementation, operation, and assessment (IFAD, 2018; World Bank, 2020). These standards are relatively recent, most enacted in the past decade, and the documents clearly consider human and natural systems as connected. In practice, however, climate and environment/natural resource management are usually treated as additional criteria that must be addressed in project design and assessment, isolated and marginalized rather than imbedded into planning.

We can consider inclusion of the natural system criteria in four phases, defined by requirements to meet the threshold to achieve a "satisfactory" rating:

1. Ignored: In this phase, environment (and climate) were rarely addressed, development was the priority and equity issues were important. ***Result: Increase in the overdraft on the natural system account.***

2. Good intentions: Environment and climate were noted in this second phase, often with what could be described as a faith-based

approach. It was not unusual to see project designs, evaluations, and supervision reports that considered commitments to compliance with donor environmental guidelines and safeguards and with national regulations to warrant a satisfactory rating. To put this in perspective, I have never seen an evaluation of an education or health intervention make a statement such as, "The design of the intervention incorporated government guidelines and a designated body has the authority to inspect and enforce, so we deem the approach satisfactory." Substitute environment for education in the previous sentence and we have a statement that is frequently made about natural resources, sustainability, and climate in supervision and evaluation documents. *Result: Increase in the overdraft on the natural system account.*

3. Do no harm: With these emerging approaches, achieving a satisfactory rating for climate and environment requires plausible design and implementation resources and responsibilities such that the intervention will not harm the environment or ignore climate. Empirical evidence is not required for a satisfactory rating but might become an expectation. Use of less harmful practices for continued resource extraction, such as climate-smart agriculture, species-specific fishing gear, protection of mangroves, forest management, and methods in road building, are deemed to not harm and so warrant a satisfactory rating. In effect, this is a type of double counting with the natural system benefits, such as improved irrigation and soil condition, required to restore production levels and support previous harmful agricultural projection practices. *Result: End of continued withdrawals on the natural system account but accumulated overdraft not addressed.*

4. Evaluation we need: In the fourth phase, evaluation for the endgame, achieving a satisfactory rating requires that restorative actions for the natural system are confirmable, central, and substantial parts of project design, operations, and adaptive management. *Result: Paying down the overdraft; learning and dif-*

fusion provide positive prospects that this will continue and accelerate.

Mainstreaming sustainability systematically locates evaluation at the nexus and is a first and essential change in the stance of evaluation; but valuing the natural system evaluation is beginning to address dominion.

Evaluation Standards Will Emphasize Achieving the Larger Goals Identified as Central to Checkmating Extinction

When the end is in sight, when the endgame is what is at play, our focus shifts from playing the game well (admirable evaluation) and from contributing to incremental improvements for beneficiaries to an absolute need to provide value to checkmating our destruction of the natural system that sustains us.

To illustrate the character of absolute evaluation standards, the International Resources Panel (IRP)[7] has shown that the planet does not have the material resources to provide for expansion of existing cities and creation of new ones resulting from urbanization, rural-to-urban migration, and population increase (Swilling, 2018). Development projects typically claim they will "contribute to" slowing rural-to-urban migration through improved rural livelihoods. Rural-to-urban migration and population growth are complicated and involve a powerful mix of push and pull factors requiring combined programmatic efforts to achieve sustainable flows and levels that will contribute to sustainable development, not undermine it. This is an illustration of a goal important for the endgame; evaluation needs to assess against achievement of that goal. If population and urban growth threaten sustainability, then the standard that needs to be applied in evaluation is achieving the goals that will remove population increase and rural-to-urban migration as important threats to sustainability. This does not mean curtailing migration and mobility, which are important to escaping severe climate

---

[7] https://www.resourcepanel.org/

and for humanitarian and economic reasons. Achieving levels of rural-to-urban migration sustainable for both urban and rural areas likely hinges on viable rural communities. And evaluation can provide value in moving from current unsustainable flows by adopting a stance that includes an expectation of verifiable achievement of important endgame outcomes that will realize specific migration goals set at sustainable levels. These goals, like the high-end climate goals of a CO_2 reduction to limit temperature increase to $1.5°C$, should be specified in absolute terms; for example, the specific sustainable population of Vancouver or Hanoi.

Standards Need to Shift to Evaluating Against Collective Achievement of Sustainability Goals, and Away from Likely Contributions by Partitioned Organizations and Interventions

Achieving the results needed to checkmate extinction requires collective and synthesized efforts; this is the required stance of evaluation for the endgame.

Partitions must be replaced by joined-up action and evaluation must adopt a collaborative focus on system achievement of the larger goals required for sustainability, regardless of whether interventions have adopted this stance. Holding interventions accountable for achieving results for which they are neither resourced nor authorized is inappropriate. However, for the endgame, evaluators should still address the needed result, what is required to achieve it, and the success and contributions of efforts toward collectively addressing this result. Setting goals that are critically important to success in the endgame is one way evaluation can observe shortcomings in collaboration and shared efforts toward achievement. It will also reveal gaps between current and needed achievements that likely span a number of individual organizational remits. This type of evaluation, focusing on what is needed, reflects the spirit of a results focus but from a collective, joined-up perspective rather than from parti-

tioned efforts. It promotes collective action and accountability for sustainability goals.

Sustainability Is Imbedded in All Evaluation Criteria Reflecting Nexus, Not Isolated as a Free-Standing Criterion

An evaluation stance recognizing the complex connectivity of human and natural systems means that all evaluation criteria should be considered from a two-system stance—sustainability and climate should not be isolated in separate and usually marginalized criteria.

Collective action means that work toward any and all of the SDGs and government and third-sector initiatives is likely to be drawing from and contributing to the sustainability of the natural system and climate. The previous element brings these into the scope of evaluation for the endgame, and this element addresses how evaluation accomplishes this. Each of the evaluation criteria and standards, e.g. the OECD DAC criteria (relevance, coherence, effectiveness, efficiency, impact, and sustainability), needs to be infused with considerations of sustainability by addressing both human and natural systems. Examples include the effect of humanitarian efforts on the physical landscape, and the many effects on the natural system of the use of plastics.

Evaluation Standards at the Endgame: Evaluating with Rapid Change and Uncertainty

Relentless rapid learning and brisk adaptation is the temporal scale required for interventions at the endgame and so must also be for evaluation.

Sustainability and climate are topics where the knowledge and practice base is improving rapidly and still features considerable uncertainty and ambiguity. Where changes in our knowledge are proceeding at a rapid pace and where considerable ambiguity still exists, longer term interventions—such as 4 or more years—will inevitably be suboptimal by the time they are halfway through their remit, perhaps highly suboptimal. Those implementing interventions must adopt vigorous adaptive management practices and be held accountable for this. We need to accelerate the pace of reflection and renewal, or else an important portion of our efforts will be applying approaches that are no longer considered efficacious at a time when we can least afford to do so. Evaluation is an important vehicle for this.

At the endgame, knowledge cycles are greatly reduced—we now think that the shelf life of some current climate knowledge is about 2 years. Severe climate events are also accelerating, becoming more frequent and severe and building cumulative effects. Two category 5 storms and resulting flooding within one month, as happened in 2020 in the Caribbean, requires very different responses than two storms of equal strength separated by 10 years.

To illustrate, consider 2030 and 2050 as forecasts of when we will pass irreversible thresholds, which make them key timings for checkmating extinction. A large portion of program and project cycles approach 7 or more years from inception to renewal, with 1–2 years for planning and funding, 1–2 years to mobilize, then operations of 4–5 years. Seven-year program cycles gives us just one program cycle until 2030 and four until 2050. The typical mid-term, end-of-term, and later ex-post evaluation approaches cannot provide information, insights, and advice in time to affect interventions in much more rapid adaptation cycles. Some evaluation approaches and methods will need to adapt rapidly and significantly to be relevant to evaluation at the endgame; fortunately, other approaches and methods are more fit for this purpose. Longer term evaluation undertakings will still provide value, such as with longer term impacts and adaptation of interventions to changing conditions, but overall, evaluation at the endgame is a new challenge for the field, requiring the evaluation stance to immediately become shorter term and employ more rapid approaches that are

relentlessly use seeking such as Rapid Impact Evaluation (Rowe, 2019a).

Evaluation for the Endgame Relentlessly Pursues Use

We no longer have the luxury to indulge the evaluation agendas and strategies that do not contribute to checkmating extinction. Our work must focus directly and strongly on the rapid adaptation and learning cycles of a proliferating landscape of actions contributing (or not) to checkmating extinction.

Conclusion: Nexus Requires New Rootstock to Grow Relevant Evaluation Functions

This chapter recognizes that we have entered the endgame of extinction and identifies what is needed for evaluation to contribute to checkmating extinction. I have sketched a trail from where evaluation is today to where it needs to be to provide value and guidance to efforts to achieve a checkmate favorable to life on the planet.

That trail first observes that evaluation at global and national levels is monastically focused on the human system and only marginally addresses the natural system. It reaches back to Judeo-Christian concepts of dominion as the origin story for our focus, and identifies narrowly framed accountability structures as an important contemporary mechanism for the exercise of dominion. Reinforcing this is institutional capture of efforts to infuse sustainability and systematically address necessary climate goals in development and associated social ambitions at all levels. The unhappy result is seen in two recent stocktaking efforts illustrating the limited contributions of contemporary evaluation to sustainability.

Evaluation at the endgame is different from the evaluation we have known and practiced up to now. Evaluation will need to take stances that will be challenging, as is any endgame effort. The six characteristics of evaluation for the endgame are:

1. The opponent at the endgame is continued destruction of the natural system by humans, meaning that both systems must be considered and addressed by evaluation at the endgame. Nexus is the required position for evaluation at the endgame.
2. When the end is in sight, our focus shifts from playing the game well (admirable evaluation) and from contributing to incremental improvements for beneficiaries to an absolute need to provide value to checkmating our destruction of the natural system that sustains us.
3. Achieving the results needed to checkmate extinction requires collective and synthesized effort, which is the required stance of evaluation for the endgame.
4. An evaluation stance recognizing the complex connectivity of human and natural systems means that all evaluation criteria should be considered from a two-system stance—sustainability and climate should not be isolated in separate and usually marginalized criteria.
5. Relentless rapid learning and brisk adaptation is the temporal scale required for interventions at the endgame and so must also be for evaluation.
6. We no longer have the luxury to indulge the evaluation agendas and strategies that do not contribute to checkmating extinction and our work must focus directly and strongly on the rapid adaptation and learning cycles of a proliferating landscape of actions contributing (or not) to checkmating extinction.

Adopting these stances at first appears to be a radical shift for evaluation, one with poor prospects for adoption. However, a growing recognition of the sustainability and climate imperative is underway. Evaluation working with biophysical knowledge partners is able right now to usefully contribute to the endgame. The hard part is recognizing that the prevailing stance of evaluation is contributing to the problem, that we need

to turn our backs on forces and institutional arrangements that have provided us comfort in exchange for complicity, and turn to a future we choose, which is to be a valued and useful contributor to checkmating extinction.

References

Adyel, T. M. (2020). Accumulation of plastic waste during COVID-19. *Science, 369*(6509), 1314–1315. https://doi.org/10.1126/science.abd9925.

Alkin, M. C. (2004). *Evaluation roots: Tracing theorists' views and influences*. Sage.

Better Evaluation. (2020). *Footprint evaluation*. https://www.betterevaluation.org/en/themes/footprint_evaluation

Braun, A., Lang, S., & Hochschild, V. (2016). Impact of refugee camps on their environment: A Case study using multi-temporal SAR data. *Journal of Geography, Environment and Earth Science International, 42*(2), 1–17. https://doi.org/10.9734/JGEESI/2016/22392.

Carugi, C., & Bryant, H. (2019). A joint evaluation with lessons for the Sustainable Development Goals era: The joint GEF-UNDP evaluation of the Small Grants Programme. *American Journal of Evaluation, 41*(2), 182–200. https://doi.org/10.1177/1098214019865936.

Chelimsky, E. (2012). Public-interest values and program sustainability: Some implications for evaluation practice. *American Journal of Evaluation, 35*(4), 527–542. https://doi.org/10.1177/1098214014549068.

Christie, C. A., & Alkin, M. C. (2008). Evaluation theory tree re-examined. *Studies in Educational Evaluation, 34*(3), 131–135. https://doi.org/10.1016/j.stueduc.2008.07.001.

Clark, W. C. (2007). Sustainability science: A room of its own. *PNAS, 104*(6), 1737–1738. https://doi.org/10.1073/pnas.0611291104.

Clark, W. C., van Kerkhoff, L., Lebel, L., & Gallopin, G. C. (2016). *Crafting usable knowledge for sustainable development*. National Academy of Sciences.

Cousteau, F. (2020, December 9). Our oceans, our future. *New York Times*. https://www.nytimes.com/2020/12/09/opinion/covid-climate-change-ocean.html

Figueres, C., & Rivett-Carnac, T. (2020). *The future we choose: Surviving the climate crisis*. Knopf.

Gilio-Whitaker, D. (2019). *As long as grass grows: Indigenous fight for environmental justice, from colonization to Standing Rock*. Beacon.

International Fund for Agricultural Development. (2018). *IFAD's social, environmental and climate assessment procedures (SECAP)* (2017 ed.). Author.

Kates, R. W. (2011). What kind of a science is sustainability science? *PNAS, 108*(49), 19449–19450. https://doi.org/10.1073/pnas.1116097108.

Liu, J. D. (2007). Coupled human and natural systems. *Ambio, 36*(8), 639–649. https://doi.org/10.1579/0044-7447(2007)36[639:CHANS]2.0.CO;2.

Nadeau, R. L. (2006). *The environmental endgame.* Rutgers.

Nadeau, R. (2008, March 19). Brother, can you spare me a planet? (extended version). *Scientific American.* https://www.scientificamerican.com/article/brother-can-you-spare-me-a-planet/

OECD/DAC Network on Development Evaluation. (2019). *Better criteria for better evaluation.* https://www.oecd.org/dac/evaluation/daccriteriaforevaluatingdevelopmentassistance.htm

Ostrom, E. (1990). *Governing the commons: The evolution of institutions for collective action: Political economy of institutions and decisions.* Cambridge University Press.

Patton, M. Q. (2020a). *Blue Marble evaluation.* Guilford.

Patton, M. Q. (2020b). Evaluation criteria for evaluating transformation: Implications for the coronavirus pandemic and the global climate emergency. *American Journal of Evaluation,* 1–37. https://doi.org/10.1177/1098214020933689.

Rowe, A. (2012). Evaluation of natural resource interventions. *American Journal of Evaluation, 33*(3), 384–394. https://doi.org/10.1177/1098214012440026.

Rowe, A. (2018). Ecological thinking as a route to sustainability-ready evaluation. In R. Hopson & F. Cram (Eds.), *Tackling wicked problems in complex ecologies* (pp. 25–44). Stanford University Press.

Rowe, A. (2019a). Rapid impact evaluation. *Evaluation, 25*(4), 496–513. https://doi.org/10.1177/1356389019870213.

Rowe, A. (2019b). Sustainability-ready evaluation: A call to action. *New Directions in Evaluation, 162*, 29–48. https://doi.org/10.1002/ev.20365.

Sustainability Working Group, Canadian Evaluation Society. (2020). *Report on stocktaking for sustainability-ready evaluation (draft).* (Unpublished manuscript).

Swilling, M. H. (2018). *The weight of cities: Resource requirements of future urbanization.* United Nations Environmental Programme, International Resources Panel. https://www.resourcepanel.org/reports/weight-cities.

World Bank. (2020). *The environmental and social framework.* https://projects.worldbank.org/en/projects-operations/environmental-and-social-policies#safeguards

United Nations Evaluation Group Working Group on Integrating Environmental and Social Impact into Evaluations. (2020). *Stock-taking exercise on policies and guidance of UN agencies in support of evaluation of social and environmental considerations, Vol I Main Report.* United Nations Evaluation Group. http://www.unevaluation.org/document/download/3712.

Uitto, J. I. (2019). Sustainable development evaluation: Understanding the nexus of natural and human systems. *New Directions in Evaluation, 2019*(162), 49–67. https://doi.org/10.1002/ev.20364.

Williams, R. (2019). Evaluation in a dangerous time: Reflections on 4 years in a central policy agency in the Government of Nova Scotia. *Evaluation Matters—He Take Tō Te Aromatawai, 5,* 41–62. https://doi.org/10.18296/em.0039.

Zak, D. (2019). How should we talk about what's happening to our planet? *Washington Post,* August 27. https://www.washingtonpost.com/lifestyle/style/how-should-we-talk-about-whats-happening-to-our-planet/2019/08/26/d28c4bcc-b213-11e9-8f6c-7828e68cb15f_story.html

Importance and Utilization of Theory-Based Evaluations in the Context of Sustainable Development and Social-Ecological Systems

Takaaki Miyaguchi

Abstract

Numerous challenges confront the task of evaluating sustainable development—its complex nature, complementary evaluation criteria, and the difficulty of evaluation at the nexus of human and natural systems. Theory-based evaluation, drawn from critical realism, is well suited to this task. When constructing a program theory/theory of change for evaluating sustainable development, concepts of socioecological systems and coupled human and natural systems are useful. The chapter discusses four modes of inference and the application of different theory-based evaluation approaches. It introduces the CHANS (coupled human and natural systems) framework, a holistic, analytical framework that is useful in evaluating such complex, social-ecological systems and resonates with the challenging elements of sustainable development evaluation.

Since the adoption of the 2030 Agenda for Sustainable Development and its Sustainable Development Goals (SDGs) in 2015, the United Nations member states and various stakeholders all over the world have been galvanizing their efforts to contribute to the achievement of the SDGs. Although the SDGs themselves were the result of international negotiation and consensus among the member states, the breadth of the partnerships and collaborations among non-state actors, including the private sector, nongovernment organizations, and nonprofit organizations, has been unprecedented.

The SDGs, of course, are not without critics. Some argue that these goals are nothing but a wish list (Hickel, 2015), while others point out the inconsistencies and incompatibility among 169 targets and their indicators and question the abilities and capacities of many states, especially those of developing countries, to adequately monitor and evaluate the current and future status toward achieving these SDGs (Leal Filho et al., 2019; Pongiglione, 2015; Stokstad, 2015).

The focus of this chapter is to look at the challenges in evaluating the status of sustainable development, which requires looking into the nexus of human and natural systems, and introduce the utility of theory-based evaluation for such purposes. The chapter introduces a holistic framework called CHANS (coupled human and natural systems), an analytical framework that is useful in evaluating such complex, social-ecological systems.

T. Miyaguchi (✉)
Kyoto University of Foreign Studies, Kyoto, Japan
e-mail: t_miyagu@kufs.ac.jp

J. I. Uitto, G. Batra (eds.), *Transformational Change for People and the Planet*, Sustainable Development Goals Series, https://doi.org/10.1007/978-3-030-78853-7_15

Challenges in Evaluating Sustainable Development

We all know that humankind should strive for sustainable development as the concept is declared and promised with the SDGs. However, evaluating the status of and progress toward "sustainable development" is extremely difficult.

Sustainable development is a concept that is not just complicated—with interventions involving multiple components, multiple agencies, and multiple simultaneous and/or alternative causal strands—but also complex, having recursive causality with reinforcing loops, disproportionate relationships with a tipping point, and emergent outcomes (Rogers, 2008). Such characteristics of sustainable development make evaluation practice all the more challenging. Rowe (2012, 2014) identified four types of challenges.

First is the challenge of *attribution*. Because the status of sustainable development is found at the nexus of human and natural systems, achieving sustainability means maintaining the integrity of the combined ecological–societal system (Kay & Boyle, 2008). One can therefore anticipate the difficulties in comparing and matching both human and natural systems against those interventions that take place from the human system (Rowe, 2012; Vaessen & Todd, 2008). Pinning down, let alone quantifying, the level of attribution (or causation) is almost impossible.

The second difficulty is one of *temporal scale*. Although temporal scales for measuring economic activities or wealth being generated can be as brief as quarterly, when we turn our attention to society, a decade or more is required for us to confirm change within any generation in that period. What presents the toughest challenge in evaluating sustainable development is related to ecological time scales. For example, to validate a change of climate through an increase or decrease of greenhouse gas emissions requires 100 years. Even 20 to 30 years is needed to witness any change in climate variability. These scales of ecological systems are beyond our socioeconomic scales.

Temporal scale also has an important subdimension: *spatial frames*. An ecological spatial frame, such as a tropical rainforest, does not respect political or societal boundaries or jurisdictions. Adequate evaluation faces a great challenge due to such ecological spatial characteristics. And our modern history offers ample evidence that such ecological timeframes or spatial frames have been blatantly ignored for short-term benefits to the economy and society.

The third challenging aspect relates to *values*—economic, societal, and environmental. What type of value we adopt is a pivotal question when evaluating progress toward achieving sustainable development. To evaluate such progress, we must identify a common type of value through which we can compare the effectiveness of the efforts toward it. One valuation type that has been overly used in our modern history has been economic, or monetary values. But one can fathom the limitations of relying solely on this dimension of value and trying to apply it to other dimensions, such as ethnic, religious, cultural, and biodiversity. The various methods developed mainly by economists allow us to put an (economic) value on natural resources (such as contingency valuation, hedonic pricing, or cost effectiveness analysis), but these are derived from and based only on the socioeconomic dimension and do not allow us to grasp the complex nature of social-ecological systems.

The fourth type of challenge is one of achieving *use* and *influence*. Numerous knowledge products and evaluation reports address sustainable development, but whether these products have been put to actual use is quite a different matter. Therefore, engaging decision makers and stakeholders in the evaluation process itself is vital so that they will put the results to use toward their decision-making processes.

In addition to these four types of challenges in evaluating sustainable development, we also see an aggregation challenge known as a micro-macro paradox (Uitto, 2014; Vaessen & Todd, 2008; Van den Berg & Cando-Noordhuizen, 2017). This refers to lack of coherence or effectiveness when many successes at a micro level do not accumulate accordingly to result in successes at a larger, macro scale. Such paradox stems from reductionism. The shortcomings of reductionism are made especially apparent when we deal with complex systems for which the whole is more than the sum of the parts (Bhaskar et al., 2010; Kay, 2008a).

Sustainable development maintains the integrity between socioeconomic and ecological systems. But more often than not, measuring, analyzing, and evaluating the status of or movement toward sustainable development has been influenced by social science disciplines rather than natural, biophysical sciences (Rowe, 2012). Such analysis leaves no doubt that all economic and social activities are based on a healthy environment and the finite resources existing on earth. Economic activity is, in effect, the conversion of material and energy from a natural resource pool as input with converted material and used energy as output. As ecological economist Herman Daly (1990) put it, there is no such thing as "sustainable growth" when every single economic activity is based on the natural resources existing on a finite planet. Although natural systems are thus the absolute foundation of all economic activities, the international discourse pertaining to sustainable development until now has been dominated by socioeconomic aspects—the human system side (Rowe, 2012, 2014).

However, the problem is not just over-reliance on social sciences; what matters is the polarization in which attempts to evaluate sustainable development happen only with either social science discipline or with natural science discipline—without their integration or synthesis. The natural ecosystems are diverse, complex, and dynamic; thus, traditional, disciplinary science is "not by itself sufficient for understanding and dealing with ecosystems" (Waltner-Toews et al., 2008, xii). In light of these current situations surrounding sustainable development evaluation efforts, we turn to theory-based evaluation and its approaches.

Theory-Based Evaluation

Before discussing theory-based evaluation and its approaches, we must clarify the term's meaning vis-à-vis other terms used in evaluation literature.

Theory-based evaluation (used by Weiss, 1997a) is, in short, a "plausible and sensible model of how the program is supposed to work" (Bickman, 1987). Other terms are interchangeable, such as logic model (Mathison, 2004), program theory (Bickman, 1990), the theory of action (Patton, 1997), theory of change (Weiss, 1997a), and theory-driven evaluation (Chen & Rossi, 1983). In this chapter, I use Weiss's terms *theory-based evaluation* and *theory of change*, which consists of *implementation theory* and *program theory*.[1]

According to Brousselle and Buregeya (2018), theory-based evaluation has emerged in reaction to current normal evaluation practice. They assert the need for a theory of change, not just for poorly formulated interventions, but especially when evaluating complex interventions. And theory-based evaluation and its approaches are "aimed at reinforcing the explanatory power of evaluations" (Weiss, 1997b).

Theory-based evaluation formulates program elements, rationale, and causal linkages. The atheoretical approach to evaluation has been characterized by "a step-by-step cookbook method of doing evaluations" (Chen, 1990). The atheoretical approach tends to focus on the relationship between inputs and effects without considering the transformational processes, referred to as "black box evaluations" (Norgbey & Spilsbury, 2014). Going beyond such atheoretical approach, theory-based evaluation takes into account the transformational processes inherent in the programs being evaluated (Chen, 1990).

Theory-based evaluation pays close attention to contextual conditions. According to Chen (1990), theory of change consists of two parts, normative theory and causative theory.[2] The causative theory "specifies how the program

---

[1] Funnell and Rogers (2011) reverses these terms so that *program theory* consists of *theory of action* and *theory of change*. Thus, somewhat confusingly, *program theory* by Weiss corresponds to *theory of change* by Funnell and Rogers. Since the terms by Weiss are used more often in international development and its evaluation field, I have adopted her terms in this chapter.

[2] According to the original terms adopted by Chen (1990), it is described as *program theory* (instead of *theory of change*, adopted by Weiss). However as explained in the previous footnote, Weiss's terminology, *theory of change*, is used in this chapter.

works by identifying the conditions under which certain processes will arise and what their likely consequences will be" (Chen, 1990).

With its focus on contextual conditions, theory-driven evaluation also shares three fundamental characteristics: (a) to explicate the theory of treatment by detailing the expected relationships among inputs, mediating processes, and short- and long-term outcomes; (b) to measure all of the constructs postulated in the theory; and (c) to analyze the data to assess the extent to which the postulated relationships actually occurred (Coryn et al., 2011; Shadish et al., 2002).

Several approaches stem from theory-based evaluation, including theory of change, realist evaluation, logic analysis, and contribution analysis. All of these approaches have philosophical and conceptual roots in a philosophy of science known as *critical realism* (Brousselle & Buregeya, 2018). And an origin in critical realism is deemed quite appropriate to evaluating sustainable development, which involves two-evaluand systems.

Critical Realism

Critical realism is a philosophy of science advocated by Roy Bhaskar. It originated as a critique of a deterministic worldview, which took the stance that if some factor X occurred—such as an intervention—then the observed result Y must follow (Forss et al., 2011). This philosophy can be understood through four modes of inference, distinction between open and closed systems, and explanatory power rather than prediction.

First, the four modes of inference are necessary to understanding critical realism. The first two, *deduction* and *induction*, are well known. Through deduction and induction inference, evaluators get to know what works (through deduction by applying a theory, and through induction with observations). The latter two modes of inference, *abduction* and *retroduction*, are less familiar. Abduction combines the deductive and inductive modes of inference and is defined as "working from consequence back to cause or antecedent" (Denzin, 2017, p. 100). In other

words, abduction means "to interpret and recontextualize individual phenomena within a conceptual framework to understand something in a new way" (Danermark et al., 2002, p. 80). In evaluation, this abduction inference is synonymous with constructing a program theory. According to Weiss (1997a), program theory refers to "the mechanisms that mediate between the delivery (and receipt) of the program and the emergence of the outcomes of interest" (p. 57). In other words, program theory is hypothesized causal linkages. In evaluation terms, then, it connotes for whom an intervention may work and, above all, how it works.

The fourth mode of inference, retroduction, provides the essence of this philosophy of science. Retroduction means to "reconstruct the basic conditions for these conceptually abstracted phenomena to be what they are" (Danermark et al., 2002, p. 80). It is one thing to talk about hypothesized (abstracted) causal linkages, but it is quite another to pay heed to the conditions under which such generative mechanisms can be triggered. Pawson and Tilley (1997), referring to this notion of critical realism, likened such conditions to a gunpower explosion that does not always take place when flame is applied, but also requires certain conditions, such the gunpower mixture being compacted, the structure not being damp and having sufficient quantity and oxygen, and heat applied long enough. Gunpower explosion functions as a generative mechanism and is synonymous with Weiss's program theory (Blamey & Mackenzie, 2007). In evaluation terms, through this fourth mode of inference, retroduction, evaluators can grasp what may work under what circumstances.

Therefore, through utilizing all four modes of inference described above, evaluators will be able to know what works, for whom, how, and under what circumstances. Theory-based evaluation and its approaches resonate quite well with this statement that is the essence of critical realism, and thus the root of theory-based evaluation.

The second component for understanding critical realism, as described by Bhaskar (2013), is the concept of the world as having three domains:

empirical (observable experiences), actual (a factual event that is generated by mechanisms), and real (the mechanisms that generate an event). These three domains establish a critical perspective in which the reality that scientists study is larger than only the empirical domain (Bhaskar, 2013).

Further understanding this concept requires a grasp of the difference between closed and open systems. A closed system is akin to an experiment in which a certain mechanism is tested in an isolated laboratory setting, allowing the mechanism to operate in isolation, independent of other mechanisms. An open system is akin to society itself, in which social events are the products of many simultaneously existing mechanisms, exemplifying the complex nature of society. Because society is inherently an open system, we must recognize that one cannot isolate a single social mechanism and do an experiment. The above-mentioned modes of inference in social science function as an experiment does in natural science (Danermark et al., 2002).

The third important element in understanding critical realism is the difference between *explanations* and *predictions*. In a closed system, explanations are synonymous with predictions, whereas explanations in an open system indicate tendencies. When attempting to seek external validity in an open system, one should seek explanations, rather than predictions or judgments (Allen, 2008), to reveal the causal mechanism hidden beneath the surface (Brousselle & Buregeya, 2018).

Importance of Theory-Based Evaluation Approaches

The school of theory-based evaluation includes approaches with different implications (Alkin, 2013). When choosing among them to evaluate sustainable development, knowing the strengths and weaknesses of the two theory-based evaluation approaches—realist approach and theory of change—is important. Evaluators need to be aware of these similar but distinct approaches and adopt the one that is appropriate to the purpose of the evaluation.

Realist approach is concerned with promising context-mechanism-outcome configurations (called CMO configurations; Pawson & Tilley, 1997). Utilizing this approach, evaluators can hypothesize various program theories to determine which are effective (or not) under certain circumstances. In other words, realist approach helps to deliver more precise and substantive program learning. At the same time, however, it is less appropriate for dealing with highly complex, multisite interventions with multiple outcomes (Blamey & Mackenzie, 2007). Theory of change, in contrast, is more concerned with overall program outcomes and helps to provide a strategic perspective on a complex program (Blamey & Mackenzie, 2007).

Theory-based evaluation approaches are appropriate for evaluating the status of and progress toward sustainable development, which is both complicated and complex. Based on the characteristics of theory-based evaluation approaches, prudent evaluators adopt appropriate approaches for different purposes. Evaluators should use the theory of change approach, for example, when evaluating the overall status of sustainable development, and choose the realist approach to hypothesize and understand certain program theories that are deemed effective for successful results within each program component. Constructing and analyzing a theory of change is an essential method for resolving the problems inherent in complex interventions (Dubois et al., 2011; Morell, 2010).

But how can we construct theories of change to apply to sustainable development evaluation? How do we assess emergent and anticipated outcomes resulting from relationships that are sometimes non-linear (Morell, 2010; Shiell et al., 2008), and how do we deal with uncertainty created by complex, self-organizing systems (Kay, 2008a)?

According to Funnell and Rogers (2011), theories of change can be constructed in three ways.[3] *Stakeholder mental model* is articulated according to how stakeholders believe a program will achieve what it is designed to do. Through *deductive approach*, a theory of change uses formal and informal documentation and research theories about a program and the needs it is intended to address. And last, *inductive approach* "involves observing the program in action and deriving the theories that are implicit in people's actions when implementing the program" (Funnell & Rogers, 2011, p. 111).

Out of these three techniques, however, there is an over-reliance on the deductive approach for theory development, with as many as 91% of analyzed cases reported to have used this approach, compared to 49% for the stakeholder mental model and 13% for the inductive development approach (Coryn et al., 2011). Predominantly, these theories are derived from social sciences. Scriven (2012) pointed out a strong tendency of professional evaluators to specialize in just one of the many branches of evaluation and only one area of human activity, further narrowing the scope of evaluation and thereby increasing difficulties in evaluating sustainable development.

This discussion of approaches has two important points. First, we find fewer cases of constructing theories of change from a natural science-based standpoint. And second, hardly any theory of change construction integrates both social science and natural science; rather, evaluators have tended one way, using either social science-based or natural science-based approaches (Rowe, 2012).

If we are to evaluate sustainable development at the nexus between human and natural systems, evaluators should integrate both social and natural sciences in constructing and hypothesizing theories of change, especially when the status of

sustainable development is about maintaining the integrity among society, economy, and environment.

Coupled Human and Natural Systems (CHANS)

Just as social sector problems and their evaluations have been dominated by the social sciences and their theories, the aspect of sustainability—especially within the context of ecological sustainability—has been equally dominated by natural, biophysical scientists. However, dealing with both social and ecological systems requires analyses that involve several components from each system, such as research on energy-water nexus and food-energy-water nexus. Despite this, studies on nexuses with three and four nodes are still very rare (Liu, Hull, Yang, Viña, Chen, et al., 2016).

One promising theoretical framework for understanding the mutual interactions and feedback mechanisms between human and natural systems has been advocated and advanced by Nobel laureate Elinor Ostrom in her pioneering work on social-ecological systems. Her research was concerned mainly with natural resources, especially common pooled resources, and provided a strong foundation to further understand the governance for successfully managing the *commons*, once considered impossible for an economic, rational, decision-maker worldview (Folke, 2007; Liu et al., 2007; McGinnis & Ostrom, 2014; Ostrom, 1990).

The essence of this so-called adaptive management and governance is about two-way interactions and feedback loops found between social-ecological systems (Evans, 2012). What Ostrom's work demonstrated was that socioeconomic entities such as fishing villages could change their way of governing themselves, adapting their decision-making rules and procedures in reaction to a situation such as a change in the ecological status of their surroundings. The related research has resulted in a general framework for analyzing sustainability of social-ecological sys-

---

[3]The original text of Funnell and Rogers (2011) used the term *program theory* instead of *theory of change*, but I have used theory of change, an interchangeable term by Weiss, to be consistent with the selection of evaluation terms and concepts in this chapter.

tems, fully taking into account both human and natural systems (Ostrom, 2009).

Stemming from Ostrom's work on adaptive management is another insightful analytical framework for understanding social-ecological systems, called the coupled human and natural systems (CHANS) framework. The primary focus of Ostrom's research was on common-pool resources in which the ecological system was either unowned or ownership was shared. However, the CHANS analytical framework goes well beyond the scale of common-pool resources and can thus provide helpful new insights that apply to the evaluation of sustainable development.

According to Liu, Hull, Carter, et al. (2016), the major barrier to effective implementation of sustainable development is the lack of sufficient knowledge about the complex relationships between humans and nature. The CHANS approach is intended "to serve as a pragmatic, heuristic tool for analyzing into relationships between people and the environment." The CHANS framework emphasizes that the human and natural components are coupled rather than separate (Carter et al., 2014, para. 6).

Among many other scholars, Ostrom has emphasized that context (i.e., not interventions themselves but the systems and subsystems that surround them, such as societal, political, and economic situations) does matter in analyzing the intricate interactions between human and natural systems. What is distinctive about CHANS is that it does not treat such contextual factors as external but as intrinsic elements within the framework. Researchers used a CHANS framework to conduct a 20-year-long study of social-ecological interactions that surround the biodiversity hot spot of the Wolong National Park of China, home to an endangered species of panda (*Ailuropoda melanoleuc*). These researchers proposed a framework that incorporates the human subsystem components such as communities and local residents, and the natural subsystem components such as wildlife and the land cover characterizing their habitat (Carter et al., 2014). The variety in the study's analyses was truly transdisciplinary. They included dedicated research on the influence and relationships within this coupled system

surrounding *Ailuropoda melanoleuc*, such as demography at household level and by distance and elevation level, education, energy transition, government policies, human dependence on ecosystem, infrastructure, livestock and livestock-panda interactions, payment for ecosystem services, scenario analysis and modeling, and spatial and tree distribution (Liu, Hull, Yang, Viña, Chen, et al., 2016).

Resonating well with the characteristics of sustainable development—complex systems involving both human and natural systems—and social and natural science disciplines, the CHANS framework "provides a platform for natural and social scientists to work together to quantify and integrate human-nature relationships at multiple organizational levels across space and over time" (Liu, Hull, Carter, et al., 2016).

Another characteristic of this framework is that it considers and treats the focal coupled system as an open system, rather than a closed system, placing the focal coupled system under specific social, economic, and political settings (Ostrom, 2009).

Why We Need a Framework Like CHANS

Especially when evaluating the complex systems of sustainable development, evaluators should consider adopting theory-based evaluation and its approaches instead of an oversimplified, one-size-fits-all, black box approach.

Among the seven traps[4] in constructing a theory of change proposed by Funnell and Rogers (2011), having "no actual theory" is on top of the list. In evaluating sustainable development, we especially need to avoid this trap by developing theories of change that are: (a) based on both social and natural sciences, (b) able to recognize

---

[4]They are: (1) no actual theory; (2) having a poor theory of change; (3) poorly specifying intended results; (4) ignoring unintended results; (5) oversimplifying; (6) not using the program theory for evaluation; and (7) taking a one-size-fits-all approach (Funnell & Rogers, 2011, p. 42).

the interactions between human and natural systems, and (c) capable of describing nonlinearity and emerging traits of complex systems and incorporating ecological temporal scale and spatial frames. Moreover, because theory-based evaluation is method neutral and suited to quantitative or qualitative methods, or both (Chen, 2005; Donaldson, 2007), the CHANS framework also offers flexibility for evaluators. CHANS can systematically guide researchers in analyzing complex sustainability issues surrounding socio-ecological systems.

Another valuable element of the CHANS framework is that it recognizes the importance of the participatory approach, or "putting researchers in the local residents' shoes" (Liu, Hull, Yang, Viña, An, et al., 2016). Many studies of social-ecological systems adopt "participatory approaches to identify, characterize, and solve management-related problems" (Norberg & Cumming, 2008, p. 238). The importance of such an approach goes beyond a specific set of rules of one method. Participatory approach is vital because complex systems cannot be captured by any single perspective and require a plurality of perspectives. Such plurality requires a variety of "forms of inquiry, inclusion of, and dialogue with persons representing different interests and different world views" (Waltner-Toews & Wall, 1997, p. 30). Because all coupled systems in question develop out of historical and cultural conditions, the future of such a system cannot have one single preferred state. As Kay (2008b) poignantly stated, researchers, if left to decide, will inquire into those aspects of the system that they themselves deem important; therefore, it is "crucial that the values, concerns, and knowledge of local stakeholders and actors be central to any inquiry" (p. 30).

Of course, this is not to claim that CHANS is the only framework through which we can evaluate sustainable development at the nexus of environment and development. However, evaluators should seek to use a framework that: (a) can encompass the complicated and complex nature of sustainable development; (b) is holistic, multilayered, and multiscaled; and (c) draws from both social and natural sciences, so that program theories develop using perspectives from both disciplines.

Appropriate Methodologies

CHANS appears to provide a useful framework for evaluating sustainable development. What can then be the appropriate methodologies and approaches for capturing such coupled systems? Evaluators have four types of methodologies to consider. First is *triangulation*, "the process of gathering scientific evidence about a system through a combination of laboratory, field, modeling, and historical investigations, facilitated by iterative and cross-disciplinary collaboration among research groups" (Plowright et al., 2008). When investigating the dynamism of complex systems, we cannot predict or reach a correct answer, because such is only possible based on a linear (irreversible, one-way) cause-and-effect worldview that excludes all influencing factors under a simple, laboratory-like system. To narrow the level of uncertainty and describe complex systems with more explanatory power, we need to shed light on the triangulation method. This method has been well practiced and its importance widely acknowledged among many evaluators (Carugi, 2016; Forss et al., 2011; Morra-Imas & Rist, 2009; Patton, 2002; Uitto, 2016).

The second type of methodology is *cross-scale/cross-layer comparison*. Complex social-ecological systems are nonlinear with reversible feedback loops, in requirement for multiple perspectives, and are multiscaled and multilayered (Waltner-Toews & Kay, 2008). Therefore, the ability to pursue several different lines of exploration at several different scales is necessary (Norberg & Cumming, 2008). For one example, analyzing or constructing simulation models only at a large, global scale (e.g., greenhouse gas emission modeling) would be inadequate; instead, the evaluator must compare different scales or layers within the systems. A local landscape is applied to a sub-watershed, which is made up of the ecological communities such as woodlots, wetlands, open fields, etc., each of which then is made up of individual species (Kay & Boyle, 2008).

To understand why certain social-ecological systems have not succeeded, we can conduct cross-scale/cross-layer comparisons and analyses at different spatial and temporal scales (Cumming, 2007; Ostrom, 2009). Evaluation already has a method that encompasses such nested nature models, called nested theories of change (Mayne, 2015; Richards, 2019; Riley et al., 2018). Although almost all the applied cases of nested theories of change in evaluation literature are found within the human (social) systems, evaluators in natural (ecological) systems can also adopt this method.

The third methodology type is *causal inference*. Even though the field of evaluation has been dominated by social scientists and their theories, the use of causal inference within natural science domains has begun to attract attention, notably in the cases of emerging infectious disease (Plowright et al., 2008) and global biodiversity scenarios and landscape ecology (Cumming, 2007). Thus, we see the utility of theory-based evaluation approaches even in the realm of natural science. Incorporating both natural science and social science perspectives in constructing theories of change is a prerequisite for starting to evaluate sustainable development; therefore, and an analytical framework like CHANS that enables such integration is necessary.

The final methodology is *cross-site synthesis* and *meta-analysis*. Because social-ecological systems are both complicated and complex, trying to identify a one-size-fits-all strategy will be in vain. At the same time, treating every single social-ecological system as a completely different and local incidence will not likely generate any externally valid insights that are generalizable to other parts of the world. Rather, to do so, "different ecological, socioeconomic, political, demographic, and/or cultural settings need to be synthesized" (Carter et al., 2014). Liu, Hull, Carter, et al. (2016) stressed the importance of seeking external validity and generalizability despite highly localized situations in each social-ecological system. They also advocated the importance of "model (social-ecological) systems," i.e., those that contain the core and essence of CHANS. By conducting cross-site syntheses or meta-analyses, CHANS researchers have been already able to identify some common aspects of

social-ecological complex systems that are applicable and spread across the globe (Carter et al., 2016).

Several CHANS sites have shared these common characteristics:

- Organizational—restoring reciprocal effects and feedbacks with nested hierarchies, indirect effects, emergent properties, vulnerability, and thresholds and resilience
- Spatial coupling—coupling across spatial scales, couplings beyond boundaries, and heterogeneity
- Temporal couplings—human impacts on natural systems, rising natural impacts on humans, legacy effects, time lags, increased scales and pace, and escalating indirect effects[5]

Evaluators are encouraged to start paying close attention to this research field on social-ecological systems and coalesce the previously separated efforts and research results from social and natural science into one, holistic framework such as CHANS.

Conclusion

Since the adoption of the 2030 Agenda for Sustainable Development in 2015, the concept and its goals have spread globally, with an increasing level of awareness and with inspiring, collaborative, multistakeholder implementation initiatives all over the world. At the outset, with 17 SDGs, the objectives seemed clear. However, beyond the political rhetoric of these goals and targets, we realize that we cannot declare achievement of sustainable development when all 169 targets are met *separately*. The essence of sustainable development is to acquire and maintain integrity among the three pillars—social, economic, and environmental. These three pillars are closely interlinked and interwoven. Accumulating each block or project successes from the micro

---

[5]For more details, refer to Liu, Hull, Yang, Viña, Chen, et al. (2016).

level will not lead to the macro-level integrity that these goals are seeking overall. Evaluators face a formidable task in evaluating sustainable development, homing in on the nexus between human and natural systems.

We face four types of challenges in evaluating sustainable development: the issue of attribution, temporal scale, the values, and achieving use and influence. At the same time, we also face an extra challenge of the micro-macro paradox. Theory-based evaluation and its approaches offer a means well suited to evaluating these complex systems that are multilayered, multiscaled, and span different time scales.

Theory-based evaluation has its roots in critical realism, a philosophy of science that emerged out of criticism against a deterministic worldview. Fully utilizing four modes of inference, critical realism can help reconstruct the basic conditions for certain phenomena to be what they are, by paying special attention to the context in which the specific generative mechanism is triggered.

Even though theory-based evaluation and its approaches are considered appropriate in evaluating complex systems, the theories of change that we develop and use tend to come predominantly from the social science discipline and be deductively constructed, instead of articulated by stakeholders or inductively constructed. When we deal with a social-ecological system, which is both complicated and complex, we need to develop theories of change that are based on well-developed principles from both the natural and social sciences—particularly ecology, economics, and political science—and we must confront this formidable task through comparative analyses of many cases (Walker et al., 2006).

This chapter introduced the useful analytical framework called CHANS (coupled human and natural systems) that is capable of addressing the issues mentioned above. This framework has a strong influence from Ostrom and her work on adaptive management and governance of resources held in common. CHANS emphasizes that human and natural components are coupled, rather than separate, and incorporates political and socioeconomic situations as an integral part

of the framework, rather than merely the external drivers of change.

By closely examining and applying the CHANS framework to ongoing and future programs concerned with achieving sustainable development, evaluators can address the four types of challenges in evaluating sustainable development. Although CHANS is not the only framework that facilitates addressing these issues and challenges, it has particular promise in supporting evaluation of sustainable development.

Knowing about a framework is one thing, but conducting actual analyses is quite another. However, the methodologies discussed here, such as triangulation, cross-scale/cross-layer comparisons, causal inference utilizing both social and natural science, and use of meta-analysis, are considered appropriate in evaluating social-ecological systems.

Although one might argue that no conceptual model exists for evaluating sustainable development with a holistic lens, using a framework like CHANS allows evaluators to construct theories of change and conduct subsequent analyses. At the same time, it supports specific analysis both quantitatively and qualitatively and utilizes both social and natural sciences.

Evaluating outcomes that a program cannot hope to influence may be impossible. However, because the CHANS framework specifically focuses on the interlinkages and mutual influence at the nexus between environment and development, it enables analysis, if not outright attribution, of a level of contribution to long-term outcomes that are seemingly outside of a program's direct scope.

With the recent increase in the level of awareness and attention to the concept of sustainable development and its goals, we should soon see more evaluations of subjects that would traditionally be considered outside the (narrow) scope of a program. Theory-based evaluation and its approaches, with the support of an analytical framework like CHANS, should be a great resource for our continuous and collaborative efforts in evaluating sustainable development.

References

Alkin, M. C. (2013). *Evaluation roots: A wider perspective of theorists' views and influences* (2nd ed.). Sage.

Allen, T. (2008). Scale and type: A requirement for addressing complexity with dynamical quality. In D. Waltner-Toews, J. J. Kay, & N.-M. E. Lister (Eds.), *The ecosystem approach: Complexity, uncertainty, and managing for sustainability* (pp. 37–49). Columbia University.

Bhaskar, R. (2013). *A realist theory of science.* Routledge.

Bhaskar, R., Frank, C., Parker, J., & Høyer, K. G. (2010). *Interdisciplinarity and climate change: Transforming knowledge and practice for our global future.* Taylor & Francis.

Bickman, L. (1987). *Using program theory in evaluation.* Jossey-Bass.

Bickman, L. (1990). Editor's notes: Advances in program theory. *New Directions for Program Evaluation, 1990*(47), 1–4. https://doi.org/10.1002/ev.1550.

Blamey, A., & Mackenzie, M. (2007). Theories of change and realistic evaluation: Peas in a pod or apples and oranges? *Evaluation, 13*(4), 439–455. https://doi.org/10.1177/1356389007082129.

Brousselle, A., & Buregeya, J.-M. (2018). Theory-based evaluations: Framing the existence of a new theory in evaluation and the rise of the 5th generation. *Evaluation, 24*(2), 153–168. https://doi.org/10.1177/1356389018765487.

Carter, N. H., An, L., & Liu, J. (2016). Cross-site synthesis of complexity in coupled human and natural systems. In J. Liu, V. Hull, W. Yang, A. Viña, Z. Ouyang, & H. Zhang (Eds.), *Pandas and people: Coupling human and natural systems for sustainability* (pp. 203–217). Oxford University.

Carter, N. H., Viña, A., Hull, V., McConnell, W. J., Axinn, W., Ghimire, D., & Liu, J. (2014). Coupled human and natural systems approach to wildlife research and conservation. *Ecology and Society, 19*(3), 43. https://doi.org/10.5751/ES-06881-190343.

Carugi, C. (2016). Experiences with systematic triangulation at the global environment facility. *Evaluation and Program Planning, 55*, 55–66. https://doi.org/10.1016/j.evalprogplan.2015.12.001.

Chen, H.-T. (1990). *Theory-driven evaluations.* Sage.

Chen, H.-T. (2005). *Practical program evaluation: Assessing and improving planning, implementation, and effectiveness.* Sage.

Chen, H.-T., & Rossi, P. H. (1983). Evaluating with sense: The theory-driven approach. *Evaluation Review, 7*(3), 283–302.

Coryn, C. L., Noakes, L. A., Westine, C. D., & Schröter, D. C. (2011). A systematic review of theory-driven evaluation practice from 1990 to 2009. *American Journal of Evaluation, 32*(2), 199–226. https://doi.org/10.1177/1098214010389321.

Cumming, G. S. (2007). Global biodiversity scenarios and landscape ecology. *Landscape Ecology, 22*(5), 671–685.

Daly, H. E. (1990). Sustainable growth: A bad oxymoron. *Journal of Environmental Science & Health Part C, 8*(2), 401–407. https://doi.org/10.1080/10590509009373395.

Danermark, B., Ekström, M., Jakobsen, L., & Karlsson, J. C. (2002). *Explaining society: Critical realism in the social sciences.* Routledge.

Denzin, N. K. (2017). *The research act: A theoretical introduction to sociological methods.* Transaction.

Donaldson, S. I. (2007). *Program theory-driven evaluation science: Strategies and applications.* Routledge.

Dubois, N., Lloyd, S., Houle, J., Mercier, C., Brousselle, A., & Rey, L. (2011). Discussion: Practice-based evaluation as a response to adress intervention complexity. *The Canadian Journal of Program Evaluation/La Revue canadienne d'evaluation de programme, 26*(3), 105–113.

Evans, J. P. (2012). *Environmental governance.* Routledge.

Folke, C. (2007). Social–ecological systems and adaptive governance of the commons. *Ecological Research, 22*(1), 14–15. https://doi.org/10.1007/s11284-006-0074-0.

Forss, K., Marra, M., & Schwartz, R. (2011). *Evaluating the complex: Attribution, contribution, and beyond* (Vol. 1). Transaction Publishers.

Funnell, S. C., & Rogers, P. J. (2011). *Purposeful program theory: Effective use of theories of change and logic models.* Wiley.

Hickel, J. (2015). The problem with saving the world. *Jacobin Magazine*, 8. https://www.jacobinmag.com/2015/08/global-poverty-climate-change-sdgs/

Kay, J. J. (2008a). An introduction to systems thinking. In D. Waltner-Toews, J. J. Kay, & N.-M. E. Lister (Eds.), *The ecosystem approach: Complexity, uncertainty, and managing for sustainability* (pp. 3–13). Columbia University.

Kay, J. J. (2008b). Framing the situation: Developing a system description. In D. Waltner-Toews, J. J. Kay, & N.-M. E. Lister (Eds.), *The ecosystem approach: Complexity, uncertainty, and managing for sustainability* (pp. 15–34). Columbia University.

Kay, J. J., & Boyle, M. (2008). *Self-organizing, holarchic, open systems (SOHOs).* Columbia University.

Leal Filho, W., Tripathi, S. K., Andrade Guerra, J., Giné-Garriga, R., Orlovic Lovren, V., & Willats, J. (2019). Using the sustainable development goals towards a better understanding of sustainability challenges. *International Journal of Sustainable Development & World Ecology, 26*(2), 179–190. https://doi.org/10.1080/13504509.2018.1505674.

Liu, J., Dietz, T., Carpenter, S. R., Alberti, M., Folke, C., Moran, E., Pell, A. N., Deadman, P., Kratz, T., Lubchenco, J., Ostrom, E., Ouyang, Z., Provencher, W., Redman, C. L., Schneider, S. H., & Taylor, W. W. (2007). Complexity of coupled human and natural systems. *Science, 317*(5844), 1513–1516. https://doi.org/10.1126/science.1144004.

Liu, J., Hull, V., Carter, N., Viña, A., & Yang, W. (2016). Framing sustainability of coupled human and natural systems. In J. Liu, V. Hull, W. Yang, A. Viña, X. Chen, Z. Ouyang, & H. Zhang (Eds.), *Pandas and people:*

Coupling human and natural systems for sustainability (pp. 15–32). Oxford University.

Liu, J., Hull, V., Yang, W., Viña, A., An, L., Carter, N., Chen, X., Liu, W., Ouyang, Z., & Zhang, H. (2016). Lessons from local studies for global sustainability. In J. Liu, V. Hull, W. Yang, A. Viña, X. Chen, Z. Ouyang, & H. Zhang (Eds.), *Pandas and people: Coupling human and natural systems for sustainability* (pp. 240–252). Oxford University.

Liu, J., Hull, V., Yang, W., Viña, A., Chen, X., Ouyang, Z., & Zhang, H. (Eds.). (2016). *Pandas and people: Coupling human and natural systems for sustainability*. Oxford University Press.

Mathison, S. (2004). *Encyclopedia of evaluation*. Sage.

Mayne, J. (2015). Useful theory of change models. *Canadian Journal of Program Evaluation, 30*(2), 119–142. https://doi.org/10.3138/cjpe.230.

McGinnis, M. D., & Ostrom, E. (2014). Social-ecological system framework: Initial changes and continuing challenges. *Ecology and Society, 19*(2), 30. https://doi.org/10.5751/ES-06387-190230.

Morell, J. A. (2010). *Evaluation in the face of uncertainty: Anticipating surprise and responding to the inevitable*. Guilford.

Morra-Imas, L. G., & Rist, R. C. (2009). *The road to results: Designing and conducting effective development evaluations*. World Bank.

Norberg, J., & Cumming, G. (2008). Complexity theory for a sustainable future: Conclusions and outlook. In J. Norberg & G. Cumming (Eds.), *Complexity theory for a sustainable future* (pp. 277–293). Columbia University.

Norgbey, S., & Spilsbury, M. (2014). A programme theory approach to evaluating normative environmental interventions. In J. I. Uitto (Ed.), *Evaluating environment in international development* (pp. 123–149). Routledge.

Ostrom, E. (1990). *Governing the commons: The evolution of institutions for collective action*. Cambridge University.

Ostrom, E. (2009). A general framework for analyzing sustainability of social-ecological systems. *Science, 325*(5939), 419–422. https://doi.org/10.1126/science.1172133.

Patton, M. Q. (1997). *Utilization-focused evaluation*. Sage.

Patton, M. Q. (2002). *Qualitative research and evaluation methods* (3rd ed.). Sage.

Pawson, R., & Tilley, N. (1997). *Realistic evaluation*. Sage.

Plowright, R. K., Sokolow, S. H., Gorman, M. E., Daszak, P., & Foley, J. E. (2008). Causal inference in disease ecology: Investigating ecological drivers of disease emergence. *Frontiers in Ecology and the Environment, 6*(8), 420–429. https://doi.org/10.1890/070086.

Pongiglione, F. (2015). The need for a priority structure for the Sustainable Development Goals. *Journal of Global Ethics, 11*(1), 37–42. https://doi.org/10.1080/17449626.2014.1001912.

Richards, R. (2019). *The value of theory of change in large-scale projects and complex interventions* (K4D Helpdesk Report). Institute of Development Studies.

Riley, B. L., Kernoghan, A., Stockton, L., Montague, S., Yessis, J., & Willis, C. D. (2018). Using contribution analysis to evaluate the impacts of research on policy: Getting to 'good enough'. *Research Evaluation, 27*(1), 16–27. https://doi.org/10.1093/reseval/rvx037.

Rogers, P. J. (2008). Using programme theory to evaluate complicated and complex aspects of interventions. *Evaluation, 14*(1), 29–48. https://doi.org/10.1177/1356389007084674.

Rowe, A. (2012). Evaluation of natural resource interventions. *American Journal of Evaluation, 33*(3), 384–394. https://doi.org/10.1177/1098214012440026.

Rowe, A. (2014). Evaluation at the nexus: Principles for evaluating sustainable development interventions. In J. I. Uitto (Ed.), *Evaluating environment in international development* (pp. 69–85). Routledge.

Scriven, M. (2012). Conceptual revolutions in evaluation: Past, present, and future. In M. C. Alkin (Ed.), *Evaluation roots* (pp. 167–179). Sage.

Shadish, W. R., Cook, T. D., & Campbell, D. T. (2002). *Experimental and quasi-experimental designs for generalized causal inference*. Houghton Mifflin.

Shiell, A., Hawe, P., & Gold, L. (2008). Complex interventions or complex systems? Implications for health economic evaluation. *BMJ, 336*(7656), 1281–1283. https://doi.org/10.1136/bmj.39569.510521.AD.

Stokstad, E. (2015). Sustainable goals from UN under fire. *Science, 347*(6223), 702–703. https://doi.org/10.1126/science.347.6223.702.

Uitto, J. I. (2014). Evaluating environment and development: Lessons from international cooperation. *Evaluation, 20*(1), 44–57. https://doi.org/10.1177/1356389013517443.

Uitto, J. I. (2016). Evaluating the environment as a global public good. *Evaluation, 22*(1), 108–115. https://doi.org/10.1177/1356389015623135.

Vaessen, J., & Todd, D. (2008). Methodological challenges of evaluating the impact of the Global Environment Facility's biodiversity program. *Evaluation and Program Planning, 31*(3), 231–240. https://doi.org/10.1016/j.evalprogplan.2008.03.002.

Van den Berg, R. D., & Cando-Noordhuizen, L. (2017). Action on climate change: What does it mean and where does it lead to? In J. I. Uitto, J. Puri, & R. D. van den Berg (Eds.), *Evaluating climate change action for sustainable development* (pp. 13–34). Springer.

Walker, B. H., Anderies, J. M., Kinzig, A. P., & Ryan, P. (2006). Exploring resilience in social-ecological systems through comparative studies and theory development: Introduction to the special issue. *Ecology and Society, 11*(1).

Waltner-Toews, D., & Kay, J. J. (2008). Implementing an ecosystem approach: The diamond, AMESH, and their siblings. In D. Waltner-Toews, J. J. Kay, & N.-M. E. Lister (Eds.), *The ecosystem approach: Complexity, uncertainty, and managing for sustainability* (pp. 239–255). Columbia University.

Waltner-Toews, D., Lister, N., & Bocking, S. (2008). A preface. In D. Waltner-Toews, J. J. Kay, & N.-M. E. Lister (Eds.), *The ecosystem approach: Complexity, uncertainty, and managing for sustainability* (pp. ix–xviii). Columbia University.

Waltner-Toews, D., & Wall, E. (1997). Emergent perplexity: In search of post-normal questions for community and agroecosystem health. *Social Science & Medicine, 45*(11), 1741–1749.

Weiss, C. H. (1997a). *Evaluation: Methods for studying programs and policies* (2nd ed.). Prentice Hall.

Weiss, C. H. (1997b). Theory-based evaluation: Past, present, and future. *New Directions for Evaluation, 1997*(76), 41–55. https://doi.org/10.1002/ev.1086.

Pathway to the Transformative Policy of Agenda 2030: Evaluation of Finland's Sustainable Development Policy

Mari Räkköläinen and Anu Saxén

Abstract

Finland has been the first country in the world to conduct a comprehensive evaluation of the national implementation the Agenda 2030. The purpose of the evaluation was to support efficient implementation of the agenda by producing information on the nation's sustainability work for all administrative branches. The evaluation results are used for coherence in the policies and long-term sustainable development activities. The evaluation produced concrete recommendations on future directions for sustainable development policy. It also proposed future evaluation approaches.

In this chapter, the authors present the evaluation approach and discuss the key results and their usage. They identify the essential elements of the utility of the evaluation in contributing to national progress of sustainable development policy. The Agenda 2030 evaluation approach was developmentally oriented and conducted in a very participatory manner. The authors reflect on the evaluative lessons learned and future options. They encourage emphasis on learning throughout the evaluation process even more in policy-level evaluations, and special attention to usefulness of the evaluation results already in evaluation design. Designing inclusive evaluation processes is a crucial precondition for evidence-informed learning and decision making in promoting transformative policy in the country context.

Introduction to the Evaluation

Finland is committed to implementation of the United Nations (UN) 2030 Agenda for Sustainable Development and its goals, with a national policy to evaluate the implementation of Agenda 2030 once during every four-year electoral term. Finland was the first country in the world to conduct an impartial and independent evaluation of the national implementation of Agenda 2030. The evaluation, called PATH2030, was funded as part the government's 2018 analysis, assessment, and research activities (www.tietokayttoon.fi/en).

In its implementation plan for the 2030 Agenda, the Finnish government was committed to a comprehensive evaluation of the national implementation efforts of the agenda. The purpose of the evaluation was to support evidence-based decision making and knowledge management. This is important because efficient

M. Räkköläinen
Finnish Education Evaluation Centre, Helsinki, Finland
e-mail: mari.rakkolainen@karvi.fi

A. Saxén (✉)
Ministry for Foreign Affairs of Finland, Helsinki, Finland
e-mail: anu.saxen@formin.fi

J. I. Uitto, G. Batra (eds.), *Transformational Change for People and the Planet*, Sustainable Development Goals Series, https://doi.org/10.1007/978-3-030-78853-7_16

implementation of the universal 2030 Agenda requires an entirely new kind of comprehensive approach and coherence in the policies of the administrative branches. By producing information on the nation's sustainability work for all administrative branches, the evaluation results could be used for coherent and long-term sustainable development policy and activities. The evaluation also considered the status of sustainable development in the foreign policy sector; specifically, the government sought to assess how Finland's foreign policy in all administrative sectors promotes the achievement of the Agenda 2030 goals (Prime Minister's Office, Finland [PMO], 2017a, b).

The objective of the evaluation was to examine the state of sustainable development in light of national sustainability indicators, key sustainable development policies and objectives, and national implementation of Agenda 2030. The evaluation was expected to produce concrete recommendations on the future directions for Finland's sustainable development policy, taking into account different timespans and levels of ambition, and proposing future evaluation approaches. Thus, the evaluation results would also provide learning.

In its timely execution, the evaluation aimed to strengthen the knowledge base for updating the Finnish 2030 Agenda implementation plan after the parliamentary elections in 2019, and to give input into the preparation of the new Government Programme. It provided content for social policy debate preceding the parliamentary elections and it produced information on the sustainability work of ministries and relevant stakeholders. Further, the evaluation could serve as an input for Finland's next voluntary national review (VNR), a component of the United Nations' implementation process for Agenda 2030, intended in part to facilitate the sharing of successes, challenges and lessons learned.

The Finnish evaluation mainly examined the national-level implementation during the period following the adoption of the 2030 Agenda in early 2016. However, it also considered other significant public instruments at the national level, such as the Society's Commitment to Sustainable Development launched in 2013 (PMO, 2013) and the supporting Commitment2050 tool, "The Finland we want by 2050" (PMO, 2016); the government report on development policy (PMO, 2017a; VNS, 2016); and the government implementation plan of the 2030 Agenda (PMO, 2017b; VNS, 2017).

Preparations

Initial discussion of the evaluation began in 2017 between ministries and other implementing parties and planning began in 2018 under the leadership of the Prime Minister's Office. The assignment highlighted the involvement of stakeholders. Accordingly, the evaluation approach was developmentally oriented and strongly participatory by nature. It emphasized comprehensive participation of key actors and stakeholders in sustainable development policy. To promote learning and sharing, evaluators collected data using interactive workshops, interviews, forums, and surveys.

The Prime Minister's Office chaired the cross-administrative steering group with representatives from the ministries of environment, finance, foreign affairs, agriculture, and forestry. A broader advisory group also was nominated to ensure wider perspective of relevant experts and stakeholders and to strengthen the use and usefulness of the evaluation results. The Development Evaluation Unit of the Ministry for Foreign Affairs had an expert role in the evaluation steering group to comment on the evaluation design, methodology, and reporting. Conducting the evaluation was an multidisciplinary team with members from three Finnish organizations: think tank Demos Helsinki, the Helsinki Institute of Sustainability Science (HELSUS), and the Finnish Environmental Institute (Syke).

The implementation of the evaluation was an intensive process, with launch taking place in August 2018 and the results published just 7 months later, in March 2019. The evaluation title, PATH2030, describes the road map toward transformative policy as put forth in Agenda 2030 (Berg et al., 2019). The publication is part of the implementation of the 2018 Government Plan for

Analysis, Assessment and Research (see www. tietokayttoon.fi/en).

Dissemination

The results of the Finnish national evaluation of Agenda 2030 delivered an overall picture of the progress and the status of implementation. Moreover, the document enhanced awareness of the role evaluation can play in the implementation of policy goals and underpinning the Agenda's Sustainable Development Goals (SDGs).

Sustainable development policy is a broad subject to evaluate, with no single, right way to produce such a vast, national-level evaluation because much depends on the context of the country and there are many variables to consider. However, Finland wanted to share its experience and serve as a motivator for other countries to produce policy-level, strategic evaluations of Agenda 2030 goals. After publication, the evaluation provided input to several proceedings and motivated further international cooperation.

We presented the PATH2030 evaluation in 2019 at the Third International Conference on Evaluating Environment and Development in Prague, Czech Republic, organized by the Independent Evaluation Office of the Global Environment Facility (GEF), the Earth-Eval Community of Practice, and the International Development Evaluation Association (IDEAS). The VNR report was presented to the UN High-Level Political Forum on Sustainable Development in New York in 2020.

In this article, we present the evaluation's approach and discuss the key results and their usage. We also reflect on the evaluative lessons learned and future options. To aid in understanding the focus of the evaluation, we also briefly introduce the coordination model of sustainable development policy in Finland.

Implementation of the Evaluation

Focus of the Evaluation: Sustainable Development Policy and the Coordination Model

The PATH2030 evaluation focused on Finland's sustainable development policy and cross-administrative foreign policy. It examined the coordination model of sustainable development in Finland, presented in Fig. 1. The model covers stakeholders, networks, and documents that sup-

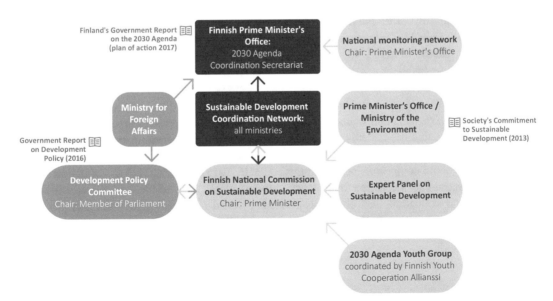

Fig. 1 Finland's sustainable development coordination model

port sustainable development policy at the national level. These are central to the coordination, management, monitoring, and continuous improvement of Finland's sustainable development policy, providing, for example, perspectives of young people and research.

Since 1993, the Finnish National Commission on Sustainable Development has acted as a coordinating body at the national level. The Prime Minister serves as chair and commission members represent broadly various sectors of society, from political decision making to ministries, research institutes, interest groups, and nongovernmental organizations. The work is supported by a secretariat at the Prime Minister's Office and by the Coordination Network of ministries. This network acts as a link between the various administrative sectors and national sustainable development policy. The Expert Panel on Sustainable Development, which has been in operation since 2013, consists of independent experts and researchers from a range of fields. In 2017, the Agenda 2030 Youth Group, coordinated by the Finnish Youth Cooperation Alliance, was established under the Finnish National Commission on Sustainable Development.

Finland's first national strategy for sustainable development was published in 1998 (Ministry of the Environment, 1998) and the second in 2006 (PMO, 2006). In 2013, the Finnish National Commission on Sustainable Development issued a new strategic statement called the Society's Commitment to Sustainable Development (PMO, 2013).

In the PATH2030 evaluation, the key concept is *sustainable development*. This concept has no single formulated definition, but it is commonly understood as development that meets the needs of the present without compromising the ability of future generations to meet their own needs. For this particular evaluation, central documents on sustainable development policy in Finland guided the conceptualization of sustainable development as a political objective; this process also helped in analyzing the coherence of concepts across those key documents (Berg et al., 2019).

Sustainable development policy, in its broadest sense, refers to all policies that affect the achievement of Finland's sustainable development targets. It may also, therefore, refer to policies that have not been included in the scope of sustainable development in previous declarations. For instance, this scope might include measures to combat climate change or to prevent the growth of societal inequality, or measures of economic policy that increase or decrease the total consumption of natural resources or the rate of employment (Berg et al., 2019).

The evaluation looked at the *operational model of sustainable development*, which refers to a comprehensive set of cross-administrative policies with the official mission of promoting sustainable development. Examples of these policies include budget reviews from a perspective of sustainable development and the integration of sustainable development as part of the strategies, measures, indicators, and evaluation of ministries (Berg et al., 2019).

The evaluation also studied foreign policy in all sectors of government. This refers to the Finnish government's aim that the nation, as a global partner, supports the sustainable development of developing countries through various means of foreign and security policy, such as trade policy and development policy (PMO, 2017a). According to the Government Programme, Finnish development policy emphasizes strengthening the business activities and tax bases of developing countries (PMO, 2015). The PATH2030 evaluation report included these government priorities, which are also linked to the UN processes that preceded the 2030 Agenda.

Evaluation Questions

The ultimate purpose of the evaluation was to create preconditions for coherent and long-term sustainable development policy and strengthen the knowledge base of the implementation plan for the 2030 Agenda. To achieve this, the evaluation needed to cover complex phenomena and the manifold policy context of the 2030 Agenda. Therefore, our evaluation team, steering group, and supporting group worked intensively to handle the comprehensive tasks required in the call for proposals and to define the main evaluation questions. With regard to assessing impact and effectiveness, we also had to take into account the short time span.

In the end, the final main evaluation questions related to:

- the state of sustainable development in Finland in light of indicators
- the main goals and means of the development policy
- challenges and strengths of sustainable development policy

In relation to foreign policy, the evaluation explored:

- the links and coherence between the different administrative branches of foreign policy and the sustainable development goals (focusing on international tax policy and trade policy)
- the different ways Finland's foreign policy can contribute to achievement of goals across all administrative branches

Carrying Out the Evaluation: Approach, Methods, and Process

The evaluation approach derived from the tradition of developmental evaluation, but it also relied on a theory-based analytical tool adapted for this evaluation. Theory-based assessment aims to understand both preconditions and mechanisms of implementation, and we paid attention to the theories of change behind the impact path-

ways (see Stame, 2004, 2006; Weiss, 1997a, b). Based on a desk study, the evaluation evolved around four central target areas:

1. The status of sustainable development
2. Theory of change behind sustainable development policy
3. Policy measures
4. Foreign policy

The 4I's framework (Brockhaus & Angelsen, 2012), in which sustainable development policy is analyzed through institutions, interests, ideas, and information, served as a key analysis structure for the evaluation. Using this framework, the evaluators analyzed how, at the institutional level, societal structures limit or promote development, how the interest of stakeholders gain a voice, and how different interest groups participate in the decision-making process. At the level of ideas and ideology, evaluators identified ideologies and explored how ideas have been accepted in politics. The evaluation team also studied the type of information that was used to support and guide policy. Table 1 presents the framework that was applied to the evaluation (Berg et al., 2019).

Table 1 Analytical framework of the PATH2030 evaluation

Category	Questions
Institutions (rules, path-dependencies or stickiness)	How do structures restrict/promote sustainable development policy? What are the issues that are hard/possible to change?
Interests (potential material advantages)	Actors' interests: Why does an actor lobby for a certain issue? Is it somehow beneficial? Are different opinions heard? Who may participate?
Ideas (policy discourses, underlying ideologies or beliefs)	What ideologies guide the action of different actors? What new ideas are emerged?
Information (data and knowledge, and their construction and use)	What kind of information is used in politics? Who has produced it?

Source: Berg et al. (2019)

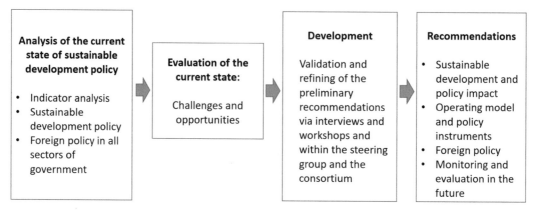

Fig. 2 The process of the PATH2030 evaluation. (Source: Berg et al., 2019)

The intensive evaluation process had three phases that formed the basis for conclusions and recommendations:

1. Analysis of the current state of sustainable development policy
2. Evaluation of the current state and its challenges and opportunities
3. Validation and development of the preliminary recommendations via interviews and workshops and within the steering group and the broader advisory group.

The evaluation team also liaised with the international evaluation community through EvalPartners and EVALSDGs to share ways of working and advice.

The evaluation explored the theory of change for Finnish sustainable development policy first and foremost by focusing on the most central documents—policy instruments for sustainable development—and the views of the representatives of ministries and other experts. The analysis sought to clarify not only the official goals but also the ways in which key actors perceive them (Berg et al., 2019). The evaluation team consulted the international SDG index (Sachs et al., 2018) and Finland's national sustainable development indicators.

The developmental aspect assured that the evaluation process was inclusive and included several participatory elements. The process gave different stakeholders in sustainable develop-

ment policy the opportunity to contribute to the evaluation and learn from each other's experience during the evaluation process. The evaluation used multidisciplinary methods and cross-sectoral data collection to acquire a wide, diverse set of material. The data consisted of both indicators and documents, with expert insight collected through questionnaires, interviews, roundtable discussions, and workshops. A total of 130 experts took part in the workshops, and we interviewed 80 bilaterally. The questionnaire produced 238 responses. The data collection process also included an international workshop held with partner organizations Stockholm Environment Institute (SEI) and Sustainable Development Solutions Network (SDSN). Figure 2 illustrates the process of the evaluation.

Key Messages from the Evaluation

The PATH2030 evaluation stated that in the future, Finland should focus on substantive issues where the nation still faces particular challenges or opportunities to progress. The evaluation found that Finland's strengths in implementation of sustainable development include societal stability, competence, and capability to mediate conflicts of interest. The biggest challenges lie in addressing climate change, the state of the environment, unsustainable consumption, and social inequality.

Although the evaluation concluded that sustainable development is broadly accepted in Finnish society, at the policy level, Finland is still missing clear common vision and a plausible plan for achieving policy goals. The evaluation also recognized that the theory of change for sustainable development needs to be clarified in terms of objectives and measures, and in use of indicators. Despite this, Finland's policy has succeeded in generating ownership and inclusiveness of sustainable development, but the evaluation noted room for improvement in policy coherence and capacity to produce transformative change. The evaluation suggested increasing proactive measures and enhancing coherence of sustainable development policy. At the institutional level, sustainable development has not yet been sufficiently integrated into all government sectors and management systems. Sustainable development is reflected quite well in strategies, but various ministries could include more management-level Agenda 2030 discussions. The evaluation also found that the systemic use of indicators and other data and knowledge in decision making and societal learning could be strengthened (Berg et al., 2019). Overall, the evaluation provides a wealth of information on the state of sustainable development and different solutions.

A key finding concerning Finland's foreign policy was that promoting sustainable development in cross-sectoral foreign policy has largely remained the responsibility of the Ministry for Foreign Affairs, especially its Department for Development Policy. Hence, the evaluation noted a need to enhance the impact and coherence of cross-administrative sustainable foreign policy (Berg et al., 2019).

Table 2 presents a summary of the evaluation's analysis of strengths and challenges (Berg et al., 2019).

The evaluation recommended that sustainable development should become the basis of future Government Programmes, that a roadmap for how to reach the goals should be created, and that, for example, the indicators and organization of sustainable development and the role of the expert panel on sustainable development should be revised. Promoting sustainable development in cross-sectoral foreign policy should be a central part of the Government Programme (Berg et al., 2019).

The evaluation also explored how the monitoring and evaluation of the SDGs could be organized in the future. It recommended strengthening the monitoring of Agenda 2030 by improving the usability of indicator data and by creating a more clearly visualized indicator system, which could serve as a broad-based, topical benchmark for discussion. More usable data would promote monitoring the achievement of goals and developmental trends. The evaluation also suggested that a systemic, cross-administrative evaluation system would help more systematically assess the impact of Finland's sustainable development policy and form a basis for the long-term follow up of the results. The systemic national level evaluation tool could increase knowledge of interconnection of activities between different administrative and policy sectors. In relation to decision making, the tool could strengthen the use of monitoring information and impact assessment as part of the policy for implementing the 2030 Agenda (Berg et al., 2019).

The Use and Usefulness of the Evaluation

At the national level, the results and recommendations of the Path2030 Evaluation were widely shared with stakeholders. Several discussions took place with political parties, four parliamentary committees reflected on the results together, and the evaluation team presented the results and organized workshops in several ministries and with the Finnish National Commission on Sustainable Development. The evaluation's timing was ideal in relation to the election and ongoing negotiation for the Government Programme in Finland. The Programme is very much built on the basis of Agenda 2030, as recommended by the PATH2030 evaluation. As a result, the Government decided to include a roadmap to achieve the SDG in its new 2019 Programme, in the form of a report on the global 2030 Agenda for Sustainable Development. The report, sub-

Table 2 Summary of analysis: strengths and challenges of Finland's sustainable development policy

Analytical pillar	Strengths	Challenges
Institutions	Finland has diverse participatory approaches to sustainable development.	Sustainable development has not been sufficiently integrated into all government sectors and their management systems.
	The pursuit of sustainable development is fairly visible, such as in the strategies of different ministries.	Government work on sustainable development is poorly resourced when the actual required workload is taken into consideration.
		Working in silos is still a core problem.
Interests	Widely shared aims and processes, such as the 2030 Agenda Government Report and reviews with a sustainable development angle (e.g., budgetary review), help mediate conflicts of interest.	Conflicts of interest (e.g., short- and long-term, different dimensions of sustainable development) decrease the coherence and transformational power of politics.
		Tightly defined commercial interests tend to outweigh sustainable development policy based on human rights.
Ideas	Sustainable development is a widely shared and mainstreamed aim.	In practice, many disagreements take place over sustainable development solutions.

(continued)

Table 2 (continued)

Analytical pillar	Strengths	Challenges
Information	A wealth of information is available on the state of sustainable development and different solutions.	The systemic use of indicators and research data in decision making and societal learning is insufficient.
		Understanding of cross-sectoral themes of sustainable development is underdeveloped and information on Finland's foreign policy aims is fragmented.

Source: Berg et al. (2019)

After publication of the PATH2030 evaluation, Helsinki hosted an international seminar on evaluation to connect the SDGs with national priorities. The event was jointly organized by the Ministry for Foreign Affairs of Finland, IIED, EVALSDGs, and UNICEF. The purpose of the joint seminar was to support country-led initiatives to evaluate national performance against Agenda 2030. Government representatives, evaluation specialists, and civil servants from more than 20 countries attended the seminar and participated in workshops on the role of evaluation in facilitating national policy dialogues and the evaluative challenge of the multidimensional characteristics of the SDGs. The main objectives were to learn from one another, share good practices, and produce an initial set of principles for SDG evaluation. Bringing people together helped identify learning needs and informed postworkshop knowledge products.

The seminar inspired emerging international cooperation that led to a coproduced handbook, *Evaluation to Connect the National Priorities with SDGs* (D'Errico et al., 2020). Although the 2030 Agenda has clear guidelines for follow-up, review, and reporting, the workshops in the seminar revealed a demand for clarity and support around how to conduct evaluations in a country-led process. The four organizations decided to meet this need together, collaborating on a guide

mitted to parliament in October 2020, serves as Finland's national implementation roadmap for achieving the objectives of the 2030 Agenda (VNS, 2020). The report presents the concrete actions the government is taking to promote the achievement of each of the 17 SDGs in Finland and globally.

to country-led SDG evaluations that would raise awareness of the role evaluation could play at the national level in the context of the 2030 Agenda. The PATH2030 evaluation was introduced and the guide was launched in New York at the UN's annual High-Level Political Forum on Sustainable Development. This event is the main intergovernmental platform for discussions about sustainable development, including the sharing of knowledge, learning, challenges, and success.

The guide has become a very timely resource for evaluation commissioners, managers, and professional evaluators who are seeking to create tailored plans and approaches to SDG evaluation in their country context. Local and national evaluators can now benefit from support in using SDG evaluations to improve policies and programs closer to home by applying tailored approaches. Rather than offering a one-size-fits-all model, the guide supports building successful evaluation around existing national context, underpinned by the principles of Agenda 2030.

When preparing evaluation approaches, commissioners, managers, and evaluation experts have to consider complex system interactions and interventions within Agenda 2030 policy implementation. Evaluation commissioners often struggle with how to prepare for an effective and useful evaluation in their country context. In the guide, experts emphasize the importance of identifying from the outset who will use the evaluation results, how they will use them, and why. After identifying this, commissioners can consider the different ways they could use evaluation to learn about SDG implementation and relationships with existing strategies, policies, and programs.

Connecting national priorities with SDGs—including 17 goals, 169 targets, and 232 indicators—is a challenge, and knowing and ensuring that the implementation is on the right track is not easy. The international expert group that prepared the guide reflected on some of the key issues in selecting the methodology and designing the evaluation setup for an SDG evaluation. They highlighted that successful evaluation usually draws on evidence from various sources. Therefore, the guide recommends integrating monitoring systems and indicators as part of the

evaluation, while giving specific evaluation methods the most robust role.

The experts observed that one of the greatest challenges is assessing integration in the context of multiple SDGs. Thus, the methodology and methods chosen must support evaluators in drawing overall conclusions from multiple findings. Selection of appropriate methodology depends on the evaluation questions, objectives, and use of the results. Participation of the key stakeholders in designing the evaluation approach also quite often contributes to choice of methodology, which increases the usefulness of the evaluation results (D'Errico et al., 2020).

The chosen evaluation approach and methodology have implications for utility. For a successful evaluation, keeping the scope manageable is important, as is limiting the number and diversity of the evaluation questions. For example, in the case of Finland's PATH2030 evaluation, the extremely complex task led to the original evaluation questions in the call for proposals being very challenging and complicated, and too numerous. After some radical revisions, the questions were clarified and simplified to better serve the purpose and usefulness of the Path2030 evaluation (Berg et al., 2019).

Below are some practical lessons learned for improving the utility of an SDG evaluation, drawn from reflections on the Path2030 evaluation.

- Plan the framework carefully and be focused.
- Keep the evaluation questions short and clear.
- Try to keep up participatory spirit—but make clear that all expectations cannot be met.
- Allocate enough time for the entire evaluation process.
- Write the report such that readers can readily understand the complex process and its results. Visualize and simplify the complex issues.
- Formulate the recommendations carefully to make them clear and easy to understand and adapt.
- Focus on the opportunities, not only the challenges.
- When communicating the results, connect to the actual challenges and focus on the most important priorities.

In national evaluations, a participatory approach enhances the usefulness of evaluation results. In organizations, designing inclusive processes is a crucial precondition for evidence-informed learning and decision making.

Evaluative Lessons for the Future

Inclusiveness and Participation

Discussing how Agenda 2030 evaluation approaches and processes can contribute to national progress on sustainable development is interesting and extremely important. The PATH2030 evaluation of Finland's sustainable development policy is an example of conducting an evaluation in a very participatory manner. The starting point for the evaluation was the tradition of developmental-oriented evaluation (see Patton, 1997, 2011), which is not a common approach in policy-level evaluations of SDGs. The evaluation focused more on systems and processes rather than only on end results. The developmental orientation was intended to serve both sharing and learning purposes and to bring together different interest groups and administrative sectors. Another important justification of the approach was that the collaborative methods would help in gathering valuable but undocumented data to identify a comprehensive status of Agenda 2030 implementation. Joint workshops revealed and produced both in-depth knowledge about how the indicators had been applied in practice and data used in the complicated context of decision making. Developmental evaluations are often long lasting, but this was not the case in Finland's exercise. Despite the intensity and short duration, the evaluation's developmental aspects were not undermined; on the contrary, we learned numerous useful lessons.

Although the Path2030 evaluation did not follow the most traditional principles of developmental evaluation, its success encourages us to apply this approach even more consistently in policy-level evaluations. The evaluation convinced us that using participatory methods consistently is possible—and worthwhile—even in a limited time span, and especially when the focus is on such a complex system. It is important to recognize the evaluation processes and structures that expand partnering, boost utilization of results, and lead to learning and transformative change. We have learned that recommendations become clearer, more concrete, and more realistic when they are formulated with the participation of stakeholders, civil servants, or policy makers who are the ultimate users of the evaluation results. Implementation of recommendations is effective if the evaluation contributes directly to ongoing reform or an organization's development process. And if the timing is not right, evaluation may not affect the policy reviews, strategies, or implementation of policies as planned.

Learning Throughout the Evaluation Process

Several proceedings and dissemination events after the publication of the Path2030 evaluation have increased the utility of the evaluation results. Enhancing learning does not mean only disseminating results. It also requires attention both to learning throughout the evaluation process and to learning from results after publication. From a learning perspective, it is more important that an evaluation is valid and fit for the purpose in the particular context than that it rigidly fulfills the requirements of comparability of the results. Locally designed evaluations (enabling ownership) to meet local conditions have proven to have a positive effect on learning and development, as long as other quality assurance elements are embedded in the evaluation system to ensure as much confidence and trust as possible. Many studies have shown that designing inclusive processes is a crucial precondition for evidence-informed learning and decision making in organizations or in country contexts. Therefore, allocating time for sharing and reflection throughout the evaluation process is important (e.g., Mayne, 2010, 2011; Palenberg et al., 2019; Räisänen & Räkköläinen, 2013, 2014; Räkköläinen, 2011; Vähämäki et al., 2011;

Table 3 Learning throughout the evaluation process

Evaluation design	Evaluative lessons learned
Methodological choices	
Contextualization	Embedding methods in context helps meet local conditions and enables recognition of tacit knowledge and know-how in order to learn why and how the results have been achieved.
Mixed-methods	Capturing the complexity by using a variety of methods and cross-sectoral data collection enhances validity, which is significant for learning.
Evaluation process	
Participation	Involvement of key stakeholders and beneficiaries with various backgrounds contributes collaborative learning and use of evaluation results, and gives opportunity for interactive implementation of the recommendations.
Inclusiveness	Structured opportunities and allocated time for reflection throughout the evaluation process contributes to evidence-informed learning.

Vähamäki & Verger, 2019; Young, 2019). Table 3 summarizes lessons learned that relate to supporting learning throughout the evaluation.

With regard to evaluation of SDGs and Agenda 2030 implementation, an important step is encouraging countries to develop their own monitoring, evaluation, and learning approaches and practical tools based on understanding of the variable needs in different contexts. Learning throughout the evaluation process should be considered beginning with the evaluation design and given special attention in the methodological choices. Concerning methodological development, one option at the local level could be to pilot the so-called real-time evaluation approach, which has its origin in developmental evaluation tradition (e.g., Cosgrave et al., 2009; Herson & Mitchell, 2005; Jamal & Crisp, 2002; Polastro, 2011). Real-time evaluation is normally associated with emergency response or humanitarian interventions because it is designed to provide immediate (real time) feedback to those in charge of interventions, programs, and projects. This feedback is usually provided during field work so that immediate improvements can be introduced and put into practice in timely manner. Real-time evaluations are often joint exercises that enable shared learning opportunities and enhance mutual accountability between different actors.

The COVID-19 pandemic has presented many obstacles to SDG evaluation—and all development evaluations. Conducting evaluations that capture the multidimensional characteristics of the SDGs is even more challenging when activities are restricted. The pandemic caused data collection problems in many countries, which may lead to a major evidence gap and make verifying the impact of the 2030 Agenda even more difficult. COVID-19 will also affect not only the way recommendations are formulated, but how they can be put into practice and linked to SDGs. The pandemic will have long-term effects on economies, jobs, livelihoods, and poverty; therefore, the evaluation approaches during this period should be strategic and forward looking. More solid national evaluation policies may also be necessary. Strengthening evaluation capacity will be instrumental in stimulating national-level ownership in evaluation and the use of evaluation results for transformative change toward sustainable development goals.

References

Berg, A., Lähteenoja, S., Ylönen, M., Korhonen-Kurki, K., Linko, T., Lonkila, K.-M., Lyytimäki, J., Salmivaara, A., Salo, H., Schönach, P., & Suutarinen, P. (2019). *PATH2030 – An evaluation of Finland's sustainable development policy. Publication series of the Government's analysis, assessment and research activities 23/2019*. Prime Minister's Office. http://urn.fi/URN:ISBN:978-952-287-655-3.

Brockhaus, M., & Angelsen, A. (2012). Seeing REDD+ through 4Is: A political economy framework. In A. Angelsen, M. Brockhaus, W. D. Sunderlin, & L. Verchot (Eds.), *Analysing REDD+: Challenges and choices*. Center for International Forestry Research (CIFOR).

Cosgrave, J., Ramalingan, B., & Beck, T. (2009). *Real-time evaluations of humanitarian action: An ALNAP guide*. Overseas Development Institute.

D'Errico, S., Geoghegan, T., & Piergallini, I. (Eds.). (2020). *Evaluation to connect national priorities with SDGs: A guide for evaluation commissioners and managers.* International Institute for Environment and Development. https://pubs.iied.org/pdfs/17739IIED.pdf.

Government Plan for Analysis, Assessment and Research for 2018. (2017). https://tietokayttoon.fi/en/government-plan-for-analysis-assessment-and-research

Herson, M., & Mitchell, J. (2005). Real-time evaluation: Where does its value lie? *Humanitarian Exchange, 32,* 43–45.

Jamal, A., & Crisp, J. (2002). Real-time humanitarian evaluations: Some frequently asked questions (EPAU/2002/05). UNHCR Evaluation and Policy Unit. http://www.unhcr.org/research/RESEARCH/3ce372204.pdf.

Mayne, J. (2010). Results management: Can results evidence gain a foothold in the public sector? In O. Reiper, F. Leeuw, & T. Ling (Eds.), *The evidence book* (pp. 117–150). Transaction.

Mayne, J. (2011). Contribution analysis: Addressing cause and effect. In R. Schwartz, K. Forss, & M. Marra (Eds.), *Evaluating the complex* (pp. 53–96). Transaction.

Ministry of the Environment. (1998). *Hallituksen kestävän kehityksen ohjelma. Valtioneuvoston periaatepäätös ekologisen kestävyyden edistämisestä. Suomen ympäristö 254.* Ympäristöministeriö, Helsinki.

Palenberg, M., Bartholomew, A., Mayne, J., Mäkelä, M., & Esche, L. (2019). *How do we learn, manage and make decisions in Finland's development policy and cooperation?* Ministry for Foreign Affairs of Finland.

Patton, M. Q. (1997). *Utilization-focused evaluation.* Sage.

Patton, M. Q. (2011). *Developmental evaluation: Applying complexity concepts to enhance innovation and use.* Guilford.

Polastro, R. (2011). Real time evaluations: Contributing to system-wide learning and accountability. *Humanitarian Exchange Magazine, 52,* 10–13. https://www.alnap.org/help-library/real-time-evaluations-contributing-to-system-wide-learning-and-accountability.

Prime Minister's Office, Finland. (2006). *Towards sustainable choices. A nationally and globally sustainable Finland. The national strategy for sustainable development.* Prime Minister's Office Publications. https://www.ym.fi/download/noname/%7B5D1F24EE-27D0-4E07-BAFE-AFCCAF451E8A%7D/97824

Prime Minister's Office, Finland. (2013). *The Finland we want by 2050 – Society's commitment to sustainable development.* https://kestavakehitys.fi/en/commitment2050

Prime Minister's Office, Finland. (2015). *Finland, a land of solutions. Strategic programme of Prime Minister Juha Sipilä's government, 29 May 2015.* Government Publications 12/2015. https://vnk.fi/documents/10616/1095776/Ratkaisujen+Suomi_EN.pdf/c2f3123a-d891-4451-884aa8cb6c747ddd/Ratkaisujen+Suomi_EN.pdf?version=1.0&t=1435215166000

Prime Minister's Office, Finland. (2016). *Society's commitment to sustainable development "The Finland we want by 2050" tool.* Adopted at the meeting of the Commission on Sustainable Development, 20 April 2016.

Prime Minister's Office, Finland. (2017a). *Valtioneuvoston selonteko kestävän kehityksen globaalista toimintaohjelmasta Agenda 2030:sta. Kestävän kehityksen Suomi – pitkäjänteisesti, johdonmukaisesti ja osallistavasti.* Prime Minister's Office Publications.

Prime Minister's Office, Finland. (2017b). *Government report on the implementation of the 2030 Agenda for Sustainable Development. Sustainable Development in Finland – Long-term coherent and inclusive action.* Prime Minister's Office Publications. https://julkaisut.valtioneuvosto.fi/bitstream/handle/10024/79455/VNK_J1117_Government_Report_2030Agenda_KANSILLA_netti.pdf?sequence=1.

Prime Minister's Office, Finland. (2020). *Voluntary national review 2020 Finland. Report on the implementation of the 2030 Agenda for Sustainable Development.* Government Administration Department Publications. http://urn.fi/URN:ISBN:978-952-287-947-9. https://julkaisut.valtioneuvosto.fi/bitstream/handle/10024/162268/VNK_2020_8_Voluntary_National_Review_Finland.pdf?sequence=4&isAllowed=y.

Räisänen, A., & Räkköläinen, M. (2013). Assessment of learning outcomes in Finnish vocational education and training. *Assessment in Education: Principles, Policy & Practice, 21*(1), 109–124. https://doi.org/10.1080/0969594X.2013.838938.

Räisänen, A., & Räkköläinen, M. (2014). Developmental assessment of learning outcomes. In S. Kalliola (Ed.), *Evaluation as a tool for research, learning and making things better* (pp. 241–266). Cambridge Scholars.

Räkköläinen, M. (2011). *What do skills demonstration reveal? The trust and confidence in the assessment process of learning outcomes. Acta Universitatis Tamperensis 1636.* Tampere University Press.

Sachs, J., Schmidt-Traub, G., Kroll, C., Lafortune, G., & Fuller, G. (2018). *SDG index and dashboards report 2018.* Berelsmann Stiftung and Sustainable Development Solutions Network (SDSN). https://sdg-index.org/reports/sdg-index-and-dashboards-2018/.

Stame, N. (2004). Theory-based evaluation and types of complexity. *Evaluation, 10*(1), 58–76. https://doi.org/10.1177/1356389004043135.

Stame, N. (2006). Governance, democracy and evaluation. *Evaluation, 12*(1), 7–16. https://doi.org/10.1177/1356389006064173.

Vähämäki, J., Schmidt, M., & Molander, J. (2011). *Review: Results-based management in development cooperation.* Riksbankens Jubileumsfond.

Vähämäki, J., & Verger, C. (2019). *Learning from results-based management evaluations and reviews.* OECD Development Co-operation Working Paper 53. https://doi.org/10.1787/3fda0081-en.

VNS. (2016). *Government Report on the 2030 Agenda for Sustainable Development. Towards a carbon-neutral welfare society.* VNS 1/2016 vp.

VNS. (2017). *Valtioneuvoston selonteko kestävän kehityksen globaalista toimintaohjelmasta Agenda2030:sta Kestävän kehityksen Suomi – pitkäjänteisesti, johdonmukaisesti ja osallistavasti.* VNS 1/2017 vp.

VNS. (2020). *Government Report on the 2030 Agenda for Sustainable Development. Towards a carbon-neutral welfare society.* Publications of the Prime Minister's Office 2020:7. VNS 3/2020 vp. https://julkaisut.valtioneuvosto.fi/handle/10024/162475

Weiss, C. (1997a). Nothing as practical as good theory: Exploring theory-based evaluation for comprehensive community initiatives for children and families. In J. P. Connell, A. C. Kubish, L. B. Schorr, & C. H. Weiss (Eds.), *New approaches to evaluating community initiatives: Concepts, methods and contexts* (pp. 65–92). Aspen Institute.

Weiss, C. (1997b). Theory-based evaluation: Past, present, and future. *New Directions for Evaluation, 76,* 41–55. https://doi.org/10.1002/ev.1086.

Young, S. (2019). How USAID is building the evidence base for knowledge management and organizational learning. *Knowledge Management for Development Journal, 14*(2), 60–82. https://www.km4djournal.org/index.php/km4dj/article/view/466.

Evaluating for Resilient and Sustainable Livelihoods: Applying a Normative Framework to Emerging Realities

Prashanth Kotturi

Abstract

Evaluation has to reflect the evolving priorities of development and measure progress on their achievement. At the same time, evaluation must also incorporate newer demands from within the field such as increasing equity focus in evaluations, gender mainstreaming, and human rights. Environment and climate change became mainstreamed into the programming of development organizations following the Rio Earth Summit in 1992 and formation of financing mechanisms such as the Global Environment Facility (GEF) in 1991. This chapter reflects on how the Independent Office of Evaluation (IOE) of the International Fund for Agricultural Development (IFAD) addressed the growing demands on the evaluation function in terms of incorporating concerns on environment and climate within existing methodological frameworks, and also adapting its methodology to meet internal and external evaluation demands. The chapter considers how evolving methodologies, methods, and tools have helped IFAD overcome these issues.

Winds of Change in Development and Response

Development is a dynamic field with new demands placed on development actors to mainstream and incorporate every few years. The Brundtland Commission report (Brundtland, 1987) brought the importance of environment to the front and center of the development debate. Similarly, the Intergovernmental Panel on Climate Change was established in 1988 to draw voices from across the globe onto a single platform to tackle climate change. However, only later did environment and climate change become mainstreamed into the programming of development partners. Environment found higher recognition first in the aftermath of the Rio Earth Summit in 1992 and the formation of financing mechanisms such as the Global Environment Facility (GEF) in 1991. Climate change also started to be reflected more explicitly in development programming in the late 2000s, in light of the global food price crisis.

Evaluation as a field has to account for these evolving trends in development. Evaluation has to reflect the emerging priorities of the development field and measure progress on their achievement. At the same time, evaluation also has to incorporate new demands emerging from within the field, such as making evaluations more equity focused, gender mainstreamed, and human rights-centric.

P. Kotturi (✉)
Independent Office of Evaluation of the International Fund for Agricultural Development, Rome, Italy
e-mail: p.kotturi@ifad.org

J. I. Uitto, G. Batra (eds.), *Transformational Change for People and the Planet*, Sustainable Development Goals Series, https://doi.org/10.1007/978-3-030-78853-7_17

This chapter reflects on how the Independent Office of Evaluation (IOE) of the International Fund for Agricultural Development (IFAD) addressed the emerging demands on evaluation function in terms of incorporating concerns on environment and climate within the existing methodological framework(s), directly or indirectly, while adapting its methodology to meet internal and external demands on IFAD evaluation. This chapter also illustrates how evolving methodologies, methods, and tools have helped IFAD overcome these issues.

Evaluation Methodology of IOE

The Independent Office of Evaluation of IFAD has had three iterations of evaluation methodology, codified in the Methodological Framework for Evaluation 2003 (MFE, 2003), Evaluation Manual 2009 (first edition), and Evaluation Manual 2015 (second edition). Each of these has built on the work of the previous methodologies. The starting point for all three are the evaluation criteria of the Development Assistance Committee of the Organisation for Economic Co-operation and Development (OECD DAC) first laid out in the Principles for Evaluation of Development Assistance (OECD, 1991) and later defined in the 2002 Glossary of Key Terms in Evaluation and Results Based Management (OECD, 2002). For the purpose of this chapter, IOE's evaluation criteria is divided into three categories: core criteria, impact criteria, and other criteria. Over the years, each of these categories has evolved to encompass different facets of sustainable livelihoods and resilience to climactic shocks. Table 1 presents IOE's evaluation methodology over time.

Table 1 Evolution of IOE evaluation methodologies

	MFE 2003	Evaluation Manual 2009	Evaluation Manual 2015
Core criteria	Relevance	Relevance	Relevance
	Effectiveness	Effectiveness	Effectiveness
	Efficiency	Efficiency	Efficiency
			Sustainability
Impact domains/ criteria	Impact on physical and financial assets	Household incomes and assets	Household income and net assets
	Impact on human assets	Human and social capital and empowerment	Human and social capital and empowerment
	Impact on social capital and people's empowerment	Food security and agricultural productivity	Food security and agricultural productivity
	Impact on environment and communal resource base	Natural resources and environment	Institutions and policies
	Impact on institutions, policies and regulatory framework	Institutions and policies	
Other criteria	Overarching factors: Sustainability Gender equality Innovation and scaling up	Sustainability	Gender equality and women's empowerment
	Performance of IFAD	Promotion of pro-poor innovation, replication and scaling up	Innovation and scaling up
	Performance of government		Environment and natural resource management
	Performance of cooperating institutions		Adaptation to climate change
	Performance of cofinancing institutions		Performance of IFAD
			Performance of government

In terms of methodology, this chapter explores three defining features of IOE's methodology and their role in evaluating environment, sustainability, and resilience to climate change. These link to a sustainable livelihood approach, constant evolution of methodology, and accumulated methodological experience through various products.

Sustainable Livelihood Approach and Evaluation Methodology

Conceptual Linkage Among Livelihoods, Environment, Resilience, and Agriculture

IFAD has the mandate to work toward enhancing the livelihood systems of rural populations through agricultural and nonagricultural livelihood options.

When discussing sustainable livelihoods, the definition from Carney (1998) reveals the various layers therein:

> A livelihood comprises the capabilities, assets (including both material and social resources) and activities required for a means of living. A livelihood is sustainable when it can cope with and recover from stresses and shocks and maintain or enhance its capabilities and assets both now and in the future, while not undermining the natural resource base. (p. 4)

This definition interweaves the ideas of livelihoods and resilience. It also lays out another important aspect of livelihood enhancement and the resilience to shocks: interaction between human and natural systems.

Central to both the Carney (1998) definition and determining the resilience of households to vulnerabilities is the idea of livelihood assets. These are the means of production available to a given individual, household, or group that can be used in their livelihood activities and have the potential to produce something that is economically desirable (Goodwin, 2003, p. 3). Natural capital, social capital, human capital, physical capital, and financial capital are the five types of assets discussed in the literature (United Nations

Development Programme, 2017), and may be tangible or intangible in nature.

According to Chambers and Conway (1991), tangible assets include food stocks; stores of value such as gold, jewelry, and woven textiles; and cash savings in banks of thrift and credit programs. This category also includes land, water, trees, livestock and farm equipment, tools, and domestic utensils. Nontangible assets include claims and access. Claims are often made in times of shocks or stress or when contingencies arise. They are made on individuals or agencies; on relatives, neighbors, patrons, chiefs, social groups, or communities; or on nongovernmental organizations (NGOs), the international community, or governments, including programs pertaining to drought relief or poverty alleviation. Access is the opportunity to use a resource, store, or service, or to obtain information, material, technology, employment, food, or income. Figure 1 illustrates the components and flows in a livelihood.

The definition of sustainable livelihoods brings to the fore the importance of withstanding shocks and uncertainties for ensuring sustainability of livelihoods, and the role that the various kinds of assets play in doing so. Some of the major shocks that the poor face include political, climactic, and economic shocks. In this context, even before climate change and environmental sustainability became more mainstreamed into development parlance and expressed more explicitly in development theory, there was an implicit recognition of the various climactic shocks and a more explicit recognition of the broader strategies to cope with them.

Over the years, evaluation criteria have evolved to encompass different facets of sustainable livelihoods and resilience to climactic shocks. The core criteria of relevance, effectiveness, and efficiency were influenced by the Principles for Evaluation of Development Assistance published by OECD DAC in 1991 and have been reflected in every iteration of IOE's methodology.

The second part of the methodology links directly to the sustainable livelihood approach

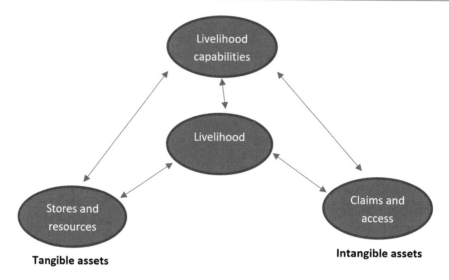

Fig. 1 Components and flows in a livelihood
Source: Chambers and Conway (1991)

through the criteria and subcriteria under the impact domain in Table 1. These criteria/subcriteria mirror the various kinds of livelihood assets—natural capital, social capital, human capital, physical capital, and financial capital—to varying degrees. Given the criteria's intricate links to the sustainable livelihood approach, IOE has been able to determine whether an intervention of a portfolio of projects is able to provide or enable its smallholder target groups with assets that help them adapt to various kinds of shocks, including climactic shocks. This goes beyond looking at physical infrastructure to considering intangible assets such as individual agency and social capital, skills that can help smallholders adapt to all kinds of shocks. However, a crucial gap in using this rubric of criteria in evaluating adaptation to climate change is that it only looks at the presence or absence of certain preconditions with a causal assumption that these will help in adaptation efforts without always placing them in the context of the climactic risks that exist in a given area.

The combination of OECD DAC criteria and those derived from a sustainable livelihoods approach form the normative framework on which IOE has built accountability and learning in rural development. This normative framework is reflected in the various iterations of IOE evalu-

ation methodology over the years (see Table 1) and has formed the basis on which IOE has built its methodology for evaluating environment and climate adaptation.

Evolution of the IOE Evaluation Methodology

As Tables 1 and 2 illustrate, each subsequent iteration of IOE evaluation methodology has undergone significant changes in terms of criteria. Livelihood assets, such as human and social assets, natural resource assets, physical assets, and financial assets, were most explicitly discussed in the MFE 2003 and are reflected similarly in the evaluation manuals of 2009 and 2015. Three broad observations illuminate the evolution of the evaluation criteria and methodology.

First, evaluation criteria have evolved from reflecting livelihood issues to a broader focus on themes and thematic thrusts. The MFE 2003 criteria focused much more on prevalence and strengthening different kinds of livelihood assets that enhance the resilience of smallholders. The strengthening and presence of these was seen as necessary for livelihoods to be resilient and sustainable. As shown in Table 1, evaluation criteria in 2009 and particularly in 2015 reflected the-

Table 2 Evolution of IOE evaluation questions

Evaluation criteria	Evaluation questions		
	2003	2009	2015
Environmental assets and natural resources	Did the natural resource base status change (land, water, forest, pasture, fish stocks, etc.)? – **access** Did exposure to environmental risks change? – **vulnerability**	Did the status of the natural resources base change (land, water, forest, pasture, fish stocks, etc.)? – **conservation** Did local communities' access to natural resources change (in general and specifically for the poor)? – **access** Has the degree of environmental vulnerability changed (e.g., exposure to pollutants, climate change effects, volatility in resources, potential natural disasters)? – **vulnerability**	To what extent did the project adopt approaches/ measures for restoration or sustainable management of natural resources (e.g., enhancement of ecosystem services, support to training and extension to foster efficient environment and natural resource management, uptake of appropriate/new technologies)? – **conservation** To what extent did the project develop the capacity of community groups and institutions to manage environmental risks (e.g., how governance-related factors are shaping the management of natural resources, influence of incentives and disincentives for sustainable natural resource use and natural resource-based livelihoods improvement)? – **governance and management of natural resources and environmental risks** To what extent did the project contribute to reducing the environmental vulnerability of the community and build resilience for sustainable natural resource management that contribute to poverty reduction (e.g., factors such as access to technologies, information/awareness creation)? – **vulnerability** To what extent did the project contribute to long-term environmental and social sustainability (e.g., through avoiding overexploitation of natural resources, loss of biodiversity, or reduction of the community's livelihoods; by empowering and strengthening the capacity of community-based natural resource management groups to ensure sustainable natural resources management; or by ensuring strong stakeholder engagement, especially of vulnerable groups, in decision making affecting natural resources use)? – **human and natural system nexus** To what extent did the project follow required environmental and social risk assessment procedures (e.g., social, environmental, and climate assessment procedures), including meaningful consultation with affected and vulnerable communities, and comply with applicable IFAD or national environmental and social standards or norms to ensure any harmful impacts are avoided or managed/mitigated through, where needed, the implementation of effective environmental and social management plans, including robust monitoring and supervision? – **safeguards compliance**
Adaptation to climate change			To what extent did the project demonstrate awareness and analysis of current and future climate risks? What are the amounts and nature of funds allocated to adaptation to climate change-related risks? What were the most important factors that helped the rural poor to restore the natural resource and environment base potentially affected by climate change?

matic priorities such as environment and natural resources, gender, food security, and, most prominently, climate change. This change in methodology can be attributed to the evolving nature of IFAD operations with a move away from integrated rural development programs to those with a theme focus, such as rural finance, value chains, market access, and integrated natural resource management. Much more important, the reflection of natural resource management as a separate criterion reflected an increasing recognition of and focus on environmental conservation and management in IFAD programming in the context of donor demands and supplementary financial resources from funding institutions such as the GEF.

Second, IFAD's evaluation focus has evolved from looking at conservation status of the natural resource base toward recognizing the importance of the natural resource base for livelihoods of rural populations and their sustainable use and management. This is most prominently reflected in the evaluation questions framed under each iteration of the evaluation methodology and the number of questions under the environment and natural resource management criteria (see Table 2). In terms of questions, the 2003 MFE had two questions pertaining to the state of the natural resource base and vulnerability of rural poor to environmental risks. The subsequent iterations of methodology in 2009 and 2015 contained more expansive coverage in evaluation questions on environment and natural resource management. The later iterations of IOE evaluation methodology have essentially focused on the sustainable interactions between human and natural systems. The questions elaborated in Table 2 are by no means exhaustive and provide only initial guidance; actual questions asked under the criteria in each evaluation may differ.

Third, climate change adaptation was mentioned as a separate evaluation criterion for the first time in 2015. As Table 2 illustrates, climate change was covered more implicitly under the criterion of environment and natural resource management in 2003 and 2009, as shown in the evaluation questions that elaborate on environmental risks. Such implicit inclusion has often

been meant to also cover climate change-induced risks. However, in the 2015 iteration of IOE methodology, the assessment of climate change is more explicit, with questions on current and future risks. Such a change took place as a result of an evolving and increasing emphasis on climate change adaptation in IFAD's strategic emphasis, corporate policies, and programmatic thrusts.

Evaluation criteria have undergone change in two ways. First, newer evaluation criteria such as climate change adaptation have been added over time to reflect the evolving criteria. Second, even when evaluation criteria have remained similar, the scope of evaluation criteria has expanded in terms of the suggested evaluation questions under those criteria. A consequence of the increase in the scope of questions is that crosscutting issues are better incorporated into evaluations. Such questions also better account for complexity. For example, the 2015 iteration of evaluation methodology recognizes the dependence of target groups on the natural resource base for livelihoods and thus looks at human and natural system interaction. Similarly, the questions under the 2015 methodology also go beyond simple conservation to incorporate governance of the natural resource base.

Accumulated Methodological Experience Through Various Products

IOE produces a wide variety of products with differing scope, focus, and purposes. In the early 2000s, most of IOE's focus was on country portfolio evaluations and project evaluations. However, IOE has increasingly moved toward a more thematic focus over the years. This started with undertaking higher plane evaluations such as corporate-level evaluations, which look at corporate and thematic priorities beyond evaluation of a country portfolio or project. This was further reinforced when, in 2011, IOE began undertaking evaluation synthesis reports on specific topics, an exercise that consolidates evaluation findings on a specific topic over a period of time.

Thus, IOE has built its experience in evaluating environment and climate change over a period of time by undertaking incrementally different kinds of analysis on the topics. For example, IOE produced evaluation synthesis reports in the past few years on topics such as fisheries, water management, and environment and natural resource management. This accumulated experience on evaluation synthesis also pointed to the need for a higher plane evaluation specifically focused on climate change adaptation. As of the time of writing this chapter, IOE was undertaking a thematic evaluation on this topic.

IOE has also been able to consolidate the findings of its numerous evaluations over a period of time to provide trends of performance on evaluation criteria in its Annual Report on Results and Impact (ARRI). Such trends are depicted on a 3-year moving average. These assessments have provided IOE with useful and contemporary insights and helped in planning evaluations that probe the underlying factors influencing performance. Figure 2 depicts the trends in IOE's rating on environment and natural resource management criteria as shown in the 2020 ARRI (IFAD, 2020).

The figure illustrates how performance saw some decline in the period of 2007–2013 before picking up again. IOE undertook the evaluation synthesis report on environment and natural resources management in 2015 to consolidate the lessons that IOE evaluations have generated on the topic and explanatory factors for performance in the area of environment and natural resource management. Thus, IOE's products have progressively built on one another to inform the debate on thematic areas of priority for IFAD. IOE has introduced newer products from time to time to meet the accountability and learning requirements of the organization. Such evolution in its products is operationalized through the evaluation methodology, which has been revised as the need arose for tackling newer challenges in IFAD operations.

Methods for Assessment Using Normative Frameworks

IOE typically has used a wide variety of methods to collect data on evaluation questions and criteria. The evolution of IOE's methods mirror the

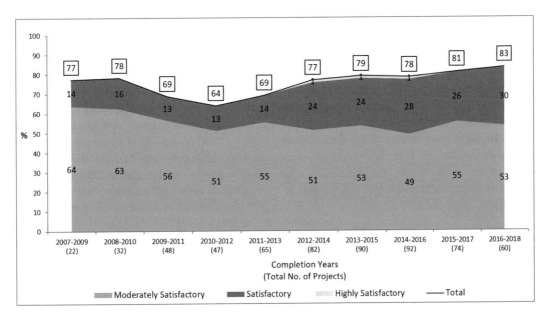

Fig. 2 Trend in ratings on natural resource management

diverse elements of the organization's evaluation methodology vis-à-vis environment and climate change adaptation. The sustainable livelihood element of the normative framework mostly employs qualitative methods.

Sustainable Livelihood Approach and Qualitative Methods

Livelihood assets such as natural, social, human, physical, and financial capital are instrumental in ensuring the resilience of rural populations. IOE methodology has historically focused on ascertaining the impact of IFAD interventions through creation of these assets. Various dimensions under these asset classes help in adaptation through different pathways and questions are asked under each of these to understand their role in helping smallholders adapt.

Physical Capital

Typical physical assets include tangible assets such as land, housing, and vehicles that are repositories of economic value. In terms of primary data, evaluators generally assess physical assets through individual interviews and direct observations of field interventions. For example, evaluators in the field enquire with beneficiaries on their state of current asset matrix, the composition of assets, and the utility of such assets in times of shocks. In terms of secondary data, IFAD's measurement and evaluation (M&E) system typically captures the state of housing, size of landholding, and household items as a part of outcome and impact surveys. Such data is used by evaluators after due diligence on the methodology in the outcome surveys. The data is also validated during field visits through interviews and direct observations.

Financial Capital

For IFAD's target groups, this category typically includes access to financial assets such as credit, insurance, savings, etc. Community-based savings groups and remittances also hold an important place in IFAD's operations in ensuring access to financial resources in the event of facing shocks, especially financial shocks. Because discussions around financial resources can be sensitive, evaluators usually deploy focus group discussions only to understand the nature of community-based and other informal and formal sources of financial services that remain at the disposal of IFAD's target groups and these sources' role in helping communities withstand shocks. For example, in many community-based savings groups that IFAD creates and strengthens, financial products exist for restarting economic activities such as restocking of livestock herds and reviving crop agriculture. Evaluators conduct interviews with individuals to understand their access to alternative avenues of income such as remittances and nonfarm income, and the resilience of those sources.

Social Capital

Social capital is defined by the OECD as "networks together with shared norms, values, and understandings that facilitate cooperation within or among groups" (2007, p. 103). Thus, assessing social capital involves looking at the social dynamics between people, the functioning of community institutions, and formal and informal bonds. This requires evaluators to explore the individual and collective social capital to understand the level of trust, solidarity, and coherence that exists in a community. This is usually done through in-depth focus group discussions with individuals who share social and geographic spaces. Such focus group discussions also attempt to understand the ability of target groups to lean on their social networks in times of shocks and the willingness of the social networks to provide avenues for relief and recovery. Interviews with individuals allow evaluators to confirm the extent to which individuals are able to rely or have relied on project- or program-facilitated social networks. In terms of secondary data, IFAD uses its M&E system from time to time to capture the state of community institutions through a composite rating or a qualitative assessment of their functioning.

Human Capital

Human capital refers to skills, knowledge, and ability to work. Building human capital usually involves training in farm and off-farm livelihood activities. To gather secondary data, IFAD's M&E system captures output data on the trainings and skills imparted. IOE also interviews participants on the nature of the skills learned and their use in the livelihood matrix. IOE pays special attention to target groups' ability to use the skills to undertake livelihood activities that promote resilience. For example, IFAD programs in some countries help build off-farm vocational skills for professions and jobs that enable people to diversify their livelihoods or migrate seasonally for income-generating activities.

Natural Capital

Historically, IFAD has relied on interviews and focus group discussions to understand the state of the common and communal natural base and the ability of target groups to access these assets. This refers not just to access as a whole but also to equity of the access for various groups. Hence, in-depth individual interviews and focus group discussions put special emphasis on interacting with a sample that is as representative as possible when assessing access to natural capital. Direct observations have also provided a general view of the state of natural resource management. However, in recent times, IOE has employed newer methods to assess interventions on environment and natural resource management, described below.

Methods for an Evolving Methodology

As mentioned previously, IOE's methodology has evolved in the past few iterations from assessing the presence of livelihood assets to a more thematic focus on climate change adaptation and environment and natural resource management. More specifically, successive evaluation methodologies have moved away from simply looking at target groups' ability to access natural resources.

Now, IOE considers the state of the natural resource base, interaction between human and natural systems, and climate change-induced risks to livelihoods and to the natural resource base. IOE has felt an increasing need to adopt newer methods because conventional qualitative methods do not sufficiently assess on these dimensions. For example, conventional methods cannot holistically assess the state of natural resources in a project/program area. A method such as direct observation might provide a general view of the state of the natural resource base at a certain point in time, but natural resources systems are highly dynamic and require a longer term view to understand program effects and natural resource systems at large. IOE assessments also need a quantitative evidence base on the evolution of natural resource systems. Similarly, anchoring the adaptation efforts of a program/ project in local climactic risks requires a temporal and scientific analysis.

With this recognition, IOE has adopted geographic information systems (GIS) analysis as one method for data collection and analysis. GIS provides IOE with two advantages over existing qualitative methods. First, GIS delivers a temporal view of IOE's area of interest. IOE can analyze the state of natural resources and other topographic features of interest across years, irrespective of whether an IFAD program was active in that area. This allows IOE to create a baseline situation in a given area even if program M&E does not have a baseline survey. It could also allow IOE to create a counterfactual for the project intervention, although this has yet to be piloted in IOE. Figure 3 illustrates an IOE recreation of a baseline situation in a given area. Second, GIS gives IOE spatial capability to evaluate beyond the mobility limitations of evaluators in the course of field visits. In simpler terms, GIS allows evaluators to assess interventions in areas they may not be able to visit.

IOE first piloted GIS in its work in 2016 as a part of an impact evaluation in Georgia, using satellite imagery to analyze changes in landscape and cropping patterns in response to irrigation system improvements. IOE retrieved satellite

Fig. 3 Sample difference in vegetation in Nepal between 2015 and 2019

imagery captured prior to the program to establish a baseline, and then from the midpoint of the program and the time of evaluation. IOE used this imagery to calculate the normalized difference vegetation index[1] and construct models to estimate crop yields before and after irrigation improvements, taking into account other variables from the program area. This analysis triangulated the findings emerging from the survey data collected as part of the impact evaluation itself.

Similarly, IOE used GIS in the context of a country-level evaluation in Nepal. Satellite imagery (such as in Fig. 3) allowed IOE to understand changes in the landscape over a period of time through visual observation of imagery. The idea was to understand the broader nature of evolution of the natural base in the general project area without constructing models, which would require more primary data on specific variables to

reach any robust conclusions. Figure 3 shows that the leasehold forestry training provided in Karnali province of Nepal did result in improved density of tree cover in one of the sites. This was triangulated through qualitative questions on the value of nontimber forest produce of the forestry groups that depend on these leasehold forests. This provided a proxy for increase in forest cover and its impact on the livelihoods of target groups.

Constraints

One of the limiting factors for using GIS in IOE evaluations is the lack of systematic streamlining geotagging of project/program interventions. This prevents IOE from capitalizing on one of the major advantages of using GIS in evaluation work: the spatial capability to evaluate beyond the mobility limitations of evaluators during field visits. Usually, one of the ways to circumvent this constraint is by considering administrative boundaries of a project site location. However, IFAD's interventions mostly target individual households at the community level and rarely work at sub-watershed/watershed level or landscape level. Thus, considering administrative

---

[1] Normalized difference vegetation index (NDVI) quantifies vegetation by measuring the difference between near-infrared (which vegetation strongly reflects) and red light (which vegetation absorbs).

boundaries would lead to weak attribution and even contribution of the program to enhancing the natural resource base. In the absence of systematic tagging of interventions as part of program/project M&E, evaluators are limited to analyzing only those sites that they have physically visited and geotagged themselves. This reduces the scope for ensuring external validity of any conclusions that evaluators may draw based on GIS analysis. Another limitation is the lack of capacity within IOE to undertake GIS-based collection and analysis of data. So far, IOE has relied on external expertise to undertake GIS analysis in evaluations.

GIS for the Future

IFAD has begun more systematic geotagging of its project/program sites and has made plans to introduce geotagging as a part of operational procedures for projects in some regional divisions. Such data is also being organized in a format that will be consistent across the organization and available on one web-based platform, IFAD Geonode. This should help IOE better use GIS to evaluate interventions moving forward. To capitalize on the opportunities provided by these new methods, IOE is exploring various avenues to build GIS-related capacity in house.

Future Perspectives on Methodology for Evaluating Climate Adaptation and Natural Resource Management

As mentioned earlier, IOE started evaluating climate adaptation and natural resource management through the framework of sustainable livelihood approach and the five kinds of capital. IOE's initial methodological approach to resilience (including climatic risks) looked at the presence of the five kinds of capital, and their presence was seen as enabling resilience of IFAD's target communities. However, resilience needs to be more firmly anchored in risks that prevail in a given context. The second and third iterations of IOE's manual emphasize more

strongly the kind of environment and climactic risks and the analysis undertaken to recognize those risks in IFAD operations.

Future methodological directions could build on the existing normative framework. Two potential lines of enquiry that could be better explored in future evaluations are appropriateness and adequacy of interventions addressing climate risks and accompanying climate adaptation needs. Appropriateness links interventions more explicitly to prevailing climate change-induced risks in a given area and seeks to identify whether an intervention is appropriate for a given risk. Although the evaluation questions in IOE's methodological iterations do implicitly recognize the risks and responding interventions in a given context, they do not do so explicitly. Adequacy of interventions pertains to whether the intensity of an intervention, the matrix of interventions, and interventions' coherence is sufficient.

However, such evolution would require a newer paradigm of evaluation methodology, methods, and specialized skill sets to understand the climactic risks in a given context. This would require evaluations to assess two aspects of a program/project. First, project design would need to include a climatic risk assessment for the program area. Second, evaluations would have to validate the robustness of the climactic risk assessment process, which would require the capacity of evaluation teams to undertake climate risk assessment through qualitative and quantitative methods.

IOE has come a long way in terms of conducting evaluations, constantly learning as an evaluation unit, and upgrading its methodology periodically. As the operations of IFAD have become more focused on the issues around natural resource management and climate change adaptation, IOE has had more occasions to build its knowledge base. In the next years, IOE will have an opportunity to undertake another revision of its methodology when it revises its evaluation manual and incorporate the new methods and experiences it has piloted.

References

Brundtland, G. H. (1987). *Report of the World Commission on Environment and Development: Our common future*. United Nations. https://sustainabledevelopment.un.org/content/documents/5987our-common-future.pdf

Carney, D. (1998). *Sustainable rural livelihoods: What contribution can we make?* Department for International Development.

Chambers, R., & Conway, G. (1991). *Sustainable rural livelihoods: Practical concepts for the 21ˢᵗ century*. Institute for Development Studies. https://publications.iwmi.org/pdf/H_32821.pdf

Goodwin, N. (2003). *Five kinds of capital: Useful concepts for sustainable development*. Global Development and Environment Institute.

International Fund for Agricultural Development. (2020). *2020 annual report on results and impact of IFAD operations*. Author. https://www.ifad.org/en/web/ioe/evaluation/asset/42126296

Organisation for Economic Co-operation and Development. (1991). *Principles for evaluation of development assistance*. Organization for Economic Cooperation and Development, Development Assistance Committee. https://www.oecd.org/dac/evaluation/2755284.pdf

Organisation for Economic Co-operation and Development. (2002). *Glossary of key terms in evaluation and results based management*. Author. https://www.oecd.org/dac/evaluation/2754804.pdf

Organisation for Economic Co-operation and Development. (2007). *Human capital: How what you know shapes your life*. https://read.oecd-ilibrary.org/education/human-capital_9789264029095-en#page1

United Nations Development Programme. (2017). *Guidance note: Application of sustainable livelihoods framework in development projects*. Author. https://www.undp.org/content/dam/rblac/docs/Research%20and%20Publications/Poverty%20Reduction/UNDP_RBLAC_Livelihoods%20Guidance%20Note_EN-210July2017.pdf

Measuring the Impact of Monitoring: How We Know Transparent Near-Real-Time Data Can Help Save the Forests

Katherine Shea

Abstract

Global Forest Watch (GFW) is an online platform that distills satellite imagery into near-real-time forest change information that anyone can access and act on. Like other open-data platforms, GFW is based on the idea that transparent, publicly available data can support the greater good—in this case, reducing deforestation. By its very nature, the use of freely available data can be difficult to track and its impact difficult to measure. This chapter explores four approaches for measuring the reach and impact of GFW, including quantitative and qualitative approaches for monitoring outcomes and measuring impact. The recommendations can be applied to other transparency initiatives, especially those providing remote-sensing data.

Global Forest Watch is an online platform that can support monitoring and evaluation of conservation projects, international commitments to reduce deforestation, and private sector zero-deforestation plans. As a partner-funded project, like other projects within the World Resources Institute (WRI), its contributions need to be evaluated and monitored. Like other open-data platforms, it is based on the idea that transparent, publicly available data can support the greater good; in this case, reducing deforestation. By its very nature, the use of freely available data can be difficult to track, and its impact difficult to measure.

The team behind Global Forest Watch (GFW) explored several methods for monitoring and evaluating the impact of open-data tools for natural resource protection. This chapter explores four approaches for measuring the reach and impact of GFW, including quantitative and qualitative approaches for monitoring outcomes and measuring impact. This chapter aims to provide a framework for other open-data platforms to monitor outcomes and measure impact to learn and iterate on the most cost-effective strategies for natural resource protection. In the case of GFW, we have found that quantitative methods are capable of measuring outcomes and impact, although they require innovative approaches. We've also found that qualitative methods are necessary to understand the mechanisms of adoption and application and produce lessons for furthering the reach and impact of open data.

K. Shea (✉)
World Resources Institute, Washington, DC, USA
e-mail: KatherineShea@wri.org

J. I. Uitto, G. Batra (eds.), *Transformational Change for People and the Planet*, Sustainable Development Goals Series, https://doi.org/10.1007/978-3-030-78853-7_18

Background: What Is GFW?

Deforestation is a critically important challenge for the global community. Forests capture 30% of the carbon emissions released each year, playing a vital role in stemming climate change. They are also home to untold biodiversity and a resource for remote and indigenous communities. But the world is losing forests at an alarming rate: A football pitch's worth of primary forest was lost every 6 seconds in 2019, according to WRI data, slightly more than in the previous year (Weisse & Dow Goldman, 2020).

Until very recently, those responsible for forests—including policymakers, protected area managers, and international commodity purchasers—had no way of knowing where deforestation was happening without visiting potential sites on foot. Tropical forests in places such as the Congo Basin are dense and inaccessible, and monitoring them can be extremely costly. Researchers recognized the lack of data as a concern and a barrier to improved management: "To enhance the efficiency of the protection, regeneration, and utilization of forest resources, information about these changes is required" (Suwanwerakamtorn et al., 2011, p. 169). In the mid-2000s, public data and advances in computing made such a system possible. When the U.S. government publicly released imagery from the Landsat satellite, the potential for such data became clear. "Satellite imagery offers an emerging source of data for analysis and a novel medium to attract greater government and public attention to domestic and international problems such as deforestation" (Baker & Williamson, 2006, p. 12). Researchers also began calling for unified and publicly available sources of data. Eventually, researchers at the University of Maryland developed an algorithm to process the vast trove of satellite imagery and discern locations of forest change (Hansen et al., 2013).

In 2014, WRI brought together the results of Hansen's new forest change analysis, the computing power of Google Earth Engine, and stakeholders in forest protection for a partnership and a platform that made forest change data accessible to anyone with an internet connection. Global

Forest Watch, as the platform was named, is a website that displays changes in forest cover as pink pixels on a map that can be overlaid with a wide range of contextual data, including protected areas or primary forest. Of the new features and data added since 2014, the most important are weekly deforestation alerts called "GLAD alerts."[1] GLAD alerts are the most spatially explicit forest change alert product publicly available, identifying areas of 30 by 30 m. GFW also carries Fires alerts, which use heat signatures to detect fires and are updated daily, but do not detect other types of clearing, such as when trees are felled for timber or to clear a road.[2] Users can also analyze areas of interest by selecting an existing area such as a state or park, or by uploading or drawing their own shapefiles. Global Forest Watch added a mobile phone application called Forest Watcher, which enables users to download data about forest change and take it offline and into the field where internet access may not be available. The theory behind the data is that users from civil society, governments, and the private sector will use this data to better manage and protect forests (see Box 1 for GFW objectives).

All of the data on Global Forest Watch can be accessed freely by anyone with an internet connection, without even the need to identify themselves. The team behind the platform recognized at the outset that by providing the data in a completely open fashion, they would lack information about the reach of the platform, and the topic was discussed in depth by developers. Eventually, the goal of complete transparency won out. The limited knowledge about users was deemed a worthy tradeoff to avoid the possibility that entering private information might become a barrier to use. This aligned well with the thinking of other researchers who argued that public access to a

---

[1] These alerts are nicknamed for the lab where they are created, the University of Maryland's Global Land Analysis and Discovery (GLAD) lab. More detail can be found at https://glad.umd.edu/dataset/glad-forest-alerts

[2] Fires alerts are drawn from the NASA Visible Infrared Imaging Radiometer Suite (VIIRS) Active Fire detection product, which operates at a 375-m resolution, to detect heat signatures of active fires.

geo-wiki—an open-access geospatial platform—would be critical for the success of remotely sensed data:

> The philosophy is that as information is aggregated in public, discrepancies will arise, but so, too, will incentives to rectify them. The public nature of the information could induce a country to provide data to correct the record, for instance. Experience with public disclosure programs suggests that not all of them work, however, so identifying conditions most conducive to success is part of the geo-wiki experiment. (Macauley & Sedjo, 2011, p. 512)

The open-data nature of Global Forest Watch created a challenge for monitoring and evaluating the impact of the platform: If donors were to fund the continued development of the tools, they deserved to understand its reach. WRI leadership recognized that the institute has a responsibility to identify the most efficient use of donor funds to protect natural resources. The institute also has a mission of investigating and sharing guidance on effective methods of natural resource management. Therefore, over the past 5 years, the team explored new and evolving ways of monitoring program progress and measuring impact. These included using Google Analytics to track the reach of the platform, requesting data from users through a login feature, gathering user stories, and a two-part evaluation.

GFW Monitors Progress

Analytics

The first step in the Global Forest Watch theory of change is getting data into the hands of users who might act on it. To measure the reach of GFW, the team turned first to Google Analytics, a tool that provides data to a website's back-end developers about the traffic that website receives. Analytics can provide data on a variety of performance indicators by breaking down website visitors into certain categories. The GFW team tracks the total number of visitors to the site, a figure that has risen steadily each year (since launch, more than three million people have visited GFW, from every country in the world). The team rec-

---

Box 1: Objectives Outlined in the Global Forest Watch Theory of Change

1. Strengthened accountability for global commitments: Accountability for implementation of global forest commitments is strengthened by credible, independent information and analysis of forest and land-use dynamics.
2. Responsible supply chains: Actors trading or financing major forest-risk commodities use smart strategies and cutting-edge information and technology tools to reduce deforestation and illegal logging in their supply chains and investments.
3. Empowered forest defenders: Civil society and law enforcement actors are better equipped to expose and combat deforestation and illegal logging.
4. A broad-based restoration movement: Around the world, communities and commercial enterprises gain access to the knowledge, expertise, and finance they need to restore degraded lands.
5. Enabling conditions for sustainable landscapes: Sound forest and land management is enabled by governance reforms, new incentives, and improved geospatial monitoring and analysis in targeted countries and landscapes.

---

ognizes that many people may visit GFW once, without integrating the information into future action or advocacy, so we began tracking *active users,* defined as those who visit more than once in a year and stay for more than 2 minutes at a time. (Repeat and longer visits suggest that the user is reviewing information over time or delving deeper into an area or topic.) We also track the number of active users disaggregated by country, and the number of users who perform certain actions such as turning on or off the GLAD alerts layer on the map. For example, in 2019 we wanted to know how easily users found the base map feature and which base map was more popular among our users to understand

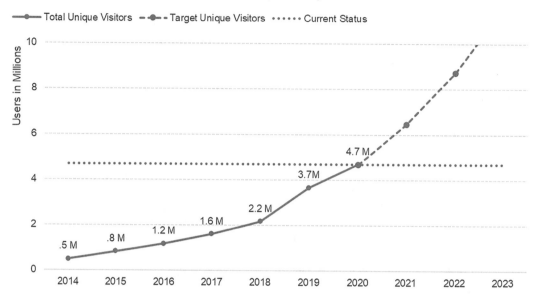

Fig. 1 Annual user data for GFW's main website
Courtesy of WRI, Stephanie Jamilla

what information they were seeking. We were able to track the number of clicks on these different features. We learned that the features were easy enough to find, since about half of the users who clicked on them were "new," compared to users who had visited the site at least once before. We also learned that planet imagery was the most popular alternative base map, suggesting that users were looking for even greater detail behind each pixel.

All this data is useful for performance monitoring and continued trends of increasing data usage imply that the site is relevant for users (see Fig. 1). Rising numbers of active users in high-forest countries, increasing time on site, and clicks on map layers suggest that the growing audience is interested in the available data. It can also tell us a bit about how to make the site more useful by highlighting which features are easiest to access, or whether users are finding new features, based on the number of clicks. Analytics data cannot, however, tell us about who those users are or how, if at all, they apply the data. It certainly cannot tell us if that application results in any action or policy that could shift the trend of deforestation.

To learn more about users and applications, the GFW team implemented one additional quantitative metric. We gathered user information through a login system, allowing all data to remain freely available while asking users who wanted additional features to provide further information. Based on conversations with users and partners, the team had identified a need for added features, particularly the ability to save areas of interest. In response, we developed the MyGFW tool, which enables users to create and save a custom set of areas and view them in a dashboard—and requires users to create a login. This provided an opportunity to gather data about users (notably, it also required the team to implement data security protocols to protect those users). Through the login, the team could ask anyone seeking to access the features a few questions about their role in forest monitoring and reason for using GFW.[3] For example, we learned that, of more than 2000 subscribers who reported on their sector, the majority worked for nongovernmental organizations (NGOs), followed by

---

[3]The login profile now asks users for their sector, role, job title, organization, and location.

researchers or students, government workers, and those in the private sector. A smaller number of users worked in journalism, community monitoring, or other categories that covered donors and United Nations (UN) agencies (see Fig. 2). We could then link that data to the locations that users were monitoring. Gathering data linking the type of user to the area they monitored would become critical to evaluating whether users applying GFW data influenced forest change.

Users and Stories

Even with the combination of analytics and MyGFW logins, the GFW team still had limited information about the way users applied data to real-world situations. This type of information could help the team, for example, make decisions about the balance between allocating resources to better data or to wider reach, or identify which type of user needed the most support, or which additional data would add the most value for users. So the team embarked on an ambitious project to gather, document, and categorize user stories for internal use. We created a user-friendly

interface and a searchable database within WRI's secure network, simply called the User Stories database. In the system, WRI staff could easily search, review, or add new user stories.

The database now contains more than 300 detailed and categorized stories about uses of GFW, painting a picture of the ways users achieve impact. Each story is categorized by variables including location, year of first use, and sector, but most important are the type of application of the data and the GFW objective to which the story contributes. We also captured stories of users who showed interest but ultimately discontinued their use of GFW. The aggregate data from the database is not representative of all users, but rather those with whom GFW staff have the most contact or those who contact GFW with questions. For example, most of these (43%) come from local NGOs, which tend be more likely to interact with GFW staff through requests for technical support, or from applications to GFW's Small Grants Fund.

Overall, the stories provide a qualitative understanding of the variety of uses and some notion of which of those are most successful. They tell the stories of journalists who used GFW

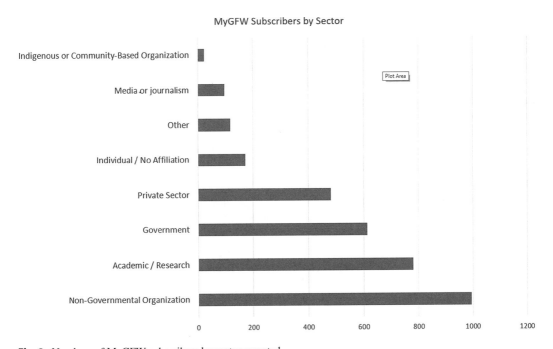

Fig. 2 Numbers of MyGFW subscribers by sector reported

to uncover illegal activities, or of teachers improving students' understanding of forests' role in climate change, or of researchers advocating for greater transparency in countries such as China.

Stories from the database also provide substantial anecdotal evidence that forest defenders have turned to GFW's near-real-time data—GLAD alerts and Fires alerts—to monitor and protect areas of interest. Data from the MyGFW system told us that users were subscribing to alerts, but stories in the database provided examples of specific successes; 84 stories came from groups tasked with monitoring and protecting forests, from government officials to civil society and even private sector users. Of those, 21 stories reported concrete results in stopping or preventing deforestation. Examples include a police chief from Brazil who manages a forest the size of Nepal with only 12 field staff. He told the team, "We effectively only started combating deforestation after Global Forest Watch. It is the principal tool of [our] police station." The chief explained how the near-real-time alerts helped him more quickly identify areas of burning and clearing, leading to at least eight arrests including company operators who were overseeing an illegal deforestation scheme. Another NGO used the data to follow up on alerts in Malaysia. They identified a logging contractor who encroached on an area that was outside a legal logging area—allegedly accidentally. The group was reported to authorities, and the logger now knows that they were being monitored and could be identified if they violate the rules again.[4] Even private sector examples emerged: In Paraguay, an investment bank noticed an alert on a farm and telephoned the farmer, who had been unaware of the fire but was able to act quickly (Guthrie, 2018).

Quantitative Evaluation

When combined with GFW's analytics data, the substantial number of anecdotes suggested that near-real-time data was having the intended impact, but could WRI provide definitive evidence? To answer this question, WRI's Managing for Results team worked with the GFW team to complete a two-part, mixed-methods evaluation funded through an internal mechanism. The evaluation aimed to explore potential methods, recognizing that definitive evidence of impact might not be within reach just yet.

The quantitative team focused on that data most likely to create a short-term, measurable impact: the GLAD deforestation alerts. Because users receive these alerts within days or months of deforestation occurring, and the alerts are specific to a 30-by-30-m area, users can act quickly to intervene. Park rangers or local communities can conduct site visits to intercede or to deter the responsible parties. In Africa, the evaluation found that subscriptions to GLAD alerts decreased the likelihood of deforestation by 18%. This impact increased when a lag time of 1 year was added between the subscription and the forest-cover measurement, suggesting that as users learn and apply the tools, they become more effective at intervening in the causes of deforestation. Indeed, these results correspond to user stories, which suggested that areas where users had subscribed to receive email alerts might see reduced deforestation.

WRI's team worked with an economist from the University of Wisconsin–Madison to design models to test the impact of the *availability* and *use* of GLAD alerts. Both models used time variation to create plausible counterfactuals. Two things changed over time: First, GLAD alerts were rolled out to different countries at different times, and second, once alerts were available, users began subscribing to alerts at different times. The research team explored a third option by attempting to build a model for *receiving* alerts. Since cloud cover can prevent the detection of deforestation, sometimes users don't receive alerts until much later, and this could affect actions. This third model proved impracti-

---

[4] All stories are courtesy of the GFW user stories database.

cal for technical reasons, but the first two yielded credible results.

In the availability model, countries that had not yet gained access to alerts provided a counterfactual.[5] This model explored the likelihood of deforestation in countries where alerts were available compared with countries where they were not, and found no discernible impact. We hypothesized that this was because adopting GLAD alerts may take time and adoption first occurs in limited areas of countries, while the sample covered many areas where no adoption occurred. It may also have included areas of forest or mining concession or other areas where monitoring was not intended to prevent clearing. The study noted that when a one-year lag effect was applied, the correlation between availability and deforestation was negative, although not statistically significant (Moffette et al., 2021).

The model assessing use of GLAD alerts explored the impact of users holding a subscription to GLAD alerts for a specified area of forest. For the use model, researchers divided subscribers into two categories, those with intent to monitor for action, such as local rangers or remote supply chain managers, and those without intent to act on monitoring, such as researchers at universities. The unit of analysis was a 1-by-1-km grid cell of forest area; the sample was limited to those cells identified as falling within the subscription areas of users with intent to monitor. Other control variables included biophysical characteristics such as slope, distance to road or ports, time-variant features such as temperature and precipitation, and macroeconomic fixed effects.

The unique approach used in this analysis hinges on the gradual uptake of new data over time. A key source of bias for understanding the impact of monitoring is the covariation of monitoring and deforestation; people tend to look where deforestation is occurring. By limiting the sample to areas that were monitored through a subscription and incorporating the deforestation rates before and after monitoring began, the researchers could eliminate this bias.

The study results showed that "the average effect of subscriptions on deforestation [was] negative, but statistically insignificant and small compared with the average yearly 2011–2016 deforestation probability (0.18)" (Moffette et al., 2021, para. 13). The analysis detected significant effects in Africa, but not in Asia or Latin America. The researchers suggested that the effect in Africa is likely because of the limited alternatives for monitoring deforestation. In South America, Brazil has a national forest-monitoring system that excluded it from the sample. Peru, which was included, developed its own forest-monitoring system during the study, and other included South American countries gained access to GLAD alerts later than countries in Africa or Asia. In Asia, alternative forest-monitoring systems have emerged that are being used by private sector actors, and subscriptions to GFW by local civil society organizations are more limited, which collectively contributed to a weaker effect.

This research offers some additional clues to two mechanisms of impact: policy incentives and experience applying the data. When data on policy mechanisms was included in the analysis, the study found that "subscriptions have a stronger deterrent effect in protected areas and in forest concessions" (Moffette et al., 2021). Further, the effect of subscriptions increased when a time lag of 1 year was applied. This suggests that as users integrate the data into protocols and systems and gain experience using alerts, they become more effective at deterring deforestation. Given that the study covered only 2 years of potential usage of alerts, we can assume the potential for

---

[5] The study covered tropical countries that gained access to alerts in four waves from 2016 to 2017. Because the timing was not correlated to deforestation, the variation offers randomly assigned counterfactuals. Wave 1, March 2016, included Peru, Republic of the Congo, and Kalimantan. Wave 2, August 2016, consisted of Brazil outside of the Amazon Biome; the Brazilian Amazon was excluded from the study because of its alternative monitoring system. Wave 3, February 2017, included Brunei, Malaysia, Indonesia, Papua New Guinea, Timor Leste, Burundi, Cameroon, Central African Republic, Equatorial Guinea, Gabon, Democratic Republic of the Congo, Rwanda, and Uganda. Wave 4, November 2017, contained Colombia, Ecuador, French Guyana, Guyana, Suriname, and Venezuela. For more information, see Moffette et al., 2021.

increased effect over a longer period, given both the lag effect and the steady increase in users.

Qualitative Evaluation

The quantitative analysis found solid evidence that GFW contributes to reducing deforestation, but questions remained about the mechanisms of adoption and the enabling factors that lead to greater success or failure in applying GFW data to forest monitoring. To answer these questions, WRI conducted a qualitative portion of the evaluation with case studies in Cameroon and Uganda. This research found evidence of impact and identified three categories of enabling factors: incentive structures, capacity, and behavior change factors. WRI staff (including this author) conducted the study with support from in-country consultants, and findings were published in a working paper by WRI (Shea & Coger, 2021).

To conduct the case studies, we chose two countries characterized by the same region and high levels of GFW usage and subscriptions but with different approaches in outreach strategy, so that we could detect contrasting results (Yin, 2014). We used the GFW theory of change to identify a set of anticipated causal pathways and targeted key informants that included users and potential users within the categories of GFW's intended objectives (see Box 1). Data from the user stories database and from subscription information contributed to identification of potential informants. We then conducted interviews and focus groups with more than 100 key informants representing a range of potential users, from national government to NGOs to community leaders in high forest-cover areas, and analyzed these results.

The results of the case studies supported evidence from the quantitative evaluation by identifying specific user groups that were applying tools in ways that directly or indirectly reduced deforestation. We found evidence that the data contributed to awareness raising and decision making for improving enabling conditions. For example, in Cameroon, government officials used the data to make decisions regarding forest titles and management plans. In Uganda, the National Forestry Authority (NFA) occasionally reviewed the data to oversee forest management. We also found evidence of GFW empowering civil society and law enforcement to more efficiently and effectively expose and combat illegal deforestation. Staff at the Uganda Wildlife Authority (UWA) used GLAD alerts to plan routes for rangers going on forest patrols; private forest owner groups used it to monitor conservation for livelihood funds; and, in Cameroon, civil society used it to independently monitor and report on illegal activities.

The most important findings from the case studies were the enabling factors. First of these was that monitoring is most effective when it links to an incentive structure. Incentives can be positive: In the case of the UWA rangers, income from tourism incentivizes the more efficient protection of forested parks. Increased access to markets such as the EU, for Cameroonian exporters, drives efforts to control deforestation at the government level, and GFW was adopted in this context even in the face of technical hurdles. Conversely, the Ugandan NFA staff had also been trained in the tools, but that agency generates income from selectively logging forested land; more than one informant mentioned a perverse incentive not to stop illegal logging, and monitoring was not adopted on NFA lands. Incentives may be financial, although these appeared to drive the most large-scale change, such as a shift across multiple agencies of government, but incentives can also be driven by other resources such as time or even reputation, both of which may have financial implications. As examples from case studies illuminated, community cohesion provides its own incentive in some forest communities, while in other cases the opportunity to express political agency and a sense of altruism may also be incentives for independent monitors.

Capacity was another key factor in the adoption and application of GFW—not only the technical capacity to receive the data, but also the institutional capacity to integrate data into process and act on that data. One significant element in successful capacity-building efforts was the role of intermediary actors, such as larger NGOs or in-country WRI staff that provided ongoing technical support and offered training sessions. Overall, key informants reported the ability to use the main functions of GFW tools at high levels, although some leaders from forest communities did report a lack of technical knowledge. Complaints about physical capacity limitations such as limited access to cell phones, computers, internet, or data plans were more common.

Institutional capacity to act on the available data is a familiar challenge for open data initiatives. "Transparency is more likely to produce the intended effects when it fulfills both the condition of 'publicity'—having relevant disclosed information actually reach the intended audience—and the condition of 'political agency'—having mechanisms where citizens can take action in response to the disclosed information" (Ling & Roberts, 2014, p. 8). In this case, some such mechanisms were barriers to the success of GFW. For example, although informants acknowledged the progress that Cameroon's Ministry of Forests had made over the years of partnership with WRI, they also noted that institutional change is slow. Changes like hiring a new unit to process and handle digital data, instituting new roles for staff, and opening new reporting channels had taken years and are still ongoing. One NGO staff member complained that independent monitors still are not able to file electronic reports; they must be printed and paper copies submitted. In his view, this slowed response times or possibly provided opportunities for special interests to intervene. Great transparency in protocols may contribute to greater effectiveness for near-real-time data.

Behavior change factors, notably trust and ownership, are also vital to adoption of new technologies and data. Trust hinges first on the validity and credibility of the data, and this may include the system for its delivery, which is also built over time through relationships. The case study revealed that in Cameroon, WRI was able to cultivate trust both through the independence of the data and by persistent engagement and collaboration with both NGOs and government. The National Forest Atlas is officially owned by the government, and this is part of its success. As Shea and Coger (2021) reported, "According to one NGO representative, the government's buy-in to a transparent system enables civil society to hold government accountable by their own standards" (p. 20). This corroborates a finding from one of the few other known studies of the impact of near-real-time forest data: "It is imperative to establish collaborative relationships with government counterparts" (Musinsky et al. 2018, p. 18). That study found that when data providers work with governments, and governments brand tools and claim ownership, systems are most effective.

In Uganda, the case studies revealed a different situation from that in Cameroon. No national agency in Uganda had adopted GFW's tools, even as field staff were using them regularly for monitoring. Little work had been done to integrate the available data and information into larger institutional contexts, or to coordinate between central and decentralized staff. This may be due to WRI's limited engagement with local actors. In Uganda, WRI worked primarily through a local NGO and did not have direct relationships with officials. One of the challenges for WRI, then, is how to continue to foster trust and ownership in other contexts where it may not have staff on the ground.

The qualitative study aimed to reduce a knowledge gap identified through a literature review finding that research examining the impact of monitoring on deforestation is scarce (Shea & Coger, 2021). It did so by identifying several successful mechanisms for the impact of near-real-time data and by illuminating some enabling factors. It also produced recommendations for the GFW team, including a more complete assessment of existing incentive structures and capacity prior to entering future engagements with partners, and establishment of a strategy for addressing these factors through intermediary actors or WRI's direct engagement. Most important, the qualitative study rounded out a picture of how GFW's data can and does achieve impact and provided recommendations to expand that impact in the interest of more effective forest protection.

Discussion and Conclusion

When it comes to the management and safe-guarding of natural resources, remotely sensed information and open-data tools can play critical roles in filling knowledge gaps, providing evidence that stakeholders need to make decisions and take action. Knowing the reach and impact of such data can help refine the tools of delivery and outreach. To do so, data providers need to monitor progress and measure impact, but the very nature of transparent, open data can make these tasks especially challenging. When anyone can access and use the tools, program managers have limited means to identify users and assess the adoption of tools or the impact of the work those users do.

Here we explored four methods—qualitative and quantitative methods for monitoring progress, and qualitative and quantitative approaches to evaluate impact—and their results. The lessons from these methods and results are broadly applicable to other platforms or tools presenting open data, especially data about natural resources.

Three lessons emerged with regard to monitoring. First, basic user data accessed through Google Analytics is useful for understanding reach and for improving basic functions and elements of the platform. Second, more specific user data is useful for understanding the potential mechanisms for impact. In particular, platform developers should explore options for two-tiered access when special features may prove useful; in the case of GFW this was a login that allowed users to save areas of interest. Logins of this type offer the opportunity to understand types of use and goals of users. More important, such data may prove critical to conducting rigorous evaluations of impact. Third, gathering qualitative stories about users' experience and application of the data in a systematic way provided the necessary complexity for additional decision making, and may contribute to evaluability of the project.

The results of the evaluations also tell us three things. First, yes, GFW has been successful in reducing deforestation, implying that other open data platforms can achieve impact. Second, the quantitative evaluation shows that measuring the impact of open data platforms is possible as long as some data about the use of those platform is available, but that innovative methods may be needed. An experimental design would be ideal for evaluating, for example, whether a more intentionally random rollout of the GLAD alerts by GFW could have facilitated a more informative evaluation. Ethical concerns about withholding access to data or technical considerations may prevent such design. In this case, using randomly varying characteristics was possible to create counterfactuals, as was the case with the staggered uptake of GLAD alert subscriptions. Third, qualitative evidence that explains the mechanisms of impact are important for understanding the results. Although this is true for any analysis, it is especially important in the situation of data that is freely accessible to any user, where information about those users may be more limited.

References

Baker, J. C., & Williamson, R. A. (2006). Satellite imagery activism: Sharpening the focus on tropical deforestation. *Singapore Journal of Tropical Geography, 27*(1), 4–14. https://doi.org/10.1111/j.1467-9493.2006.00236.x.

Guthrie, A. (2018, March 7). *Latin American banks hone socially responsible lending policies*. Latin Finance. https://www.latinfinance.com/magazine/2018/march-april-2018/latin-american-banks-hone-socially-responsible-lending-policies

Hansen, M. C., Potapov, P. V., Moore, R., Hancher, M., Turubanova, S. A., Tyukavina, A., Thau, D., Stehman, S. V. S., Goetz, S. J., Loveland, T. R., Kommareddy, A., Egorov, A., Chini, L., Justice, C. O., & Townshend, J. R. G. (2013). High-resolution global maps of 21st-century forest cover change. *Science, 342*(6160), 850–854. https://doi.org/10.1126/science.1244693.

Ling, C., & Roberts, D. (2014). *Evidence of development impact from institutional change: A review of the evidence on open budgeting*. Policy Research Working Papers, the World Bank. doi:https://doi.org/10.1596/1813-9450-6968.

Macauley, M. K., & Sedjo, R. A. (2011). Forests in climate policy: Technical, institutional and economic issues in measurement and monitoring. *Mitigation & Adaptation Strategies for Global Change, 16*(5), 499–513. https://doi.org/10.1007/s11027-010-9276-4.

Moffette, F., Alix-Garcia, J., Shea, K., & Pickens, A. H. (2021). The impact of near-real-time deforestation alerts across the tropics. *Nature Climate Change, 2021*. https://doi.org/10.1038/s41558-020-00956-w.

Musinsky, J., Tabor, K., Cano, C. A., Ledezma, J. C., Mendoza, E., Rasolohery, A., & Sajudin, E. R. (2018). Conservation impacts of a near-real-time forest monitoring and alert system for the tropics. *Remote Sensing in Ecology and Conservation, 4*(3), 189–196. https://doi.org/10.1002/rse2.78.

Shea, K., & Coger, T. (2021). *Evaluating the contribution of forest monitoring tools: Two case studies on the impact of Global Forest Watch*. World Resources Institute.

Suwanwerakamtorn, R., Pimdee, P., Mongkolsawat, S., & Sritoomkaew, N. (2011). The application of satellite data to monitor the encroachment of agriculture on forest reserve in the Phu Luang Wildlife Sanctuary, Loei Province, northeast of Thailand. *Journal of Earth Science and Engineering, 1*(3), 169–176. http://www.davidpublisher.org/index.php/Home/Article/index?id=6185.html

Weisse, M., & Dow Goldman, E. (2020, June 2). *We lost a football pitch of primary rainforest every 6 seconds in 2019*. Insights: WRI's Blog, June 2020. https://www.wri.org/blog/2020/06/global-tree-cover-loss-data-2019

Yin, R. K. (2014). *Case study research design and methods* (5th ed.). Sage. https://doi.org/10.3138/cjpe.30.1.108.

Application of Geospatial Methods in Evaluating Environmental Interventions and Related Socioeconomic Benefits

Anupam Anand and Geeta Batra

Abstract

Environmental interventions underpin the Sustainable Development Goals (SDGs) and the Rio Conventions. The SDGs are integrated and embody all three aspects of sustainable development—environmental, social, and economic—to capture the interlinkages among the three areas. The Rio Conventions—on biodiversity, climate change, and desertification, also intrinsically linked—operate in the same ecosystems and address interdependent issues, and represent a way of contributing to the SDGs. Assessing the results of environmental interventions and the related socioeconomic benefits is challenging due to their complexity, interlinkages, and often limited data. The COVID-19 crisis has also necessitated creativity to ensure that evaluation's critical role continues during the crisis. Satellite and other geospatial information, combined with existing survey data, leverage open-source and readily available data to determine the impact of projects. Working with geospatial data helps maintain flexibility and can fill data gaps without designing new and often expensive data tools for every unique evaluation. Using data on interventions implemented by the Global Environment Facility in biodiversity, land degradation, and climate change, we present the application of geospatial approaches to evaluate the relevance, efficiency, and effectiveness of interventions in terms of their environmental outcomes and observable socioeconomic and health co-benefits.

Introduction

Environmental interventions are important mechanisms for delivering the objectives laid out in the Sustainable Development Goals (SDGs) and the United Nations Rio Conventions. The SDGs are integrated and embody all three aspects of sustainable development—environmental, social, and economic—with the intention of capturing the interlinkages among the three areas (United Nations Environment Programme [UNEP], 2013). The three Rio Conventions—on biodiversity, climate change, and desertification, also intrinsically linked—operate in the same ecosystems and address interdependent issues, and represent a way of contributing to the SDGs. The various activities related to sustainable development are linked through feedback mechanisms, resulting in both benefits and tradeoffs. This is where a systems approach that recognizes the dynamic, interdependent complexity of real-

A. Anand (✉) · G. Batra
Global Environment Facility Independent Evaluation Office, Washington, DC, USA
e-mail: aanand2@thegef.org; gbatra@thegef.org

© The Author(s) 2022
J. I. Uitto, G. Batra (eds.), *Transformational Change for People and the Planet*, Sustainable Development Goals Series, https://doi.org/10.1007/978-3-030-78853-7_19

world contexts across different scales, and recognizes dynamic shifts over time, is helpful in addressing environmental issues (Kass, 2019).

The Global Environment Facility (GEF) was set up in 1992 as a financial mechanism for the Rio Conventions. The GEF supports the implementation of projects in five focal areas—biodiversity, climate change, land degradation, international waters, and chemicals and waste—through 18 implementing agencies. Since 2010, the GEF has moved toward integrated programming that seeks to bring about changes in the multiple domains necessary to achieve the desired long-term transformation. These programs consider causes across the environment and different realms of human activity, generate benefits in two or more GEF focal areas, and generate social and economic benefits. This recent emphasis on multifocal and integrated programming presents its own sets of challenges, primarily those of evaluating the results and measuring other related benefits.

Drawing on GEF projects and programs in biodiversity, land degradation, and climate change, this chapter presents the application of geospatial approaches to evaluate the relevance, results, and sustainability of GEF interventions in terms of their environmental outcomes and their socioeconomic and health co-benefits. The first section of the chapter includes an introduction to geospatial data and analysis, the trends in its use, and the reasons behind the increase in the use of these approaches. The next section illustrates the usefulness of geospatial data in evaluation using examples of specific applications by the GEF Independent Evaluation Office (IEO). The final section discusses insights from geodata applications in environmental evaluations.

Geospatial Approaches and Methods

Geospatial data is unique because it contains spatially explicit information. The data can be collected from various sources such as remote sensing platforms, geotagged photographs, and ground sensors, or from survey data sets that include such information. Geospatial methods include the creation, collection, analysis, visualization, and interpretation of geospatial data.

Thus, geospatial data and methods can provide spatially explicit, synoptic, time-series data for various earth system processes, and have been used in the monitoring of environmental processes over the past 40 years (Awange & Kyalo Kiema, 2013; Melesse et al., 2007; Spitzer, 1986). Its application in environmental evaluations has gained traction in the last 2 decades. Evaluators initially used geographic information systems mainly to visualize and detect change in combination with other evaluation data (Renger et al., 2002). Others in evaluation have recognized the usefulness of spatial data for determining baselines, outputs, and monitoring of results over time (Azzam, 2013; Azzam & Robinson, 2013). Evaluators have employed quasi-experimental designs (Andam et al., 2008; Buchanan et al. 2016; Ferraro & Pattanayak, 2006) using geospatial data in impact evaluations of biodiversity and forestry interventions. Geospatial analysis has also been used recently in randomized control trials (Jayachandran et al., 2017).

Drivers of Increased Use

Recognition of the role of geospatial science by intergovernmental agencies and major environmental and development policy frameworks is growing as countries move toward more evidence-based policy decisions (Lech et al., 2018). The United Nations Convention to Combat Desertification (UNCCD) has recommended using indicators obtained from remote sensing to monitor progress toward reversing and stopping land degradation and desertification (Minelli et al., 2017). The United Nations Framework Convention on Climate Change (UNFCCC) and the Convention on Biological Diversity (CBD) have also endorsed the use of objective indicators, many of which are derived through geospatial methods.

Other factors have influenced the increased use of geospatial data and analysis. First, we have seen an unprecedented flow of spatial data from multiple sources, including satellite data.

Moderate and coarse resolution data is free, and high-resolution data is becoming less expensive and more widely available. The recent developments in data science have influenced the availability and cost of data. The infrastructure and tools to work with large quantities of geospatial data or big geodata have increased substantially. The availability of application programming interfaces (APIs), cloud-based services, and browser-based development environments have allowed access to geospatial data and analysis without the need for significant computational infrastructure (Lech et al., 2018). Traditional statistical tools are often incapable of dealing with the volume and variety of geospatial datasets, thus paving the way for machine learning and artificial intelligence algorithms in the last 5–7 years.

Use of Geospatial Data and Analysis in Monitoring and Evaluation

Evaluators often encounter methodological challenges and data issues during the course of evaluations, including lack of baseline data, sampling bias, difficulties in selecting appropriate counterfactuals, and accounting for the impact of multiple scales and contexts on processes and interventions. Geospatial approaches and tools can effectively address these gaps and can be applied to evaluate environmental and socioeconomic outcomes, to measure environmental change. We can also combine them with existing qualitative and quantitative methods and the results of interventions over time, while recognizing and accounting for the complex interrelationships across the various factors.

Application of Geospatial Approaches by the GEF IEO

The GEF IEO has been one of the earliest adopters of geospatial methods to answer important evaluation questions on the relevance, effectiveness, efficiency, and sustainability of GEF interventions. This section presents examples from GEF IEO evaluations.

Assessing the Relevance of GEF-Supported Interventions to Combat Land Degradation and Desertification

As the financial mechanism of the UNCCD, the GEF uses land degradation focal area strategies consistent with the UNCCD global priorities, including its focus on combating desertification in Africa, emphasis on drylands and non-drylands, and achieving land degradation neutrality. The GEF is gradually moving toward integrated approaches in this area to deliver global environmental benefits in multiple focal areas while generating local environmental and development benefits. A 2017 land degradation focal area study conducted by the IEO analyzed 618 land degradation projects or multifocal area projects with a land degradation component. The study looked at the relevance and performance of GEF's investments in addressing land degradation.

The IEO used geospatial analysis to assess the relevance of GEF interventions at global, country, and site levels. The analysis involved a feature overlay of the GEF-supported land degradation projects with the areas of land degradation severity. The analysis showed that the GEF implemented interventions to address land degradation in all developing regions of the world (Fig. 1) with Africa appropriately receiving the highest share of land degradation focal area project financing (37%), followed by Latin America and the Caribbean with 24% (GEF IEO, 2017). Africa has the largest share of land with extreme degradation in semi-arid areas (United Nations Environmental Programme [UNEP], 2002). Degraded soils are also found in regions undergoing deforestation such as Indonesia and Brazil and areas with high population pressure such as China, Mexico, and India (UNEP, 2002).

The study also noted that India, Mexico, Brazil, Indonesia, and China received the majority of land degradation financing from the GEF and the majority of national projects focused on forest and agricultural lands and rangelands. Overall, the results from this study showed that the GEF was supporting land degradation projects where most needed and relevant (GEF IEO, 2017).

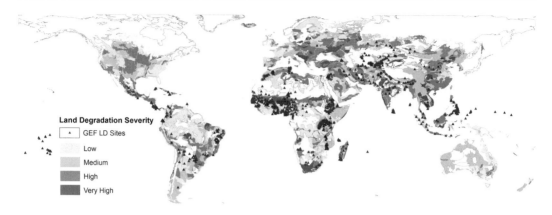

Fig. 1 Land Degradation Levels and GEF Interventions

Assessing Impacts in GEF-Supported Protected Areas

Protected areas are among the critical strategies for biodiversity conservation (DeFries et al., 2005). Global commitments and targets recognize the importance and role of protected areas in biodiversity conservation, ecosystem services management, poverty reduction, and generation of economic benefits. As the financial mechanism for the United Nations Convention on Biological Diversity (CBD), the GEF applies a strategy consistent with the CBD's strategic plan, reflected in its support to protected areas over the last 26 years. Between 1991 and 2015, the GEF provided $3.4 billion in grants to 618 projects involving protected areas, matched by $12.0 billion in cofinancing, to help protect almost 2.8 million km[2] of the world's non-marine ecosystems (GEF IEO, 2016). These figures exclude the support provided by GEF outside of the protected area systems.

Assessing the effectiveness and impact of GEF-supported protected areas is challenging mainly due to the scale, different timelines, and difficulty in collecting primary data due to the remoteness of protected areas. The IEO addressed these challenges using remote sensing data in the Evaluation of GEF Support to Protected Areas and Protected Area Systems (GEF IEO, 2016).[1]

At the global level, the evaluation used observations from satellite data, conducting geospatial analysis for the period 2001–2012 in GEF-supported protected areas and their buffers at 10 and 25 km to compare the extent of forest loss in these areas. The study examined forest change for 37,000 protected areas in 147 countries using a global dataset derived from satellite data analysis (GEF IEO, 2016).

Results of satellite data analysis of GEF-supported protected areas demonstrated that these protected areas experienced less forest loss than their surrounding 10 km buffer zones (see Fig. 2). In the 2001–2012 period, GEF-supported protected areas had up to four times less forest cover loss than the overall respective country averages, and at least two times less than protected areas not supported by the GEF in the same biomes and countries.

Using analysis through the biome lens, the evaluation found the greatest loss in protected areas in tropical and subtropical moist broadleaf forests, followed by tropical and subtropical conifers and tropical and subtropical dry broadleaf forest biomes (see Fig. 3). The findings confirmed the global trend of the most extensive forest loss in the tropics, followed by boreal and subtropical forests. The percentage loss of forest cover was highest in temperate conifers and temperate grassland, followed by tropical and subtropical grasslands, savannas, and scrublands; and then tropical and subtropical dry broadleaf forests. These results are consistent with global

---

[1] The evaluation was conducted in collaboration with the Independent Evaluation Office of the UNDP.

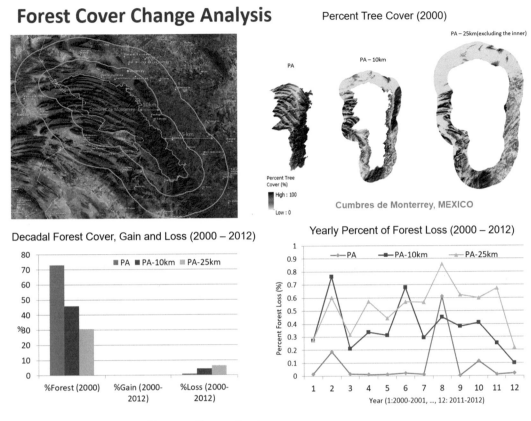

Fig. 2 Forest Loss in Protected Areas vs. Surrounding Areas

trends of tropical and subtropical forests exhibiting the most significant loss, followed by temperate and boreal forests (Hansen et al., 2013). These findings indicate the GEF's relevance to protected areas and affirm that the funding is going to areas experiencing a significant loss in protected forests—an important indicator of ecosystem integrity.

The results also showed that the median percent forest loss in GEF-supported protected areas was 1.2% percent while averaging 4.1% in the countries. The highest net percent forest losses were seen in Côte d'Ivoire (14.72%), South Africa (6.75%), and Guatemala (5.37%), and the highest net area forest losses were observed in Nicaragua (2528.76 km²), Honduras (1592.78 km²), and Bolivia (1072.29 km²).[2]

---

[2]Details of the approach, data, methods, and key findings of this evaluation can be found in the original evaluation report (GEF IEO, 2016).

Assessing Socioeconomic Co-Benefits

Despite widespread interest and extensive research into the socioeconomic impacts of environmental interventions in the past few decades, evidence remains inadequate and inconsistent (Awange & Kyalo Kiema, 2013; Melesse et al., 2007; Spitzer, 1986). Studies that have attempted to generate analytical insights and evidence have faced challenges, including the varying nature of co-benefits attributable to environmental initiatives, the typology and breadth of implementation approaches, and the data and methods used to assess co-benefits. The difference in methodology, data, and temporal and spatial scales also makes it tricky to draw overarching insights from these studies (Alpízar & Ferraro, 2020; Naidoo et al., 2019).

Studies have begun applying satellite and other spatial data sources to assess the co-benefits of development initiatives. These studies have

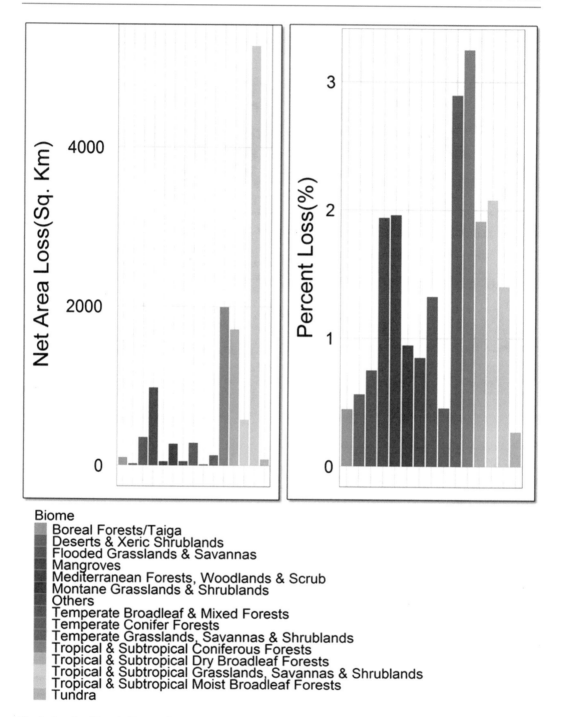

Fig. 3 Levels of Loss in Protected Areas

demonstrated how to leverage satellite-based data sources for evaluating environmental outcomes. Building on the recent developments in research and impact evaluation, the GEF IEO conducted a study to estimate the global and local-level contributions of GEF environmental initiatives and their related benefits.

IEO evaluators used a geospatial approach to determine the socioeconomic benefits associated with GEF-supported sustainable forest management (SFM) interventions. The GEF has a long history of providing support to improve the sustainability of forestry resources to increase environmental benefits and deliver socioeconomic co-benefits. This evaluation assessed the impacts of GEF-supported SFM interventions on biophysical and ecological variables and co-benefits measured in terms of socioeconomic indicators, and estimated monetary values of ecosystem services using the principle of natural capital accounting (Runfola et al., 2020). To examine the socioeconomic effects, the study used both a portfolio-wide approach (based on night light activity[3]) and a recent case study from Uganda, which was the first attempt to combine geospatial data with other survey data. To detect the impact of GEF projects on proximate (within 50 km) households, evaluators used the World Bank's Living Standards Measurement Survey of in-country household information (see Fig. 4).

The evaluation used the geographic locations of GEF SFM projects and data on the measurements of environmental outcomes based on suggested indicators from the CBD (2016) and UNCCD (2015). Night lights are a frequent proxy for socioeconomic outcomes, and the study used satellite-based measurements of nighttime light intensity over time. It also utilized a quasi-experimental approach to analyze GEF interventions' effectiveness along both environmental and socioeconomic dimensions. Details of this evaluation's methods and approaches are available in the original evaluation report (GEF IEO, 2019).

The portfolio-level, global scope analysis of economic and social co-benefits of GEF SFM projects indicated a small, positive impact on socioeconomic benefits as indicated by nighttime light intensity. The study found that projects implemented since 2010 showed a positive effect on nighttime lights (+0.24), a proxy for economic

development, that had not been observed in prior years. The study noted that, in the absence of precise geographic location information, these findings could have been an underestimate of the actual impacts across the GEF SFM portfolio. The study recognized that results from the nighttime lights at the portfolio level were not evident and expanded the analysis to include local-level data. The local-scale case study in Uganda using survey data helped fill the portfolio-level analysis gap and further explore the impact of GEF SFM projects on socioeconomic outcomes. The results showed that GEF SFM projects were associated with an increase in household assets. By matching the longitudinal survey data locations from the World Bank household survey that were close to GEF interventions to those farther away from GEF intervention sites, the evaluation found that GEF SFM projects were associated with increased household assets between $163 and $353 (within 40–60 km, respectively). The Uganda case study showed that households proximate to a GEF implementation site tended to experience average improvements in assets of approximately $310 (within 50 km) as compared to those that were not close to a GEF implementation site. Although results from a single case study cannot be considered representative of the entire portfolio, the study provided useful insights that help in understanding the main dynamics taking place in these areas,

Assessing Health Co-Benefits

GEF-supported projects and programs seek to influence positive environmental outcomes across critical areas such as biodiversity, land degradation, and climate change by generating global and local environmental benefits. It is well understood that improved environmental outcomes such as cleaner air, water, and soil undisputedly contribute to better living conditions and health. The COVID-19 pandemic has compelled a reexamination of the consequences of environmental destruction and its direct implications for human health. Anthropogenic activities leading to land use mismanagement,

---

[3] Studies have demonstrated that nighttime light levels are highly correlated with economic activity, population, and establishment density (Mellander et al., 2015).

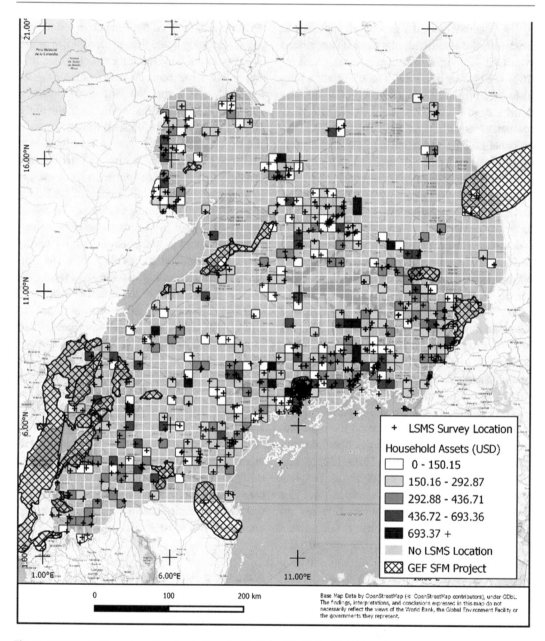

Fig. 4 GEF Project Impacts on Proximate Households Using World Bank's Living Standards Measurement Survey

fragmentation and destruction of natural habitats, and overexploitation of wildlife have fundamentally created more opportunities for the spread of infectious and zoonotic diseases (Liu et al., 2013; Olivero et al., 2020). Besides zoonotic diseases, poor water and air quality are still the leading causes of mortality worldwide. Therefore, human health issues cannot be separated from the environmental agenda and actions.

Building on the earlier work on SFM in Uganda, the GEF IEO undertook a further analysis to examine and quantify the association between GEF interventions and health co-benefits. The study looked at the health conditions of children under the age of 5 in Kenya, focusing on health measures including the prevalence of diarrhea and coughs (Fig. 5). The study explored whether improving environmental and socioeconomic co-benefits through GEF-supported projects led to improved health outcomes. It utilized the health survey dataset from the Kenya Department of Health Services (DHS, 2014), which contained 1594 survey clusters, each of which represented 19–25 households. The analyzed projects were drawn from the GEF's biodiversity, land degradation, climate change, and sustainable forest management focal areas and programs. Only projects implemented before 2014 were considered for the analysis. Evaluators used a quasi-experimental geospatial interpolation (QGI) method on Kenya's health data to quantify the association between GEF interventions and children's health conditions. The QGI method has three parameters: sample density, upper distance bound, and maximum matching difference. It uses a propensity-matching approach to pair treated and controlled survey clusters based on covariates. Runfola et al. (2020) provide more details on the QGI approach.

The study observed localized associations in both variables tested, with a 17% reduction in the occurrence of coughs within 10 km of the GEF intervention areas, and a 9% reduction in the occurrence of diarrhea within a distance of less than 3 km. Besides these direct measures of health outcomes, GEF-supported projects also had positive impacts on water access, including the access to source water in dwellings and the presence of water at hand-washing facilities. The results were found to be stronger in clusters closer to GEF interventions (see Fig. 6).

Assessing Outcome Sustainability in Fragile and Conflict Situations

Assessing the sustainability of outcomes is challenging because, in most cases, projects do not have the resources or the mandate to look at the project results after closure. Examining outcome sustainability can be more of an issue in fragile and conflict-affected situations due to logistical

and safety concerns. Geospatial analysis using remote sensing data can help in such situations where field visits and primary data collection are not possible.

The IEO used satellite-based data to assess the sustainability of environment-related project outcomes, part of the evaluation of GEF support in fragile and conflict-affected situations (GEF IEO, 2020). The IEO analysis looked at the trends in the change of forest cover in Sapo National Park, Liberia (Fig. 7). Evaluators compared the loss in forest cover for different periods (before, during, and after the project) to those periods in areas outside the protected areas and to trends in the overall national forest cover loss.

Sapo National Park is Liberia's only national park and a biodiversity hotspot within the Upper Guinea Forest ecosystem. It has faced long-standing threats from illegal farming, hunting, logging, and mining. In postwar Liberia, GEF-supported programming illustrates its catalytic potential in situations affected by conflict and fragility. The project Establishing the Basis for Biodiversity Conservation on Sapo National Park and in South-East Liberia, approved in 2004, marked one of the earliest GEF-funded projects in postwar Liberia. The World Bank implemented the project, and Flora and Fauna International (FFI) executed the project in collaboration with the Forestry Development Authority (FDA) of Liberia. The World Bank's re-engagement in Liberia started after the Second Liberian Civil War ended in 2003 (Independent Evaluation Group, World Bank [IEG], 2012, p. xiii.). Taking place from 2005–2010, the project was deemed successful, and project documents noted that "implementation occurred within a period of profound governance, environmental, institutional and societal changes in Liberia following a decade and half of the civil instability" (FFI, 2010, p i).

Since then, the GEF has supported various projects in Libera in different focal areas. Two other relevant GEF-funded projects— Consolidation of Liberia's Protected Area Network, from 2008 to 2012, and SPWA-BD: Biodiversity Conservation through Expanding the Protected Area Network in Liberia

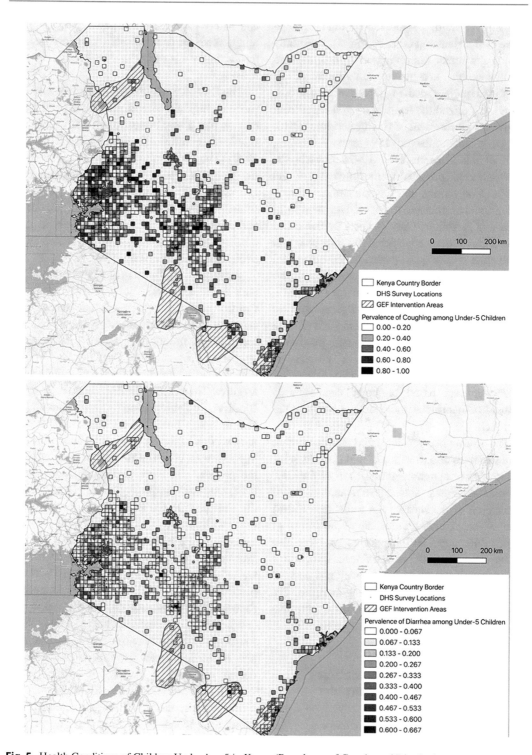

Fig. 5 Health Conditions of Children Under Age 5 in Kenya (Prevalence of Coughs and Diarrhea)

(EXPAN)—followed the first project and were also implemented by the World Bank. The Forest Development Authority of Liberia executed the projects. Both of these projects were "built on

Fig. 6 Health
Co-Benefits of
GEF-Supported Projects

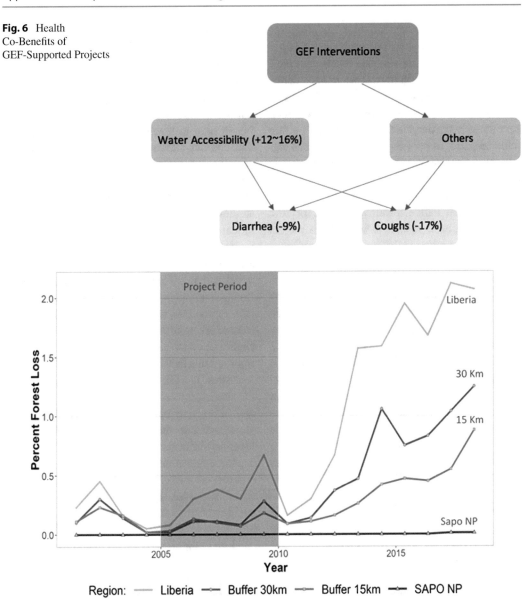

Fig. 7 Trends in Forest Cover in Sapo National Park, Liberia

successful GEF investments in Sapo NP" (World Bank, 2007, p. 4) and focused on biodiversity conservation, protected area management, community participation, and reducing rural dependence on forests and wildlife in Liberia.

Drawing on these projects' lessons, the World Bank continued its engagement with the forests and protected area interventions in Liberia, expanding the protected area systems and strengthening capacity to maintain them.

Ultimately, the Government of Liberia received grant funding ($37.5 million) through the World Bank from the Government of Norway for the cost of the Liberia Forest Sector Project, 2016–2023, which expanded substantially on the initial GEF projects (World Bank, 2016). This project supports priority investments to strengthen the on-the-ground management of Sapo National Park, including physical demarcation, provision of vehicles and equipment, and updating the park's management plans (World Bank, 2016).

The remote sensing analysis results in Fig. 7 indicate minimal forest loss, close to zero deforestation within the park boundary (flat dark line). This could be explained by the prohibition on all economic activities, including mining, within national parks, as per Liberia's National Park legislation. Legal mining concessions are present in the buffer zone.

The results illustrate how efforts to protect Sapo National Park's resources during the first project have been sustained beyond the project duration and supported through subsequent interventions. This trend inside the park contrasts with the phenomenal increase in forest loss outside the park borders (see Fig. 8) and in Liberia as whole, mainly driven by illegal activities such as mining and logging for sustenance in the post-war nation.

The Liberian economy is highly dependent on natural resource exports from the mining, forestry, and rubber sectors. According to an International Monetary Fund (2008) study, the small-scale mining sector for gold and diamonds in the country was estimated to involve as many as 100,000 artisanal miners in 2008, but only 48 artisanal and small-scale mining (ASM) miners (Small & Villegas 2012).

The two dips in the forest loss outside the Park (around 2005 and 2010) shown in Fig. 7 coincide with the eviction[4] of illegal gold miners and settlers in Sapo National Park (FFI, 2010). The lack of financial, technical, and human resources, and lack of capacity and conducive legal environment in Liberia to effectively monitor ASM sites and other illegal activities also explain forest loss in the Sapo National Park's buffer zone (World Bank, 2020).

Conclusions

This chapter demonstrates the utility of geospatial approaches and data to evaluate complex environmental interventions and assess their socioeconomic and health co-benefits. Geospatial analysis can help answer key evaluative questions on relevance, effectiveness, and sustainability of outcomes.

Geospatial methods can save financial and human resources and be very useful when working in hard-to-reach areas, especially in fragile and conflict situations or in a limiting context such as the COVID-19 pandemic. Geospatial analyses are also scalable and provide a cost-effective and efficient approach for meaningful studies at the project site, portfolio level, and global level. The results generated through these methods provide objective evidence and thereby aid transparency. The analyses can also reveal patterns that are not obvious and help in understanding complex processes. Geospatial methods and approaches work well in a mixed-methods framework and assist with common evaluation challenges such as lack of baseline, finding the right counterfactuals, and addressing accessibility issues. The GEF IEO has used geospatial tools for sharing evaluation results through 2-D maps and interactive maps and visualizations. These tools facilitate the communication of complex ideas and information.

As environmental programming becomes complex as it interlinks with other economic and social variables, and the demand for globally consistent and locally relevant data keeps growing, geospatial data and analyses offer an efficient and complementary approach to evaluators to explore new and increasingly complex questions and topics. Whether applied on its own or in combination with other complementary data and processes, geospatial approaches and methods have undoubtedly opened up new avenues for use in evaluation, and are here to stay.

Acknowledgments The authors gratefully acknowledge the role of IEO partners with whom we have collaborated for these studies, including faculty at the University of Maryland, NASA, and AID DATA, who contributed to the various evaluations mentioned in this chapter. The material presented in the case studies is based on these completed evaluations. We particularly thank Min Feng, Dan Runfola, and Jiaying (Tina) Chen for their contributions.

---

[4]The Liberian Government used the term "Voluntary departure" for the 2010–2011 removals.

SAPO NP, Liberia - Forest Loss

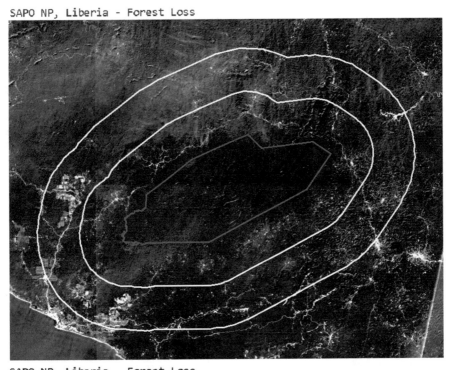

SAPO NP, Liberia - Forest Loss

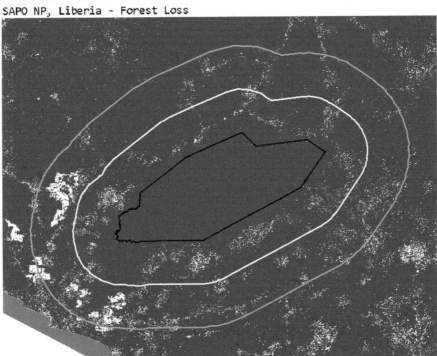

Fig. 8 Forest Loss Within and Outside of Sapo National Park, Liberia

References

Alpízar, F., & Ferraro, P. J. (2020). The environmental effects of poverty programs and the poverty effects of environmental programs: The missing RCTs. *World Development, 127*(March), 104783. https://doi.org/10.1016/j.worlddev.2019.104783.

Andam, K. S., Ferraro, P. J., Pfaff, A., Sanchez-Azofeifa, G. A., & Robalino, J. A. (2008). Measuring the effectiveness of protected area networks in reducing deforestation. *Proceedings of the National Academy of Sciences, 105*(42), 16089–16094. https://doi.org/10.1073/pnas.0800437105.

Awange, J. L., & Kyalo Kiema, J. B. (2013). Environmental monitoring and management. In J. L. Awange & J. B. Kyalo Kiema (Eds.), *Environmental geoinformatics: Monitoring and management* (pp. 3–16). Berlin Heidelberg: Springer. https://doi.org/10.1007/978-3-642-34085-7_1.

Azzam, T. (2013). Mapping data, geographic information systems. *New Directions for Evaluation, 2013*(140), 69–84. https://doi.org/10.1002/ev.20074.

Azzam, T., & Robinson, D. (2013). GIS in evaluation: Utilizing the power of geographic information systems to represent evaluation data. *American Journal of Evaluation, 34*(2), 207–224. https://doi.org/10.1177/1098214012461710.

Buchanan, G. M., Parks, B. C., Donald, P. F., O'Donnell, B. F., Runfola, D., Swaddle, J. P., Tracewski, Ł., & Butchart, S. H. M. (2016). *The impacts of World Bank development projects on sites of high biodiversity importance.*. College of William and Mary AidData. http://docs.aiddata.org/ad4/files/wps20_world_bank_biodiversity.pdf

DeFries, R., Hansen, A., Newton, A. C., & Hansen, M. C. (2005). Increasing isolation of protected areas in tropical forests over the past twenty years. *Ecological Applications, 15*(1), 19–26. https://doi.org/10.1890/03-5258.

Ferraro, P. J., & Pattanayak, S. K. (2006). Money for nothing? A call for empirical evaluation of biodiversity conservation investments. *PLoS Biology, 4*(4), e105. https://doi.org/10.1371/journal.pbio.0040105.

Flora and Fauna International. (2010). *Establishing the basis for biodiversity conservation on Sapo National Park and in south-east Liberia* (Project implementation report). Author.

Global Environment Facility Independent Evaluation Office. (2016). *Impact evaluation of GEF support to protected areas and protected area systems* (Evaluation 104). https://www.gefieo.org/sites/default/files/ieo/evaluations/files/BioImpactSupportPAs-2016.pdf

Global Environment Facility Independent Evaluation Office. (2017). *Land degradation focal area study.* https://www.gefieo.org/evaluations/land-degradation-focal-area-ldfa-study-2017

Global Environment Facility Independent Evaluation Office. (2019). *Value for money analysis of GEF interventions in support of sustainable forest management.* https://www.gefieo.org/evaluations/value-money-analysis-gef-interventions-support-sustainable-forest-management-2019

Global Environment Facility Independent Evaluation Office. (2020). *Evaluation of GEF engagement in fragile and conflict-affected situations.* https://www.gefieo.org/evaluations/evaluation-gef-engagement-fragile-and-conflict-affected-situations-2020

Hansen, M. C., Potapov, P. V., Moore, R., Hancher, M., Turubanova, S. A., Tyukavina, A., Thau, D., Stehman, S. V., Goetz, S. J., Loveland, T. R., Kommareddy, A., Egorov, A., Chini, L., Justice, C. O., & Townshend, J. R. G. (2013). High-resolution global maps of 21st-century forest cover change. *Science, 342*(6160), 850–853. https://doi.org/10.1126/science.1244693.

Independent Evaluation Group, World Bank. (2012). *Liberia country program evaluation: 2004–2011.* https://ieg.worldbankgroup.org/sites/default/files/Data/reports/Liberia_cpe.pdf

International Monetary Fund. (2008). Liberia: Poverty reduction strategy paper. *IMF Staff Country Reports, 08*(219), 1. https://doi.org/10.5089/9781451822984.002.

Jayachandran, S., de Laat, J., Lambin, E. F., Stanton, C. Y., Audy, R., & Thomas, N. E. (2017). Cash for carbon: A randomized trial of payments for ecosystem services to reduce deforestation. *Science, 357*(6348), 267–273. https://doi.org/10.1126/science.aan0568.

Kass, G. (2019). Systems approaches for tackling environmental issues. Institution of Environmental Sciences. https://www.the-ies.org/analysis/systems-approaches-tackling

Kenya Department of Human Services. (2014). *Kenya demographic and health survey 2014.* https://dhsprogram.com/publications/publication-fr308-dhs-final-reports.cfm

Lech, M., Uitto, J. I., Harten, S., Batra, G., & Anand, A. (2018). Improving international development evaluation through geospatial data and analysis. *International Journal of Geospatial and Environmental Research, 5*(2) https://dc.uwm.edu/ijger/vol5/iss2/3.

Liu, X., Rohr, J. R., & Li, Y. (2013). Climate, vegetation, introduced hosts and trade shape a global wildlife pandemic. *Proceedings of the Royal Society B: Biological Sciences, 280*(1753), 20122506. https://doi.org/10.1098/rspb.2012.2506.

Melesse, A. M., Weng, Q., Thenkabail, P. S., & Senay, G. B. (2007). Remote sensing sensors and applications in environmental resources mapping and modelling. *Sensors, 7*(12), 3209–3241.

Mellander, C., Lobo, J., Stolarick, K., & Matheson, Z. (2015). Night-time light data: A good proxy measure for economic activity? *PLoS One, 10*(10), e0139779. https://doi.org/10.1371/journal.pone.0139779.

Minelli, S., Erlewein, A., & Castillo, V. (2017). Land degradation neutrality and the UNCCD: From political vision to measurable targets. In H. Ginzky, I. Heuser, T. Qin, O. Ruppel, & P. Wegerdt (Eds.), *International yearbook of soil law and policy 2016* (pp. 85–104). Springer. https://doi.org/10.1007/978-3-319-42508-5_9.

Naidoo, R., Gerkey, D., Hole, D., Pfaff, A., Ellis, A. M., Golden, C. D., Herrera, D., Johnson, K., Mulligan, M., Ricketts, T. H., & Fisher, B. (2019). Evaluating the impacts of protected areas on human well-being across the developing world. *Science Advances, 5*(4), eaav3006. https://doi.org/10.1126/sciadv.aav3006.

Olivero, J., Fa, J. E., Farfán, M. A., Márquez, A. L., Real, R., Juste, F. J., Leendertz, S. A., & Nasi, R. (2020). Human activities link fruit bat presence to ebola virus disease outbreaks. *Mammal Review, 50*(1), 1–10. https://doi.org/10.1111/mam.12173.

Renger, R., Cimetta, A., Pettygrove, S., & Rogan, S. (2002). Geographic information systems (GIS) as an evaluation tool. *American Journal of Evaluation, 23*(4), 469–479. https://doi.org/10.1177/109821400202300407.

Runfola, D., Batra, G., Anand, A., Way, A., & Goodman, S. (2020). Exploring the socioeconomic co-benefits of Global Environment Facility projects in Uganda using a quasi-experimental geospatial interpolation (QGI) approach. *Sustainability, 11*(18), 3225. https://doi.org/10.3390/su12083225.

Small, R., & Villegas, C. (2012). Artisanal and small-scale mining in and around protected areas and critical ecosystems project: Liberia case study report.

WWF-World Wide Fund for Nature and Estelle Levin. https://www.levinsources.com/assets/pages/ASM-Liberia-Final.pdf

Spitzer, D. (1986). On applications of remote sensing for environmental monitoring. *Environmental Monitoring and Assessment, 7*(3), 263–271. https://doi.org/10.1007/BF00418019.

United Nations Environment Programme. (2002). *Global environment outlook 3: Past, present and future perspectives.*. Earthscan.

United Nations Environment Programme. (2013). *Embedding the environment in sustainable development goals* (UNEP Post-2015 Discussion Paper 1 Version 2). https://sustainabledevelopment.un.org/content/documents/972embedding-environments-in-SDGs-v2.pdf

World Bank. (2007). *Consolidation of Liberia's protected area network implementation completion memorandum (ICM)*. https://www.thegef.org/project/consolidation-liberias-protected-area-network

World Bank. (2016). *Liberia – Forest sector project, project appraisal document*. http://documents1.worldbank.org/curated/en/385131468184765418/pdf/PAD1492-PAD-P154114-Box394888B-PUBLIC-LFSP-PAD-FINAL.pdf

World Bank. (2020). *Liberia Forestry Development Authority: An institutional capacity assessment*. https://www.profor.info/sites/profor.info/files/Liberia-Forestry-Development-Authority-An-Institutional-Capacity-Assessment.pdf

Printed in the United States
by Baker & Taylor Publisher Services